Up from Dragons

Up from Dragons

The Evolution of Human Intelligence

John R. Skoyles
Dorion Sagan

McGraw-Hill

New York   Chicago   San Francisco
Lisbon   London   Madrid   Mexico City   Milan
New Delhi   San Juan   Seoul   Singapore
Sydney   Toronto

Library of Congress Cataloging-in-Publication Data

Skoyles, John R. and Sagan, Dorion.
 Up from dragons : the evolution of human intelligence / John R. Skoyles and
Dorion Sagan.
   p. cm.
 Includes bibliographical references and index.
 ISBN 0-07-137825-1
 1. Intellect. 2. Genetic psychology. 3. Human evolution. I. Sagan, Dorion
II. Title.
 BF431 .S558 2002
 155.7—dc21   2001007857

McGraw-Hill

A Division of The McGraw-Hill Companies

1 2 3 4 5 6 7 8 9 0 DOC/DOC 0 8 7 6 5 4 3 2

ISBN 0-07-137825-1

Printed and bound by R. R. Donnelley and Sons Company.

This book is printed on recycled, acid-free paper containing a minimum of 50%
recycled de-inked fiber.

To the memory of
Dorion's curiosity- and Carl-nurturing grandparents,
Sam ("Lucky") and Rachel Sagan,
and John's own mind- and thought-nurturing parents
Edward ("Ted") and Beryl Skoyles

Contents

Preface

Reembarking on an Intellectual Voyage

A quarter of a century ago, in 1977, when one of us—Dorion—was eighteen, his father asked us to look at a book in manuscript. That book went on to win the Pulitzer Prize, and remains to this day one of the most thought-provoking and interdisciplinary examinations of the evolution of human intelligence. Its writer was a scientist who had explored whether there was life on Mars, and speculated on whether it might exist outside our solar system. Here, he had written upon an even bigger scientific issue: why he—and the rest of the human species—possessed the intelligence to ask such questions. As Darwin had once put it, our minds originated from a "lowly organized form," yet we now possess a "godlike intellect." How did this astonishing event—the rise of mind and intelligence from the material world, an event as incredible as the existence of matter or the rise of life— occur? Did consciousness come into existence once, and only here, or has it done so on many worlds? Is our "godlike intellect" alone in the infinitude of space, or might we one day reach out and talk to extraterrestrial intelligences?

Where was this godlike power to reason, to think, to do science taking us? Was it leading to hell, the nuclear-precipitated, species equivalent of an adolescent suicide? Or would it deliver us to the stars, to encounters with other, more intelligent species from whom we might learn the tricks of cosmic survival? Remember, this was 1977, at the height of the escalation of the nuclear arms race. What the twentieth-century Mars gazer scribed about the evolution of intelligence and its extraterrestrial possibilities quickly topped the best-seller lists and won a Pulitzer Prize, the first ever for a science book. This and his later books and television series had by the mid-1980s made him the one scientist's face to which every American could attach a name. Before Dorion, on the title page of the pile of paper, were the words: *The Dragons of Eden: Speculations on the Evolution of Human Intelligence* by Carl Sagan.

Twenty-five years later, that eighteen-year-old is now at roughly the age Carl was when he passed on those speculations for a proof check, and like him he is a successful science writer—that torch has been passed on. Alas, Carl is no longer with us to carry his exploration of the history of the human mind into the new millennium. While they have mostly stood the test of time, his facts and arguments look like 1970s museum pieces—not so much by what they say, as by what they leave out. In the years after Carl Sagan wrote his ideas, brain science took off; nearly ten times as many papers were given at the 2000 Society of Neuroscience Conference as in 1977. Technology that in 1977 appeared only in science fiction—cameras photographing the living brain in bread-slicer-thin sections, taking snapshots of thoughts and feelings—by the mid-1990s had become mundane reality, another inventory item of brain lab and hospital radiology departments. In 1977, there were thought to be only about a dozen neural transmitters; today, we know of many dozens more, each with up to eighteen receptors dovetailing — as yet mysteriously — to concoct our dreams, cognitions, and emotions. In 1989, in recognition of the takeoff of the neurosciences, the U.S. Congress decided to call the 1990s the *Decade of the Brain*.

Carl, a space age Columbus who, if he didn't go to the stars, interpreted the meaning of the spaceships that did, had also plunged into that inner no man's land, the brain. If back in 1977 only the outlines of the vast, if condensed, cranial continent were known, twenty-five years later we have the equivalent of satellite photos and landers, and we have sent intricate mapping devices deep into the brain's interior. The modern reader, reading Carl's speculations now, might wonder why he did not mention this fact or that fact—only to realize that the particular piece of knowledge in question was discovered ten or twenty years after the book was written. Which brings us to the present: What new version of *The Dragons of Eden* would Carl have written had he decided to embark, with the range of new tools and results now available, upon the twenty-fifth anniversary of the original investigative voyage?

Carl had in fact already given us clues to where he would have gone. At the start of *The Dragons of Eden*, he raises the question of our extended childhood "in which young humans are dependent on adults and exhibit immense plasticity—that is, the ability to learn from their environment and their culture." As Carl noted, that meant the human species could change on much shorter timescales than allowed for by natural selection. As he put it, "We have made a kind of bargain with nature: our children will be difficult to raise, but their capacity for new learning will greatly enhance the chances of survival of the human species." In other words, while other animals depend upon information prewired into their brains, we use information picked up from others and our own experience. Moreover, humans exploit what Carl called *extrasomatic knowledge*—information such as writing that gets passed down from one generation to the next. In the The *Dragons of Eden*, he starts to fill in that story. Indeed, his first proper chapter, "Genes and Brains," is a well-thought-out preparation. But the science was only beginning to provide the facts needed to narrate the story he planned to tell. His sketch was rough in places; he did what he could. Like an astronomer armed only with binoculars but looking for new solar systems, *The Dragons of Eden* senses mysterious shapes, wonders not yet revealed by the science of the time. Hungry for deeper connections, it speculates.

The *Dragons of Eden* is thus two books—the one Carl wrote, and the one he promised but which was, in fact, too early given the neuroscience at the time. Brain science in the 1990s came up with the required facts in spades. Those cameras looking under our skulls showed us a brain actively rewiring itself in ways that a mechanical view of evolution had never anticipated. It showed us a proactive brain tugging apart the threads of its genetic leash. It showed us a brain which, using the ever more popular metaphor of the computer, was not just programmed by natural selection but could rather program itself. This book, then, *Up from Dragons*, tells the story that the original *Dragons* couldn't tell, even though it wanted to, because it was before its time—the story of that world-changing supercomputer, the human brain.

The original *Dragons* contains many facts that now need updating. And to that which is faded in it, we apply a new coat of scientific paint. For example, few of us now in 2002—after the Cold War—would emphasize the arms race as Carl did in 1977—though Carl also shows a remarkable anticipation of modern concerns such as abortion and animal rights. In short, Carl wrote the book he could, but provided the sketch for another for which there was not yet the requisite science. *Up from Dragons* is thus a push deeper, using new tools, into the mysterious realm outlined by Sagan's mix of science and informed contemplation; it is a parallel, not a copy. In the first book, Carl left the possibility of a new voyage open. We believe this alone justifies this sequel: a sequel not as reworking of the original but as radical revision, setting off again as if for the first time on the intellectual odyssey for which science was not ready when Carl first began his daring expedition.

John Skoyles
Hampstead, London and the Centre for Philosophy
of Natural and Social Science, London School of Economics,
July 2002

Dorion Sagan
Manhattan, July 2002

Acknowledgments

The authors express their debt, not only to each other but to the mindlike computer technologies that allowed them to collaborate across the Atlantic (like neurons in a global brain) and to the many individuals who inspired and supported them. The single most direct inspiration was that of Carl Sagan on John Skoyles; without the breadth and depth of focus of the original *Dragons of Eden*, this project, *Up from Dragons*, with its additional synthesis of decades of recent brain science, might never have found its final form. We also must thank, from our coauthorial heart, the indefatigably enthusiastic Howard Bloom, who introduced us. We thank also our courteous and responsive agent, Richard Curtis, and our highly intelligent and professional editor, Amy Bianco. As usual, Lynn Margulis ("Professor Marvelous") earned her nickname. For server facilities for the www.upfromdragons.com Web site, thanks to David McFadzean. For intellectual feedback John thanks with deep appreciation Rodney Cotterill, Nick Humphrey, Patricia Goldman-Rakic, Michael Posner, Robert Steinberg, Brian Butterworth, Chris Frith, Alwyn Ackle, and many others too numerous to individually name. Dorion wants especially to thank his son Tonio and his sweetheart Jessica.

1

The Cosmic Mirror

L ook in a mirror. What do you see? A human?

Self-confident, we tell ourselves that we are a gifted species that can see further than any other. And, because of our big brains, we are probably right. But smart as we are, we have a lot to learn. Science tells us that our own individual perspective is limited. A quick look outside and the Earth seems flat. It is not. Mountains and oceans may appear vast, but they are only tiny specks and coatings on a smallish planet orbiting a medium-sized star in the arm of a modest-sized, spiral-shaped galaxy of which the cosmos has 125 billion.[1] Out there, life forms could exist as far beyond us as we are beyond an ant or a microbe. Religion may have gotten many things wrong, but one thing it got right was humility. Humility—the word comes from the same Indo-European root as human—is something of which, science and religion can agree, we can never have too much, if we want to see our true place.

As vast and daunting as space is, time holds our breath with its mystery. Museums of science and of natural history show us the evidence and the story. We stretch our necks in awe at the brontosaurus skeleton in its frozen walk. But a less well known museum exists in which to find awe: ourselves. Look again into that mirror. The history of the universe is also the history of you. From the Big Bang to the origin of life, reptile ancestors to mammals, primates to modern humans, you are not a bystander but part of the storyline. Your face hides the relics of creation. Look in the mirror again.

See that wetness in the corner of your eyes, or the dampness of your tongue? You are made of water, as are your tears and your saliva, two parts hydrogen, one part oxygen. The hydrogen nuclei there (ignoring a few created by later radioactive decay) are Big Bang relics. The universe's first few

1

seconds made nearly all the hydrogen you see in water, proteins, and the like. If you thought you were 16 or 60, think again. You—the hydrogen particles of your body—have been around a lot longer; the present best estimate of your true age is 12.5 billion years![2]

And the rest of you is only a little younger. Take proteins, such as those making up the keratin of your skin and the crystallin of your eye's lens. Their elements—oxygen, carbon, and nitrogen—formed a billion years after the cosmos began, when hydrogen in stars had burnt into heavier nuclei. Stars kept fusing elements into heavier ones until their cores were iron, the metal that makes hemoglobin and your lips in the mirror reddish. Still heavier nuclei were then made when stars exploded in supernovas that spewed out vast fluxes of neutrons. These neutrons upped the atomic weights of the already formed elements. You can see some of the resulting atoms in the silver behind the glass of the mirror. The gold on your finger or the fillings in your teeth might have been created from the collisions of neutron stars, the eventual cosmic leftovers of supernovas.[3] You are many, many things, but the most basic is star matter, ancient star matter. The matter of your body has been borrowed for the umpteenth time, recycled through the births and deaths of stars for about 12.5 billion years.

Feel your hair and see how it flops or hangs. If you cry, tears fall upon your cheeks. We cannot ignore gravity, yet its constant presence blinds us to its influence over our every move. It is a frustrating companion and yet, if we think about it, also a friend. We feel gravity from only one source, and it is moderate considering what is out there. It is unimaginably intense in some parts of the universe, such as in black holes. When the solar system formed, 4.5 billion years ago, 10^{22} tons of mass came together and generated the gravity weighing you down before the mirror. Planet Earth was born.

Matter should not do it, but it does: It reproduces, evolves, and grows even more complex. It has a superstrange power we call life. Whether this trick was a one-of-a-kind event here on Earth, or whether it has happened millions of times elsewhere in the universe, we do not know. We only know it happened, and it happened early. Space debris left after the formation of the solar system bombarded the early Earth with a 500-million-year cosmic hailstorm that pitted it constantly with Hiroshima-scale impacts. One smasher was particularly huge. The size of Mars, it hit with such force that it threw off chunks that came together as the moon. Hellhole Earth had pretty much ceased, however, by about 4 billion years ago.[4] And within only a 100,000 years, 3900 million years ago, life had started (if it did not come from elsewhere in the universe, in which case there exists an even more remarkable story). DNA-based (an earlier RNA-based form is thought to have come first) life started off eating hydrogen and hydrogen sulfide from volcanoes and vents in Earth's bubbling surface. And then it hit upon a trick that changed the very nature of air. Around 3 billion years ago, life started to use light, and it released a by-product—oxygen.[5] It was one of the first of many tricks that cells perfected. For example, initially DNA freely floated around cells, as it still does in bacteria. But this did not allow it much control over a cell's chemistry. At least 2700 million years ago—we know by the trace molecules they left—cells discovered how to organize DNA in a nucleus.[6] The thread of life now became long: If you stretched one from any of your cells, with its 33,600 genes, it would most likely reach higher than you—2 meters.

What you are really seeing in the mirror is nearly a hundred thousand billion cells. You are a fantastic multicellular organism. The first and the most simple of multicellular organisms were not so fantastic—bacterial "trees" and multicolored expanses growing in puddles and by the ocean in living tissues called *microbial mats* and layered stones called *stromatolites*. About 2 billion years ago, a change as dramatic as any in the history of life took place when bacteria merged genomes to become cells with nuclei, the ancestors of all fungi, plants, and animals.[7] Then around 1 billion years ago, these amoebalike beings themselves came together. Their multicellularity was a miracle of cooperation. It took a long time to arise, but when it did, strange things happened.

We are not what we appear. Those hundred thousand billion cells of you each have taken on a role laid down in your DNA by Hox genes, which provide the blueprint by which your cells multiplied, some to become your head rather than your legs, and so on.[8] If all you could see were Hox genes, you would not notice much of a difference between you and insects, corals, jellyfish, or hydra. Beneath our appearances, we are not such distant kin with these slimy and creepy-crawly organisms but almost like close family. Not that insects, corals, jellyfish, or hydra Hox genes are exactly like yours; while most invertebrates have only one cluster, you have four clusters of these genes—more for enabling more complexity. But nonetheless, their DNA is much the same: If you swap a fly Hox gene for a mouse one in a mouse embryo, it will still organize the embryo to grow as a mouse. A big bang of biology happened when Hox genes arose and provided animals with body plans. It turbocharged the complexity of life. Thus, from apparent nothingness, suddenly body forms of greatly subtlety and variety—worms, crustaceans, sea cucumbers, snails, and early fishlike creatures—burst onto the scene at virtually the same geological moment, in what paleontologists call the *Cambrian explosion*.

Hidden away, beneath what you see in the mirror, you have a spinal cord, a skull, a brain, an encased heart, and gill arches (they appear in the fetus and underlie the development of the thymus and part of the pituitary). No insect or snail has such things, since they are arthropods and mollusks, not vertebrates. And yet, extraordinarily, your inner parts started off in some tiny filter-feeding larvae.[9] If you went back 545 million years in a time machine and looked at such a larva with a microscope, you would see a rough sketch of the humans to come. Its thousands of cells would start to unfold as an embryo much as you did in the womb of your mother. Differences emerge quickly as it grows into a different creature, but unless you had a specialist's knowledge, you could not tell its launch into life from your own.[10] As useful as these traits are, however, they are not the secret of the vertebrates' success. That secret lies behind what is expressed in your features. It's not your skin but key cells in it, the pigment cells; not your teeth but the cells that create its dentin; and so on for your eyes, skull, and facial nerves. The formation of each is directed by neural crest cells. These are a kind of add-on, a sort of biological super clay, unique to vertebrates, that enabled them to create the sophisticated and clever tissue that is before you.[11] Really, neural crest cells should be called miracle cells.

That mirror is vertical. We hold our head erect; our body hangs on a vertebral column, the backbone. Your face also hangs with a sculptured shape; it is not jelly held loosely in a bag. Bone was once thought to have had its origins in armor-plated fish skin, but it now seems more likely that it arose from teeth in creatures called conodonts around 510 million years ago.[12] Until bones gave the strength of calcium to skeletons, bodies used cartilage. Fortunately, evolution gave us something stronger; otherwise we could neither stand nor run. All vertebrates also have gill bars. In the mirror you can see yours, except you do not call them that; they have another name when mineralized into bone—*jaws*. One of your gills became a jaw around 450 million years ago with the first jawed fish, also ancestors of sharks.

You recognize yourself in the mirror. You can live without doing that, but you could not survive without the inner recognition of "self" that your immune system makes possible. With this hidden bodily self-awareness, the immune system spots proteins that are foreign and thus alerts the body to fight viruses, bacteria, blood parasites, and even cancers. Your immune system and its T cells and antigen-presenting histocompatibility molecules arose 450 to 500 million years ago.[13] Whereas mineralization of cartilage evolved to protect against big attackers, the immune system evolved to protect against the unimaginably small inner ones.

We have been asking you to look at your *head*. As you did this, we bet, you wobbled it around. That is odd. Fish do not have necks. They do not need them. If they want to move their heads they shift their bodies. That is easy because they are in water. You may not be an amphibian, but for being able to move your head so freely we have them to thank. When our backboned ancestors made the move to land and air, they needed to be able to neck around and spot food, places of safety, and predators.

Examine your skin: It's dry. But skins were not always dry: Obviously, fish have wet ones, as do the amphibians into which some fish evolved. Amphibians still breathe through their skins. But around 300 million years ago, reptiles evolved, and by 250 million years ago they were taking over the world. You are looking at one of the innovations that let them do this: dry skin. Theirs is scaly, and yours is not totally dry: As the antiperspirant ads keep reminding us, it sweats. As we will see in the next chapter, mammals rewrote the script of cold-blooded animals and took control of their body temperature. They needed hair to keep out the cold and sweat glands to cool off. And one innovation led to another: From the glands for sweat came the mamma glands with which mothers feed milk to their young. Mamma are why biologists call us *mammals*.

Look at your jaw again. It is a single bone. Reptiles have a less efficient crusher made of four. Yours is also hinged to your skull in a more efficient manner. Raise your tongue and you feel your palate; it lets you eat and breathe at the same time—a small innovation of immense utility, which, again, reptiles never had. Open your mouth a little and you can see a great variety of teeth: molars, premolars, incisors, canines that tear, crush, and slice food. Chew something. Notice you chew on one side of your mouth, then the other—another mammalian trick to best pulp what's in your mouth. A reptile has one set of noshes and you two, one for ingesting your mother's food and milk and another for eating adult food when your head is larger.[14] Mammals developed this extraordinarily efficient food processor to squeeze every extra calorie out of what they eat. Unlike reptiles, they had to keep warm and raise their offspring.

When you look in the mirror, both your eyes see your reflection. They are frontally paired primate-style so that their overlapping fields of vision let you see depth. Stereoscopic vision helps primates live in trees. Other mammals live there—squirrels, sloths, pine martens—but they are not so brainy with their hands, touch, and sight. Primate hands often have opposable thumbs and remarkable dexterity. With sight they can coordinate their fingers, aided by pads on their tips. The replacement of claws by nails in primates helps. Moreover, with eyes in the front, they did not need to move their heads from side to side to sense distance—which is useful for keeping balance at the edges of branches and for spotting juicy insects without being noticed.[15]

You see your face in the mirror in all the richness of its color. That ability—which depends on retinal cones sensitive to the difference between red and green—evolved first in the group of primates that included Old World monkeys and apes. Minds have been seeing in color for perhaps 35 million years. Color vision arose not, as you might think, to spot ripe fruit but rather to locate fresh tropical foliage. The extra eye pigment that lets us see the beauty of a Monet or a stained glass church window started from something as mundane as a need for primates to find young (they are slightly red) edible leaves against a background of mature leaves.[16] But then, to a hungry stomach, nothing is more beautiful than food.

Touch your forehead and raise your hands forward up above your head. Circle them around. Not a very useful skill but a relic of what made you an upright monkey—an ape. Unlike monkeys, our ancestors could hang, swing, and climb around in trees with a near-upright posture that made a tail a nuisance. That's why you don't have one. With their more horizontal posture, monkeys needed the tail as a "fifth hand." Vertical life brought other changes, including a strengthening of the abdomen. If you are proud of your flat stomach, thank your ape ancestors.[17]

Being upright changed the face, making it more easily seen by others. Your face is an emotional billboard. Fur was cleared away to make the face even better for expressing emotions. Now naked, the face could show pain, happiness, anger, fear, and shock very clearly. Try grimacing, smiling, sneering, frowning in the mirror. How wonderful to have a face! Chimps can express most of these emotions, too (though not disgust and surprise). But you are better at it. Not, as you might think, because you have more facial muscles—you do not. But feel your skin. You can pull it around, and that elasticity does not exist in chimps. Our lips protrude out and are highlighted red. Our irises are dark – from a distance nearly black, surrounded by white. Only humans expose so much of their eyeballs. As a result others can see from a distance where and at whom or what we are looking, and what we feel, much more easily than with a chimp.[18]

Look to the edge of your face. Where has your body fur gone? You are a naked ape. Sweat cools, but if there is too much hair, sweat glues it into a wet mess, as it cannot evaporate. When our ancestors came down from the trees, they generated heat in amounts and for periods that no other mammal had done, so they needed to sweat efficiently. Humans are superbly designed for running, not in a big cat's sprint, not in the cheetah's dash, but for grueling long distances, far longer even than marathons. As we will see in Chapter 16, this has a price—heat. Marathon runners can lose 5½ liters of water as sweat. Imagine them wearing a chimp's thick coat instead of shorts and T-shirts! On two legs, outrunning prey, we had to be naked—except for certain patches of hair. Shave off your eyebrows. Thought they were cosmetic? No doubt they sharpen the visual appearance of the face. But do some hard work and it becomes obvious that they serve as facial sweat bands, stopping perspiration from stinging the eyes.

Look at your chin. Elephants have them, but, oddly, early species of humans did not. The Neanderthals, for example, were basically chinless. Why? Perhaps it has to do with our having flatter faces, or maybe it is related to chewing as the jaw develops. But human chins are observable in embryos of five months. There are puzzles in that mirror.[19]

Another puzzle is skin color. Look under a gorilla's fur and it is black, under a chimp's and it is white (though their naked faces are black). Like us, apes can have a wide variety of skin colors. We do not know the color you see in the mirror, but among our readers, we know there will be every shade. In the mirror we may see one color, but we know that all the different shades of other people's skins are as much part of our humanity as ours is of theirs. It is possible that as few as only three (though possibly as many as six) genes out of 33,600 control skin color—visually a big difference but genetically hardly noticeable.[20] Why does human skin color show geographic variation? Frustratingly, this question lacks a complete or simple answer. Intensity maps of ultraviolet (UV) radiation provide a partial clue: Where UV is most strong, people tend to have darker skin. That does seem biologically sensible. Melanin protects against UV light both directly, as a blocker, and indirectly, as a molecular scavenger of the harmful oxygen radicals it generates. That protection could save the unborn from spontaneous abortions and defects such as spina bifida, which is caused by a deficiency of folic acid. UV light destroys folic acid molecules as they flow in blood vessels under the skin. All women of childbearing age should take folic acid supplements; it reduces the chances of spina bifida by up to 75 percent. Folic acid is also needed for healthy sperm. It is thus reasonable to suggest that dark skin helps people if they lack folic acid in their diet, especially if they live in a UV-intense part of the world.[21] What factor selected for light skin? UV light aids the production in the skin of vitamin D. Because they would have been more lightly clothed and thus more exposed to the sun, people in warmer climates could afford to have black skin, even though it produces vitamin D at a sixth the rate of white skin. In colder climates people had to bundle up, and white skin maximized their production of vitamin D without exposing them to too much UV radiation. There are gaps in our understanding of that face you see in the cosmic mirror. Much can be traced, but not all. No story is completely satisfactory.[22]

Now a question for men. If you are typical of most Western men, such as boxer Mike Tyson or President George W. Bush, you try to make yourself look like an adolescent boy. Modern men shave their facial hair and thus stop themselves from looking like natural, bearded men, looking instead more like young teenagers and women. Ancient Greek men did not shave and viewed men who did as effeminate. (We can tell from a joke by the ancient playwright Aristophanes.[23]) Why the difference? In one word, fashion, that "despot" which, as Ambrose Bierce, the nineteenth-century writer, observed, "the wise ridicule and obey." The fashion now is to look like youths, so we obey. And that tells us something profound about ourselves: We care deeply how others judge our appearance.

You have seen your face in photos and videos. You can easily recognize yourself in a snapshot or home movie. This is a trivial thing for us but something most people who ever lived never experienced. Reflections existed before modern times in wetted slates, polished metals, and still waters; but they were limited. Even mirrors cannot provide the opportunity to study our appearance as seen by others, as photos do. That is a privilege we gained only 150 years ago, with the invention of the daguerreotype.

Stop! There is something you are not seeing in the mirror. Imagine the face most humans would have seen. Imagine lines on your face from the year when you lost a fifth of your weight because the harvest failed. Imagine smallpox disfiguration. Imagine your face unwashed for several months. Your face is privileged by living in the time of industry and science. Food is cheap and available, many diseases have been conquered, taps give hot water, and shops offer us a bounty of beauty products to make us look good. These things have come about only in the most recent tiniest fraction of time during which people have existed, the snowfall of history on the iceberg of prehistory. We may hate our industrial, high-tech society in some ways, but we should thank it for letting us look much more young and healthy than our ancestors did. The difference can be seen in portraits. You have been grinning at yourself in the mirror—proud to show your teeth. But portrait artists until recently were careful never to reveal people's teeth, since they were so often rotten. The first self-portrait to show teeth, by Louise Elizabeth Vigée-Le Brun, was painted only in 1787. Before then (and long afterward), a toothy smile would show blackened teeth if any. Aristocrats of past ages had their skins painted white to hide the pockmarks of smallpox. If time machines were possible and you stepped out of one a few centuries ago, you would be shocked at people's haggardness, their emaciation, and they at your health and seeming ageless youth.

If that face in the mirror has changed in the last few hundred years, how will it change in the next few hundred? In this book, by looking at the past—by understanding our changing face in the mirror of time—we can also, we believe, offer a peek into our human future.

2

Up from Dragons

"To sleep, perchance to dream." Shakespeare wrote it, and we do it. But why? Reptiles never dream. But you do—and this puzzle is as good a place as any to start our journey up from dragons. For in our dreams exist the brain waves of reptiles. It is as if at night our distant ancestors—long gone reptiles—reassert themselves and creep back, making us return to their forgotten world. Hidden in dreams, science now finds an ancient compromise that allowed reptilian brains to become your mammalian one. To understand this compromise, we first need to look at the reptilian script of life that the first mammals tore up.

Reptiles strut the stage of life as actors reading the part given them by the sun and the weather. They are ectotherms, which means that they cannot generate heat internally. As a result, their pattern of activity is determined by the sun. We sometimes mistakenly call this cold-bloodedness, but reptiles can be as warm as you. What makes you different is that your body automatically—whatever the weather—accurately keeps much the same temperature. We have *homothermy*, from the Greek for "same heat." Reptiles, in contrast, can undergo wide variations in body temperature depending on the weather—*poikilothermy*. Not that reptiles do not try to control their temperature. They do, but, critically, through their actions rather than with sophisticated body adaptations.

Those were the innovations of mammals (and birds). We mammals have hair, sweat glands, insulating fat under the skin, and cells densely packed with power-generating *mitochondria*. The temperature may differ among species—wolves have the highest at 30.5°C (105°F) and echidnas the lowest at 23.2°C (73.7°F)[1]—but whatever the set point, these adaptations act to keep it stable. This inner control of temperature evolved originally in mammals so they could rule the night. In the dark, they could not be seen, and in the cold, poikilotherms might be sluggish. It was an opportunity to hunt invertebrates all night and steal the eggs of those that ruled the day—dinosaurs.

The reptiles' inability to control their own temperature does not make them less successful than mammals. There may be no polar iguanas or Arctic sea snakes, but reptiles can easily be as numerous in many places as mammals—between 1 and 10 billion lizards live in Italy alone.[2] Indeed, a big advantage exists for cold-bloodedness. Keeping one's blood warm within narrow limits requires eating vast amounts of food. Even at an ideal air temperature, a mouse uses up to thirty times more energy than a similar-sized foraging lizard. If the temperature drops to 68°F, the mouse uses thirty-three times more, and at 50°F, its use of energy shoots up 100-fold. Mammals, as a result, need to eat at least ten times as much as reptiles.[3] Warm-bloodedness is thus not always worth the effort; sometimes it is better to dine light and live cheap on borrowed warmth than to create your own internally.

But cold-bloodedness is a dead end for the great story of this book—the evolution of intelligence. Certainly reptiles could evolve huge sizes, as they did over vast sweeps of Earth as dinosaurs. But they could never have evolved our quick-witted and smart brains. Being tied to the sun restricts their behavior: Instead of being free and active, searching and understanding the world, they spend much time avoiding getting too hot or too cold. Some lizards, for example, bask in the morning sun to warm up their blood, cold from the previous night. Others restrict their foraging only to the early hours of darkness, in which the sun's heat still lingers.[4] Some nocturnal reptiles emerge after sunset out of cool daytime burrows and bury themselves in sun-baked sand. Others, such as some snakes, will wait for the right night or day air temperature. Australian nocturnal snakes, for example, go out at night only when the temperature is within certain narrow limits.

Accordingly, most reptiles lack an active life—at least compared to mammals that scamper and scurry through the cold of the night. Because of endothermy, a mammal is free to do what it wants, when it wants, and where it wants. Mammals do not stop being curious merely because the night has cooled or the first frosts have appeared. We need only to look at ourselves to appreciate the value of this freedom. While we are rarely our best at night, if we arrange our sleep, we can be active at any hour. Indeed, one in five of us, from night security personnel to astronomers, does shift work that involves resetting our biological clocks, sleeping in the day, and shifting our consciousness to the night. There is another advantage for a constant body temperature—it increases the information processing potential of the brain. Neural circuits depend on many temperature-sensitive processes that need to be fine-tuned to work together, and that fine-tuning depends on their working at a constant temperature. Get that temperature wrong—as in fever, heat exhaustion, or hypothermia—and our neurons cease to work efficiently, and we become demented.

Only mammals and birds dream. Yet dreams offer no clear advantage for survival. If anything, dreaming would even seem to be an evolutionary handicap. A toothless lion has an obvious survival problem compared to one with a mouth full of prey-ripping canines, slicing incisors, and grinding molars. But why should a lion that never dreams be any worse off than one that does? It is not that we remember our dreams; we usually do not unless we are awakened. Nor are dreams needed for sleep, for much of sleep happens without them. Yet they are clearly vital, since if we are deprived of dreaming sleep, our ability to learn new things stops; we become less able to think; we start making errors and having accidents. Our immune system weakens, our language becomes less inventive, and our voices become a monotone—we become zombielike.[5] And the dream sleep we miss, our brain makes sure we make up later.[6] Why are dreams so important?

Our need to dream turns out to be the price we have to pay for breaking with the reptilian script of sun-determined rest and activity.[7] When we look in the mirror, we do not see a reptile; but of course, a bit of the reptile remains, as one might say, under the hood, in our brainstem. Evolution works by tinkering with what already works, with the result that new parts get built onto older ones even if the newer ones work in entirely novel ways. And that can create problems.

Mammalian Revolution

For hundreds of millions of years, natural selection had tinkered with neural circuits in cold-blooded bodies so they could be paced directly or indirectly by the warmth of the sun. Then the evolutionary opportunity arose for active night animals with a new kind of mental awakeness. That ideally would require dumping the old parts of the brain tied to cold-bloodedness and designing the brain afresh. But evolution could not do this, at least not in full. Brains—particularly areas in the brainstem—even if they had their own internal tempo—had been wired in all sorts of ways to the external rhythm of the world. Endocrine, immune, growth, stress-response, and other essential maintenance systems that were firmly in place controlled many vital functions already integrated with established patterns of being asleep and being awake.

A strong need existed, however, to get rid of reptilian awareness without interfering with all the systems with which it had been integrated. The problem was that the reptilian alertness in the brainstem was not smart enough for the new way of life. Reptiles lack the ability to associate across their senses. A snake uses its eyes to strike its prey, and then touch and smell in sequence to open its mouth and swallow. Reptiles do not have the mammalian ability to associate what they see with what they hear or smell. The warm-blooded animals that evolved from reptiles to live in the dark needed something better. They needed a mind sharp enough for a split-second, on-the-ball way of mental life. After all, their constant need for food required them to be relentlessly learning new things—what food was safe, where predators lay, how best to open a beetle or a snail. Fortunately, the reptilian brain had some smarter circuits in its roof, which mammals expanded to create the cerebral cortex. They could not afford to be too handicapped by the old reptilian brain and its limited ability to think and learn—however, neither could the old brain be dumped. Yet mammals had to find a way to switch wakefulness to this new brain part.

The answer was to shift the workings of the reptilian brain into a period in which their processes did not interfere with those of the new mammalian consciousness—sleep. Mammalian sleep, as a consequence, is a compromise that mixes the ancestral remains of reptilian sleep with reptilian wakefulness.

Dragons of Dreamland

Curiously, given that it takes up nearly a third of every human existence, the odd, imperfect nature of sleep was overlooked until 1953 and its discovery by Eugene Aserinsky and Nathaniel Kleitman.[8] Scientists studying brain activity until then had assumed that sleep was a bland monotonous state, so no one had bothered to examine its brain wave activity for more than a few minutes. But this pair, a PhD student and his supervisor, did. What they found shocked them, and gave us new words: rapid eye movement (REM) sleep and slow-wave sleep. Aserinsky and Kleitman discovered that sleep went through a roughly 90-minute cycle of alternating shifts of neural gears. First people entered into a sleep full of slow waves and then into one in which their brain activity resembled "awakeness." During this "awakelike" sleep, people's eyes under their lids were busy in movement. Aserinsky and Kleitman awoke sleepers during this stage of sleep and asked them what was happening. They said they were dreaming.

Later it was found that dreams also occur during slow-wave sleep (science is never simple[9]), but it is in REM sleep that we hallucinate the most.

Dreams are puzzling. Each of our brains hallucinates wildly in them for over an hour each night, and usually recalls not a trace. If dreams were boring, that would be not surprising. But the events we experience in dreams, if they had happened during the day, would be unforgettable. However, unless we are awakened, we rarely remember them. Why is sleep studded with such intense but instantly forgotten hallucinations? Reptiles, after all, have a much simpler kind of rest: They do not dream. It is difficult to imagine what we might gain from experiencing several periods every night of consciousness akin to LSD trips.

There are other mysteries. During REM but not slow-wave sleep, the regularity and depth of breathing and the heartbeat become variable.[10] Most skeletal muscles in the body undergo atonia, a kind of muscle paralysis. A clinical condition called "rapid eye movement sleep behavior disorder," in which atonia fails to kick in, shows us the importance of this paralysis: People with the condition attack their bed partners as they live out their dreams.[11] (We should point out that this condition has nothing to do with the night terrors and sleepwalking that affect one in seven of us at some time; these happen during slow-wave sleep, when the body is not paralyzed.[12]) To add to the puzzlement, not only are we paralyzed during REM sleep, but we also lose control of our body temperature—we revert to poikilothermy. REM dominates the sleep of the newborn, and fetuses spend nearly 17 hours each day in REM sleep. If we need to do a lot of learning, we have more REM.[13] Also, if REM sleep is disturbed, it affects the memory of the skills we tried to acquire the previous day.[14] The length of the REM/slow-wave sleep cycle depends upon the size of the animal, with larger animals having longer cycles of alternation. REM sleep is most intense toward the end of a night's sleep.

Science is like solving a jigsaw puzzle where most of the pieces are still in the box, lost under the carpet or eaten by the family pet. Not all these phenomena are understood, but they fit together to suggest a rather odd idea—that REM sleep is built on reptilian sleep. That idea might seem a little curious, but let us first look at slow-wave sleep and an idea that might be thought even more bizarre—that slow-wave sleep (which tends to lack dreams) started out as reptilian consciousness.[15]

Alert reptiles and alert mammals have different patterns of brain activity. This is not surprising, since one comes from the brainstem, the other from the cerebral cortex and a bit below and closely working with, the thalamus. What is unexpected is that the pattern of brain activity in alert reptiles looks like that in mammalian slow-wave sleep; both states are, as might be guessed, full of slow waves.[16] If that were all, we might suspect it was just coincidence, but there's more. When humans process new information (such as a noise in the night), spikes occasionally appear in their slow-wave sleep. These are called K-complex waves.[17] Reptiles have such spikes when they process information during their wakeful state.

Moreover, during "awakeness" reptiles regulate their warmth by behavior, but during their sleep, such regulation is turned off (with that of the heart and breathing). Similarly, our internal temperature regulation is working while we are in slow-wave sleep (as in wakefulness) but turned off during REM. Such nocturnal poikilothermy ("cold-bloodedness") can be dangerous. It is to prevent this poikilothermic state from going on too long that small animals split up their REM sleep into shorter periods than large animals do. Thermal inertia cannot protect their bodies from changing temperature rapidly, so small animals must get out of REM as quickly as they can. That turning off of regulation during sleep might also explain why our breathing and heartbeat become more variable during REM sleep.

Further, reptiles are active during their period of alertness and temporarily torpid—they will not stir from their slumber—while in sleep. Our paralysis in REM sleep parallels this reptilian torpor. In contrast, we can move during slow-wave sleep—and many do when sleepwalking. Such activity is normally blocked in our brainstem, but not always successfully. As Rubén Rial, a Spanish neurologist, notes, "Sleep-walking could be a failure in the switch blocking the intrusion of reptilian activity during sleep. In fact, the absent-minded behavior observed during somnambulism could perhaps receive the qualification 'reptilian,' and particularities of the visually driven behavior of sleepwalkers is reminiscent of the 'blind vision' which seems to be solely due to lower visual centers after lesions in the visual cortex."[18]

The burying of the reptile brain and its sleeping and awake states in our sleep does not explain why we dream, however. Surely mammals could have shunted the old reptilian sleep into their new one without having to hallucinate. But the buried reptilian sleep has a function, at least during some periods, that favors brain activity even when asleep. Neurons, when they are actively engaged in living, are not in the best circumstances to permanently lay down the patterns of information they are acquiring. They need a maintenance break free of competing activities that might interfere with such stabilization. Memories and new skills need to be consolidated when nothing else will disrupt that consolidation process. This is why when we learn something, we know it better after we've slept on it.[19] The consolidation of memory and skills started with reptiles and their primitive learning abilities. Mammals, however, with their new brains, greatly expanded such abilities, including their consolidation in the "off-line" state of sleep. Legacy determined which part of sleep was going to be most important for such consolidation—the part that had originally evolved to underlie reptilian memory and skill consolidation. At least this would be the case for skills using brain areas still tied with circuits to old parts of the reptilian brain (such as the hypothalamus for emotions and the basal ganglia for movement). New parts, free of such connections, might use the other part of sleep. That leads to a prediction: The laying down of skills and memories that depend upon the cerebral cortex's working with the old parts of the brain, such as those for emotion and motor control, should be disrupted by interruption of REM sleep. In contrast, cerebral cortex skills that do not have such reptilian links, such as the ability to associate words, should be linked to the part of sleep unconnected with reptilian sleep, and thus avoid competition with the consolidation processes active during REM. This turns out to be the case—at least according to some recent research.[20] When REM sleep is disrupted, emotional and procedural memories are lost, but if slow-wave sleep is interrupted, the learning of associated pairs of words suffers.[21] This need to consolidate also would appear to explain why the sleep of infants contains so much REM time: They are busy consolidating the information-processing capabilities of their growing brains.

But why experience dreams only to instantly forget them? First, the processes needed to consolidate memories require the neural circuits of the brain to shift into a specialized mode. In this consolidation mode, neural circuits store what was experienced the previous day, but not the activations taking place during consolidation.[22] That does not itself explain the consciousness we have during dreams. That comes at least in part from fragments of actual previous events that the brain relives in sleep and their associations with other previously stored ones. Using electrodes implanted in experimental animals, neuroscientists have found that during sleep, neurons mimic the patterns of activation they showed during important learning events the previous day.[23] This area of research, still in its early days, concerns slow-wave sleep and does not involve human subjects. However, reliving of earlier events is known to happen during dreams as well. About 50 percent of dreams contain entities and circumstances that relate to what occurred in the preceding day.[24] People playing the computer game Tetris report that experiences of the game intrude into their dreams as they go off to sleep.[25] Indeed, this is a personal experience we all have when we recall our dreams—they are bizarre because they collage familiar past events. Such reliving of the past could be linked to learning and memory consolidation.

Second language learning indeed strongly suggests this: Not only do English speakers learning French start to incorporate French into their dreams, but how early in the course they do this, and the intensity with which they do it, is linked to how well they pick up the new language.[26]

This may be why dreams are so incoherent. If only isolated aspects of the former events are being relived, there is no need for coherence. They might be picked out at random and processed in a jumble. Further, if we look at dreaming brain activity with functional imaging, it differs radically from that of daytime consciousness. During your waking hours your frontal areas are active and coordinate activity in the back parts of the brain involved in sensation. But during sleep, the front part of the brain is deactivated,[27] and, in terms of patterns of brain waves, the front-back coherence is lost, presumably due to the breakdown of the logic such a link usually imposes on the rest of the brain.[28] The brain is busy but does not care that what it is doing has no order.

This still gives us only the beginnings of an explanation. Another possible component lies in those rapid eye movements. According to David Maurice, of Columbia University, there may be a physiological reason for them.[29] We mention his theory as it is interesting, even though we have doubts about it. In this theory, the eye's cornea lacks blood vessels (it has to be transparent) and so is in danger of "suffocating" when covered with the eyelid (the web of blood vessels under the latter's unexposed side helps to minimize this). Its main source of oxygen at such times is the aqueous humor behind the cornea, but in order for oxygen to be carried across this fluid, it needs to be occasionally shaken, as happens when the eyes move. This seems to be the function of the rapid eye movements of REM—at least according to David Maurice. He argues that if we did not roll our eyes during REM sleep, we would risk going blind. Cold-blooded animals do not need to move their eyes when they sleep, since the need for oxygen of the cornea, like that of any tissue, is less at low temperatures. During slow-wave sleep the body moves around, and this helps churn the aqueous humor. But during periods of paralysis in REM sleep, those eye movements need to be triggered in some way. The cerebral cortex, shut down in sleep, cannot by itself do this. So, Maurice argues, eye movement is generated instead by *ponto-geniculo-occipital potentials* that originate in the brainstem and then, after entering the cerebral cortex, activate the motor area that in turn sets off eye movements.[30]

For whatever reason they exist, such potentials provide the cerebral cortex with an important source of stimulation during sleep. This stimulation, and the activation they trigger in the areas that control eye movement, creates mental experience—at least the fleeting one of dreaming. This is known because the more your eyes move in REM sleep (and so the more intense your ponto-geniculo-occipital potentials are), the richer is the imagery of your dreams.[31] This correlation between eye movements and dreams also happens between the occasional eye movements and the imagery of dreams in slow-wave sleep.[32] This suggests that our consciousness in dreams is linked not to REM sleep itself but to the mental activation in the brain that is linked to eye movements.[33] Further, ponto-geniculo-occipital potentials also trigger in the brain "gamma synchronization,"[34] something that, as explained in Chapter 5, is thought to underlie consciousness. Gamma synchronization, as we show in this chapter, is in part responsible for the binding of sensations that underlies experience. This raises the question of what the gamma synchronization triggered by ponto-geniculo-occipital potentials might be binding, and so turning into consciousness. First, we should note that sleep eye movements, like any eye movements, are acts of looking; the brain moves the eyes in order to attend to something. At the same time that the eyes are moving and causing gamma synchronization, the cerebral cortex is being activated by memory and skill consolidations. Gamma synchronization therefore could be binding whatever content is being consolidated at that moment in the cerebral cortex and this act of attention. In effect, what would be bound together would be a bizarre world in which the haphazard reliving of previous events was glued together around the random eye movements of an onlooker.

The Gift

Evolution is blind, but it had a kind of foresight, since it gave mammals a gift that ultimately led to you—the ability to survive cosmic disaster. Sixty-five million years ago an asteroid or a comet smashed into the Yucatán area off the Mexico coast. Not only did its impact kick vast amounts of dust into the upper atmosphere, but it started worldwide fires that belched out smoke clouds of Armageddon proportions. As a result, the Earth was lightless for several years, vegetation died, and there were mass extinctions, including the vanishing of all the dinosaurs. Some mammals, of course, survived; otherwise we wouldn't be here to tell the tale. There had been pockets where things were not so bad. While living at night would not protect mammals in the Yucatán peninsula, it gave at least some of them elsewhere a better chance of finding a way to survive following the maelstrom. At such a dark moment in the world's history, it paid to be preadapted to living in the night, in the cold, and to be possessed of an intelligent brain. Also, an animal that lives in the night has to find well-hidden places to sleep in the day. It cannot just lie down and rest. Thus, our scurrying predecessors just before the disaster hunted out nooks and crevices into which to hide to escape daytime predators and awoke to find themselves safe and sheltered from global apocalypse. And ironically, they were provided with the biggest dinner party ever—the bodies of the last dragons.

We can imagine the gruesome scene. Any herbivorous dinosaurs not killed by shock waves and fire would have slowly starved in the cold dark. Carnivorous ones would have at first dined on their flesh, but they in turn would have died when the meat of the herbivores went rancid. All the while, worms, insects, and other small creatures would have been having their feasts. And on these invertebrates, mammals would have enjoyed a survivor's banquet. Here was the inaugural party, the kickoff event for millions of years of further evolution of mammals. In many places, the recycling of flesh would not have lasted until the light and greenery returned, and with it new life. But in some areas it did. That was enough to let mammals emerge into a dinosaur-free world. The brain and body revolution that had allowed our ancestors to steal dinosaur eggs now allowed them to take control over their future.

Mammals thus entered evolution's center stage. Nighttime life had given them the ability to regulate their own warmth and be active when nearly every other creature was asleep; more important, it had prepared them in other ways. It had required mothers to feel emotions for their young, their children. Mammals had evolved into devoted parents, not only letting their offspring nestle and huddle into their fur for warmth but also suckling them with milk from their mammary glands (the most primitive mammal, the echidna, does not have teats; its young suck on hairs, which suggests that nursing has its origins in that snuggling up to a mother's warmth). Reptiles had no such care. These were beginnings from which smartness could evolve into more smartness. The emotional attachment of mothers to children enabled the rise of other attachments, such as friends to friends. And such emotions, as we shall see, did much to drive the expansion of the brain. Rising up from dragons, mammals had created a brain that—after tens of millions more years of evolution—could become yours.

Carl Sagan made much of the *triune model*, in which there was a reptilian brain, an old mammalian "limbic" brain, and a new mammalian "cortical" brain.[35] Here we see the start of the cortex's story: The reptile brain gave rise to the old mammalian parts of your limbic system—parts like the hippocampus and amygdala that gave nocturnal mammals their advantage over reptiles. Later, after the Yucatán event, the visually oriented cerebral cortex greatly expanded as mammals, free from the competition of dinosaurs, became active during the day. Sight forces the brain to intensively process the associations within its senses and their links to motor control. It invites us to plan more; in the night obstacles could not be seen ahead, but vision allows just that. A tree full of ripe and unripe fruit requires a strategy as to how to most efficiently pluck the sweet and not the sour. The dark of night had protected mammals from predators; now they had to look ahead and anticipate them. And this was only the beginning, since they soon found they needed to plan how to gain advantage not only over the physical world but also over the social one they made with one another. The cerebral cortex as a result greatly increased in size, as animals were forced to become smarter and smarter.

But as Carl Sagan understood, there is a limitation. New parts of the brain do not just get added on; they also change—within limits—the function of earlier ones. The evolution of intelligence lies in how the new changes the old. And it's not just physical additions to the brain that create change. We will argue that the history of our intelligence—and indeed its future—lies beyond genes. Hinted at but left only partially explored in *The Dragons of Eden*, our "extrasomatic inheritance" enables us to expand the capacities of our brains by what we learn from each other. We call this extrasomatic inheritance *mindware*. Centering around the abilities of language, it includes thinking, speaking, and writing as well as using technology. And if the move from dragon to mammal brains was not a one-way influence but involved changes in both directions, so mindware is not a simple addition from outside the body. It is an unexpectedly powerful tool that reshapes all that has come before, reintegrating it toward an exciting future—*braintech*.

Dragons have vanished. But not without a trace. They still linger in our dreams. Ancestral beings do not totally disappear. They leave trails that can help us reconstruct who they were and how they became us.

The Hunter-Gatherer in the Mirror

Galaxies formed, life evolved. Cells in the ocean came together, laying down calcium as bone in fishes that wobbled on fins that became the stubby legs of amphibians. These evolved into reptiles that became smarter mammals hiding under cover of the night, later readapting and evolving some big species of their own in a world without dinosaurs. One relatively small line of mammals became big-eyed scurriers across the forest floor that clawed their way up tree bark and evolved into early primates. Some became monkeys with tails, and some of these lost their tails and became apes. The apes were at home in the forest and could swing from the trees, but some became adept at moving on two feet, freeing up their hands to extend their bodies with loose branches and rocks. They used these to gang up on larger mammalian beasts and perhaps on others of their own kind. They became talkers and killers, berry gatherers who made love and ate meat. Yet they might have escaped the notice of an intelligent alien looking at our section of the galaxy. They had not yet expanded across the planet. Then, in the evolutionary story of life, we arrived.

But not yet you. Look in the mirror again. The person you see is a problem for modern science. Simply put, none of us resembles the first human hunter-gatherers. In a mere blink of evolutionary time—120,000 years—a two-legged hunter-gatherer ape adept at life in the African savanna turned into you, a smartly dressed twenty-first-century book reader. The transformation was amazingly swift. And science has little idea how it took place. It is the big remaining unanswered question about that person you see in the mirror.

The first people biologically like ourselves appeared between 100,000 and 200,000 years ago—the estimates center around 120,000 BC.[36] These first members of our species lived what to us would have been odd lives. They had no mirrors, no books such as this, not even some of the very inventions we associate with "primitiveness." Remember what it was like as a kid telling stories and cooking hot dogs or toasting marshmallows around a campfire, a simple stone hearth aided by air-intake ditches? It seemed the height, if one can call it that, of low tech. And yet for at least half the time that we humans have been on this Earth, such fires were beyond our technological reach; evidence of their use appears only after 60,000 BC.[37]

Or perhaps you picture the earliest humans with a bow and arrow? But no: People have been shooting arrows for only a third of our existence—from around 40,000 BC.[38] What of simple wick lamps? Again, no: Such lamps arose only around 40,000 years ago, many millennia after we did.[39] Might these people then have had at least clay pots and cups? Humans have been firing the simplest unglazed crockery for less than 10 percent of our history. It goes without saying that our earliest ancestors could not read, drive cars, or write computer programs. Instead, they suffered the bites of fleas, the rumbling inside of intestinal worms, the death of nearly every child, recurring fatal hunger and thirst every few years with famines and droughts, and a blackness they could not brighten with a candle, which had not been invented, let alone an electric lightbulb. They knew not health care, maps, media celebrities, flight, roads, supermarkets, schools, jobs, pensions, TV, or cash. Compare yourself with people from the dawn of our species, and they seem primitive almost beyond belief.

And yet one fact stands out in striking contrast to all these marked differences: Their genes were, for all practical purposes, the same as ours. Humans of 120,000 years ago and those of today belong to the same biological species, *Homo sapiens sapiens*—the name was given to us by the Swedish biologist Linnaeus, and the Latin means "the wise, wise humans." So 120,000 years ago the human body—except for a few details such as skin pigmentation—was already biologically "finished," including the human brain. Bakers, hunter-gatherers, candle or market makers, or any of the varieties of human life in between possess the self-same anatomy. From conception, humans share the same potential, at least in theory. This means that your genes, which happen to take the form of a twenty-first century citizen, could just as easily have belonged to one of those first hunter-gatherers. Take a time machine with 100,000 BC on the dial and transport a modern baby into the hands of one of the first human hunter-gatherer mothers, and she would raise it to be like any other hunter-gatherer. By the same token, adopt one of their infants into a modern family, and he or she would grow up like any other present-day kid, with dreams of being Michael Jackson, Michael Jordan, or a future president. Our genes prepare us equally well for both modern and ancient life.

Why should the same genome that codes for a naked ape digging for roots in the African grassland also code for us? No law of nature required that hunter-gatherers come endowed with the extra potential to become us. Evolution 120,000 years ago did not anticipate the future—certainly not our high-tech lives amid cell phones, communications satellites, and computers at the beginning of the third millennium. Natural selection produced hunter-gatherers that were not too different technologically from some great apes. Yet in giving our ancestors the wherewithal to bang bones and carve rock, genes for brains were selected that 120,000 years later were to become Shakespeare, Mozart, Curie, Einstein, and you.

This endowment, this changeling nature, this plasticity, makes us unique among animals. Other species, for the most part, remain as they were when they evolved; we instead broke that older pattern of nature and went on to discover new forms of life, thought, and enjoyment. No other animal species before us has traveled so much evolutionary territory in so little time. And yet the ticket for this immense trip is not genetic; genetically, when you look in the mirror what you see could be a hunter-gath-

erer who was living at the beginning of our species' existence. Another kind
of ticket existed that was to purchase the journey on this great evolutionary odyssey. What was it? In our view, this question is one of the greatest
scientific mysteries. It can perhaps be answered simply, even glibly, with
notions such as "culture" or "ideas." But until now science has not been able
to provide any sort of detailed answer. Here we shall explore what happened to the hunter-gatherer in the mirror. How did such people get here?
What had evolved earlier in their brains to help them? And how did they
get here so damn fast?

3

Neurons Unlimited

A revolution has hit brain science, and it goes by the name of *neural plasticity*.[1] To understand what this entails, we need to know two "geographical" facts about the brain.

First, if we could peer through our skulls to look at the two cerebral hemispheres, we would find them covered with areas that can be described as *neural maps*—lots of them. These maps correspond in detail to our bodies and to our sense organs, such as the eye's retina and the inner ear's cochlea. On each hemisphere they are numerous: More than 10 maps exist for the retina (plus 22 more helping to process vision); 8 correspond to our inner ears.[2] Seven other *homunculi* (literally, "little men") correspond to the touch and feel of our bodies, and others exercise control by mapping our limbs and muscles and their movements. At the beginning of this century, the anatomist Korbinian Brodmann counted and marked nearly 50 such areas on each hemisphere. Later in Germany, Oskar and Cécile Vogt, a husband-and-wife team, using detailed anatomical differences, subdivided each hemisphere into over 200.[3] Per Roland at the Karolinska Institute, Stockholm, has also observed some 150–200 smaller "connectivity" areas in each hemisphere.[4]

These maps can be detailed. On one part of the brain an electrode can detect a tiny touch made on a single finger. Moved slightly, the electrode will pick up a similar touch made on another finger. The maps exist on the surface 2.8 mm (on average) of our cerebral cortex, tightly folded into itself to make a walnutlike shape that has roughly the area of four *Time* covers.

The maps, of course, are only the icing, as it were, on the cerebral cake; below them lies a nest of complex wiring, of subcortical nuclei, axons, and nerve fibers that link our brain with our eyes, skin, limbs, and organs.

While such subcortical parts of our brains make up only a fifth of it, they are just as important as the above-mentioned cerebral cortex, that takes up the rest.

"It seems likely each brain is as individual to its possessor as his face."[5] This comment by one anatomist hints at a second important geographical fact: Our brains differ as much as our bodies do. Indeed, they differ more. One part of the brain, the *anterior commissure* (the smaller of the two links between the two cerebral hemispheres) varies sevenfold in area between one person and the next. Another part, the *massa intermedia* (whose function is not known), is missing entirely in one of every four people.[6] The primary visual cortex can vary threefold in area.[7] Our amygdala (responsible for our fears and loves) can vary twofold in volume—as can our hippocampus (involved in memory).[8] Most surprisingly, in people with normal intelligence the cerebral cortex varies nearly twofold in volume.[9] Such differences are found throughout the brain. How far these differences affect who we are is not known, though they no doubt are of considerable consequence in shaping us and our abilities. Even though we are in the dark about their effects, it is important to note that they exist. The brain is often discussed in generalities, but we should remember that each mind, and individual brain, is unique.

Neural Plasticity

These two geographical facts—the existence of neural maps and their variability person to person—raise some questions. Are brain maps as variable as the brain? Do maps alter with experience? Are they a kind of neural "software" with which we may be able to "write" ourselves, or are they more like hardware, physically fixed in the brain? In short, is the brain's geography variable?

Before the mid-1980s, the answer was that maps were hardware. Their outlines appeared first in the womb, and then for a short while after birth their connections went through a fine-tuning process. After this critical period, they were hard-wired. David H. Hubel and Torsten Wiesel in the 1960s and 1970s had shown that a critical time existed for creating such maps in the primary visual cortex. These maps combine input from both eyes in alternating neural units called *columns*. However, if input from one eye is blocked—for instance, by childhood cataracts—then the columns for receiving input from that eye do not form. This happens only at a critical period; after the visual cortex has matured, cataracts in adults have no such effect. What was true of the primary visual cortex was thought to be true of the whole brain. For discovering the critical period in brain development, David H. Hubel and Torsten Wiesel won the 1981 Nobel Prize for Medicine.

But the formation of columns in the visual cortex for joining the inputs from the two eyes is unusual.[10] It is now known that most maps on the cortex never become fixed but retain lifelong plasticity. If you lose a finger, its map on the brain shrinks and the maps controlling the other fingers expand to take its place. Use one finger more than the others, and its map increases at the expense of those not being used as much. Use it or lose it: There appears to be a continuous feedback between the neural map and the usefulness of the parts it represents. The area of the brain devoted to the right index (reading) fingertip of Braille readers, for instance, has been found to be larger than that for their nonreading fingertip, and larger than that for the right fingertips of non–Braille readers.[11]

Large shifts happen when the brain adapts to limb amputations; the maps of parts that survive invade areas no longer receiving inputs. Thus, part of the brain that would have reacted to, say, a pat on the arm now reacts to a pat on the face. Touch the face of amputees and they may also "feel" it on the lost arm and fingers. Douse their face with warm water, and they may feel water dripping on their phantom arm and fingers; touch their cheek, and they may feel touched on their phantom thumb; touch their chin, and they may feel it on their missing pinkie; and so on. It sounds strange, but you do not have to take their word for it. Magnetic resonance imaging (MRI) scanners see facial sensations lighting up the area that used to map hands.[12] The hand maps have been invaded by those for the face, an encroachment that may cover up to 3.5 cm! Cortical maps also move in response to the slow destruction of the brain by tumors. In such cases, motor maps may shift by up to 4.3 cm, venturing into the premotor area and even the parietal cortex.[13] After Hubel and Wiesel's discovery of the critical period, no one had expected the human brain to show such dramatic rewiring.

Living Cartography

Neural plasticity has become hot science, and it looks to become even hotter. For one thing, it is the key to progress in the treatment of many medical problems, including injuries and birth defects.

Those with injured spines were once thought condemned to remain in wheelchairs. But now, we can manipulate neural plasticity in the spine below their injury using body-lifting hoists and treadmills to create walking reflexes. In a German study, 14 of 18 wheelchair users given this treatment became "independent walkers," but only 1 person in 14 did so when given conventional physiotherapy.[14] Those with damaged nerves in their arms can have their arms reinnervated with the nerves that normally go to chest muscles; the motor cortex shifts from the control of breathing to that of arm movement.[15] A 65-year-old art director suffered the misfortune of having most of his external penis removed due to cancer. The erotic areas in his brain, however, transferred from his glans to his testicles, his anus, and the area surrounding his penis and thus enabled him, with the aid of his wife, to continue having orgasms through massage during lovemaking.[16] He even started having double orgasms. Similar transfer of erotic areas occurs in transsexual men following their sex-change surgery.[17]

Then there is *syndactyly*, in which people have a *club hand* of webbed, shortened fingers. Not only are the fingers of their hands fused, but the cortical maps of their individual fingers also form a club hand. The fingers can be surgically divided to make a more useful hand. Surgeons did this at the Institute of Reconstructive Plastic Surgery in New York to a 32-year-old man with the initials O. G., but scientists wondered if the hand in his head would change. They tested for this by carefully touching O. G.'s fingers before and after surgery while using MRI brain scans. Before the surgery, the fingers mapped onto his brain were fused close together; afterward, the maps of his individual fingers did indeed separate and take the layout corresponding to a normal hand.[18]

Neural plasticity can heal, but it can also cause problems. It has radically changed our understanding of pain. First, neural reorganization has been found to have a link with the pain found in those with phantom limbs.[19] Such links have also been found in those suffering chronic back pain.[20] Once people with chronic back pain were told to rest, but now the best advice is to keep active.[21] Rest can encourage reorganization through neural plasticity, which leads to increased pain; activity suppresses such reorganization.

Repetitive strain injury is another case where neural plasticity causes suffering. Nancy Byl, Michael Merzenich, and William Jenkins at the Keck Center for Integrative Neuroscience at the University of California put monkeys through a daily routine of repetitive actions, and like humans doing repetitive actions, they lost control over their muscles. Neural plasticity and the maps in the brain were to blame, as the representation of hand and fingers in their brains dissolved and blurred when deprived of differentiated sensory feedback.[22] Auditory maps can reorganize and produce the distressing phantom sounds known as *tinnitus*.[23] Neural plasticity can thus not only heal but also cause problems; understand how, and we are closer to prevention and better treatments.

Brain scientists found another vivid demonstration of the plasticity of the brain when they experimented on developing ferrets, diverting the inputs that would normally go to the visual cortex to the auditory cortex instead. The result was that where the brain would have heard, it now saw—the auditory cortex, with different input, grows up into "visual cortex."[24] Nature has performed the opposite experiment on the blind mole rat. This African rodent, which has been described as looking like a saber-toothed sausage, lives underground in tunnels in the desert. Like sighted animals, the blind mole rat has a fair-sized visual cortex, but it does not go to waste. Instead, the mole rat uses it for hearing; evolution has redirected auditory input into it.[25]

Our brains are set in gray matter, not in stone; their parts are predisposed but not absolutely preset for particular functions. They are built of general-purpose bioprocessors that, after being formed, become specialized in response to their inputs and outputs[26]—not of preevolved, rigidly specialized processors. There may be a protomap specifying which is to happen, yet this is easily rubbed out. Thus, neural abilities may be fated, but they are not determined.

The Unhandicapped Brain

The modern evidence of brain cells' switching between sight and sound suggests that the visual cortex of individuals blind at birth should not wither away[27] but take on tasks other than sight. Indeed, strong new evidence suggests that it does. The visual cortex is active in people born blind—more active, in fact, than in blindfolded sighted people.[28] When sighted people locate the positions of sounds, electrical activity is found roughly in the area of their ears. This is not so in those born blind. Scientists now find that in them the location of sound lights up the visual cortex. The scientists doing this work have even suggested that the area activated is as far back as the primary visual cortex.[29] This suggests that the maps on their brains have been rearranged, with the part of the brain that would normally have been seen turning into a hearing area. And not just hearing: There is evidence that the primary visual cortex also helps people blind from birth to visualize textures felt on their fingertips and even to read Braille.[30] Indeed, strokes that destroy a blind person's visual cortex destroy the ability to read Braille.[31] Thus, the part of our brain that usually sees reworks itself to do other things in the blind. Physical handicap does not mean brain handicap.

The blind have better hearing abilities than the sighted. Consider blind cats. Meet one blind from birth, and you will never guess it. With an improved sense of touch and hearing, they so effectively make up for lost sight that only careful tests can show that, in fact, they are blind. This extra sensitivity comes in part from using the areas in the brain that would have been devoted to vision.[32] The visual areas retune their binocular sense of depth to aid spatial hearing. Do similar changes in humans give them better hearing? It is possible. The blind can hear speech against background noise better than those with sight (in some circumstances twice as well).[33] They have faster reaction times to sounds.[34] This is not because they have better physical hearing; they are no better at hearing pure sounds. Instead, it seems to concern a better ability to use what they hear.

Another example is the ability of blind humans to hear their surroundings. Blind people feel lost in rooms that fail to return the reverberations that let them sense their position. The white cane is not only a protective probe; its tapping helps them locate where they are going. It is suspected that the sound of their own footsteps even lets them "hear" their surroundings. Such self-made sounds can powerfully aid perception. Using the sound they make, blind people can "hear" a 4-inch change of position when a 1-foot disk 2 feet away is moved. This is about as good as, or possibly even better than, the distance accuracy of a sighted person using one eye.[35] They can locate the source of a sound better than sighted people using only one ear.[36] Blind children can ride bicycles aided only by clicking sounds made by their mouths—true human echolocation.[37] Consider Burns Taylor, a Texan who lost his eyesight at age 3. Locals call him "the cricket." As he rides around on his bicycle, he makes clicking sounds with his mouth. He uses widely spaced clicks to hear, through echo delay, the location of buildings. He uses softer, more rapid clicks to hear nearer and more troublesome objects, such as overhanging obstructions. These rapid clicks work by the Doppler effect: Taylor can hear the slight change caused by movement in the frequency of the returning sound. He says it is easy. He may have lost his eyes, but his brain was born with the ability to adapt.

People who are born deaf may experience the same rearrangement, but research on them is less advanced. Their auditory cortex, like that for vision in the blind, does not wither away.[38] It assumes new functions. As noted above, work on ferrets suggests the auditory cortex has the ability to "see" if input from the eyes is routed to it in the womb. Indeed, evidence exists that in the brains of cats deafened at birth, the area that would have heard learns to see.[39] Deaf people who learn to talk with sign language "hear" these visual signs in the auditory cortex.[40] Deaf people see differently from the way hearing people do. They are better at focusing on things in the periphery of their vision. This improved ability of deaf people to see is centered on activities taking place in the visual cortex in the left hemispheres of their brains. In hearing people, by contrast, peripheral vision links to neuronal activity over the top of the brain in the parietal area and on the right side of the brain.[41]

The Secret Swiss Watchmaker

Having focused on the flexibility of the brain, we should not leave you with the impression that all the brain has such plastic properties. Far from it— much of the development of the brain's circuits is time-tabled to a plan. Neural plasticity arises against the backdrop of some very fixed processes. For instance, the nerve cells (or *neurons*) and their axon inputs that make up the neural circuits of the cortex are not plastic as to where they lead.

To understand this, let us look at how these circuits arise and what they are made of. In common language, the brain is made in part of gray matter, or gray cells. (Actually your gray matter is gray only after you are dead. When alive, blood pumping through it makes it pinkish tan.) Gray matter makes up only 60 percent of our cerebral hemispheres; the other 40 percent is white matter, axon fibers covered with myelin sheaths linking up neuronal circuits in different areas.

This gray (or pinkish tan) matter, seen under the microscope, looks like a thick canopy of interlocked neurons. Each neuron has a long extension called an *axon*, which connects to other neurons at communication junctions called *synapses*. Mainly, these synapses occur at the receiving cell's *dendrites*, which extend out of the main body of a neuron like weedy bushes. Axons can have amazing reach. Most axons in the cortex stay local, but many travel to the other hemisphere or sink below to link with neurons in subcortical nuclei or deep in the spine. And other axons exist that enter the cortex, traveling up in the opposite direction. It all makes for a very crammed biochip.

Neurons are not the only inhabitants of the brain; indeed, their cell bodies make up only around one-fortieth of its volume.[42] Brain cells without axons also exist; these are *glia*. It used to be thought that glia acted only to "support" neurons, but it is now known that they also aid in communication. They come in many kinds, and their numbers vary, so in some parts of the cortex they may be outnumbered by neurons 3 to 1 and in others they outnumber them 13 to 1.[43] On average, however, approximately nine glia exist for every neuron in the cortex. Each person's cerebral cortex holds something like 15 billion to 20 billion neurons, and thus 135 billion to 180 billion glia.[44] These neurons and glia are not alone: Enmeshed with them are tiny capillaries supplying a constantly changing flow of blood. It is estimated that not less than 350 miles (560 km) of them keep our biochip of neurons and glia alive.[45]

The cortex, made up in this way, is a many-layered sandwich—on average a thickness of just under 3 mm divided into six layers of "brain wiring" and brain cells. (Anatomists, one should add, find layers within these layers, and even sometimes layers within the layers within layers.) Textbooks used to give a rigid story that the various layers in the cortex specialized in sending and receiving inputs and outputs to and from different parts of the brain. For instance, according to the textbooks, the thalamus sends its axons mainly to layer IV. But this appears to be true only for primary sensory areas, such as the primary visual cortex. Elsewhere on the cortex there are no such easy generalizations.[46] This canopy of circuits on top of our brains is still being explored. But we know something about how it is made, and it is not the story you would expect.

For a start, the cortex's neurons are made outside it. From a cell-birth zone that is deep within the brain during the middle part of pregnancy (10 to 18 weeks), neurons migrate along "guidelines" to the cortex.[47] Along the way they do a kind of "dance." At one stage, for instance, the axons from the thalamus stop and wait for the arrival of the cells with which they will later link. Their movement is choreographed: The cells for the six layers arrive in reverse order—the cells for layer VI, at the bottom, first, and those at the top last. But it is not a dance that strongly shapes different areas in the brain's cortex, as it is not perfect. Many of the "dancers" leave their guidelines. A visual input axon or one of the cells to which it links might go badly off course, ending up in the auditory cortex rather than the visual one.[48] But they are not free, as wherever they end up, they must go to layer IV. Brain development is not totally flexible. Our brains represent a fine balance between order and plasticity.

There is a hidden wisdom here. The brain cannot make itself like the Swiss make watches using precision parts—neural cogs and sprockets. It needs neural plasticity to form properly. Like microchips, which often have manufacturing flaws, brain development is rarely perfect. It must occur in a way which minimizes the impact of such flaws. Neural plasticity helps it do just that.

Another reason for this plasticity is that the brain's genes start off blind as to what the brain will later do. Our bodies differ greatly.[49] On the lookout for variation, Charles Darwin was one of the first to catalog this vast diversity, which exists inside even seemingly similar or identical bodies. The differences are not just in appearance but in function. Anatomists find our stomachs can vary eightfold in size; some human retinas contain three times the number of cone cells in their foveae than others have.[50] Molecular genetics suggests that some women have four, not three, types of retinal photopigments (and as a result, richer visual experience).[51] The sinuses of children vary 20-fold in volume.[52] Blood temperature is 98.6°F in only 1 out of 12 of us; the rest of us are somewhere between 96°F and 100°F, the most common being 98.2°F.[53] That red line on the thermometer is a medical myth. Our hands differ: The muscles that straighten our index fingers may or may not attach to a second tendon for an adjacent finger (many a pianist's career has been ruined by this variation in human anatomy). The palmaris longis, one of the muscles that flex our wrist, does not exist in 1 of 8 people; 1 in 100 have two of them. (This and other differences between our hands are responsible for making our signatures so different and hard to copy.) The pectoral muscles, which let us hold and throw things, attach to different ribs in different people. Yours may attach to ribs 2, 3, 4, and 5 or just to 3, 4, and 5; to 2, 3, and 4; or to 3 and 4.[54] People can be born with more than five fingers. In one individual with seven fingers, his brain adapted and let him exploit his extra digits; indeed, he claimed, they "gave him some advantages in playing the piano."[55] (Nowadays, such supernumerary fingers are removed shortly after birth.) One anatomist quoted by Darwin noted that out of 36 people examined, not one was found "totally wanting in departures from the standard descriptions of the muscular system given in anatomical textbooks."[56] Medical school libraries contain two kinds of anatomy books: one showing the body as it ought to look, another a catalog of human variation detailing what medical students actually find.[57]

All these differences among our bodies affect what our brains process. The brain cannot do anything before the body forms. The brain is therefore shaped by its body—something that can be seen in embryo transplants. For instance, if you graft an eye from a large species of salamander into a small one, the number of axons in the optic nerve is increased and so are the number of cells in the visual part of the brain.[58]

The growing brain thus does not try to anticipate what sort of hands or eyes it will link up with. Rather, it lets its links with the body guide and determine its neural tasks. This flexibility shaped the "design philosophy" of the brain during evolution. Rather than laying down specific functions in advance, the brain evolved to build itself out of initially plastic brain units. Thus, it is developmentally open for its inputs and outputs to shape what it does.[59] Visual input from the eye directs the development of the visual cortex. The visual cortex, therefore, does not have to be predesigned for sight;

it is only optimized for it with more synapses.[60] Auditory input likewise makes the cortex that receives it hear. But both cortices, if given different inputs, could have done different things.

No law of nature says that this plasticity should stop when the brain has grown. Flexibility is, after all, useful later in life if the brain needs to adapt to, say, losing a limb or learning a new skill. While we have already discussed work on those born blind, research suggests that changes also occur in the brains of people who go blind later in life; in certain hearing tasks the distribution of their brain activity parallels that of those blind early in life.[61] Indeed, as the brain ages, neural plasticity allows mental skills in areas that cease to function so well together to be taken over by other areas, thus minimizing cognitive problems.[62] Neural plasticity thus never entirely goes away. Indeed, we should view the brain not as fixed but as constantly shifting what it does through neural plasticity.

A whole family of processes, both cortical and subcortical, keep the brain flexible. Some affect the brain early in life, others throughout life; some happen nearly instantaneously, others take many weeks or months. One early process is the elimination of excess axons and synapses following their initial overproduction. But even after this early pruning, cells in the brain are constantly created to form structures such as glia, capillaries,[63] and even possibly neurons.[64] A quick neural plasticity results from the uncovering of suppressed connections. A slower process involves neurons growing larger in size; developing a bigger, bushier spread of dendrites; and sprouting new axons.[65] In rats such changes can expand the thickness of the cortex by 3 to 6 percent.[66] In humans such size changes can occur through long use and practice: If they start young, professional keyboard players undergo an expansion in that part of the cerebral cortex connected to hand movements.[67] These processes give the human brain its greatest flexibility in the first 10 years of life,[68] but the brain retains much of its innate ability to change throughout life.

Genes and Brains

Some are skeptical about neural plasticity, believing that if it were significant, it would prevent our genes from shaping our brains. But clearly, this is not so. People with identical genes—identical twins, who are genetic clones—have near-identical brains and near-identical EEGs (electroencephalograms), despite their differences in experience. Their patterns of blood flow at rest and when doing IQ tests and their evoked potentials when spotting the differences between sounds are remarkably similar.[69]

Genes work with—not against—neural plasticity to shape the brain. Instead of programming the cerebral cortex, genes seem to bequeath some general directions. Their shaping is indirect. The relationship between genes and brains is rather like the way a government builds roads and railways and then lets towns grow around them. The genes give no direct or final "blueprint" that lays down a design, say, for the auditory cortex to compute sound rather than sight. What our genes do is lay down the basic links. Change those links, however, and we change what the brain does. In sum, the brain is not programmed to hear or to see, but to adapt. It is, you might say, programmed to get us programmed.[70]

Neural Plasticity and Injury

Another objection is that neural plasticity suggests that the brain should be self-repairing. But the tragic effects of brain injuries—car and gun accidents and strokes—suggest that the brain cannot heal itself. The problem here is not with the brain itself but with our perception of it; we do not vaunt its successes. People have small "silent" strokes without noticing them.[71] Brain tumors can grow as large as a plum—5 cm—before people notice them, simply because the brain compensates so well for the damage being done by the tumor. The average symptomless tumor grows at a rate of 2.4 mm a year.[72] Moreover, people tend to find they have them not because of deteriorating mental faculties but because the tumor triggers seizures. Surprisingly, the young brain is sufficiently plastic for the mental skills of one side to be taken over by the other if one of them atrophies (including even some aspects of motor control).[73]

While the brain can recover from minor strokes and other injuries made in small stages, the same is not true for a sudden major injury.[74] And most brain injuries, unfortunately, are sudden. Brain cells at the center of injury die through lack of oxygen, but in their death they release excitotoxins that kill nearby brain cells. So a small "brain bruise" can end up killing a large part of the brain.[75] Moreover, traumatic injuries hurt the brain badly. When the head is hit, the brain wobbles like jelly, tearing itself against bony protrusions at the bottom of the skull. Further, the various parts of the brain are linked to work together as a whole, so damage to one part of it can disrupt the working of far-off parts. Scientists using brain scanners thus see brain injuries putting parts of the brain out of action that have not directly been damaged.[76]

Nonetheless, even after sudden injury neural plasticity plays a key role in recovery. You can lose nearly two-thirds of the area of your brain devoted to mapping your fingers and retain sensitivity of touch equal to that of those who have not been injured![77] If none of the original area survives, the maps dealing with our fingers can move into areas that had formerly mapped only the face.[78]

The above examples of neural plasticity concern our primary sensory and motor cortices, which are linked with our sense organs and body. This puts limits upon the freedom of neural plasticity to do new things; input from a sense organ, for example, will strongly control what happens in the cortex to which it projects. But much of the brain links only to other parts of the brain, and these are the parts that have expanded most in our evolution. Although our visual and motor cortices are no larger than what would be expected for an ape our size, our brains are packed with cortices whose links are only with other parts of the brain.[79] Research on neural plasticity thus raises a major question: If what is restricted by genetically wired inputs and outputs is now found to be "soft," what about our brain's new unrestricted parts? Freed from links that could tie down what they do, they must be even more able to map new things. This is where evolution has made us different. It radically challenges how we look at our origins.

But first we need to observe another aspect of brain flexibility.

Mind Skills

Is a grapefruit bigger than an orange? Our brains think, and can answer the question. But how? Scans of people imagining a visual pattern show much the same brain areas activated that are used in sight. For instance, the more the primary visual cortex is activated, the quicker people can imagine a letter (tested by their quickness in judging whether it contains any curves—like "u" but not "v").[80] Indeed, in some studies the primary visual cortex takes up more blood when imagining something than when actually seeing it; perhaps it should really be called the "imagery cortex."[81] The same is true of our brain's control of our bodies. Our imagining doing something and our actually doing it overlap. When we imagine ourselves running, our bodies react as if we were—our heart rate goes up.[82] When we move our fingers in a sequence of movements, we activate part of our motor cortex known as the supplementary motor area. When we imagine making the same finger movements, the same area of the brain is also activated.[83] People with missing hands also activate their motor cortex when they imagine tapping their phantom fingers.[84]

The mind is working even when we are not conscious of thinking or orchestrating motor movements. Reuse of visual and motor skills underlies thought. When deciding whether a grapefruit is bigger than an orange, we use the part of our brain with which we judge whether a fruit we see is a grapefruit or an orange.[85] Motor rehearsal skills also get used in problem solving: Expert Chinese abacus operators do mental calculations in their heads. They see an abacus in their mind's eye—in their visual cortex—and move its beads around.[86] Concert pianists can work out how best to do the fingering on a piece by playing it in their heads. We know this from the experience of the pianist Paul Wittgenstein (brother of the philosopher Ludwig Wittgenstein), who lost his right arm in World War I.[87] He continued to play with only his left hand, and was famous for commissioning Ravel to write his Piano Concerto for the Left Hand. But he did not lose his expert ability to work out right-hand fingering when tutoring students. His mind and brain were able to think through the execution of a piece of music on fingers that no longer existed, except in his head.

Motor and sensory imagination are plastic: They can come together, borrow from each other, swap and delegate tasks. Suppose you are asked to say quickly whether a picture of a hand is of the right or left hand. On the face of it, this seems a straightforward sensory task—but it is not. The time taken to make the decision depends upon where your own hands are. If they are in the same position as the picture you see, then your judgment is quicker than if they are not. Indeed, the more you need to mentally revolve the pictured hands to position them like your own, the longer the judgment takes. *Positron emission tomography* (PET) scans show us this task links to the brain's own control of the hands. If the picture of the hand matches your right hand, it activates areas on the left side of your brain, the side that processes nerve impulses to and from your right hand. If instead it matches your left hand, it activates the same areas, but on the other side. Our sense of left- and right-handedness thus connects with our brain's own processing of its left and right sides; we do silent work.[88] And not only are we doing this by an imaginative process, but so deep is it to the act of thought that we are doing it unawares.

Read our lips: An inner imagination is at work when we speak. Inner speech has long been linked with outer speech, since when we concentrate, we move our lips and tongue slightly.[89] Now brain scanners can see the same parts used during outer speech, Broca's and Wernicke's speech areas, working when we use inner speech.[90] And this is true not only of speech: When we imagine the words of songs, we activate the areas (the superior temporal cortex in both hemispheres) with which we hear them.[91]

Before we speak, we buffer a little of what we are going to say. It gives us space to check what we are saying for errors and make sure that we are saying what we really want to communicate. The same is true when we think to ourselves: We use our inner voice as a sketch pad to ruminate ideas.[92] Indeed, we go further: Like a memory-starved computer we reuse other buffers.[93] We can hear the words we say in our head and thus store them in the buffers normally used to store overheard speech. We know that the brain does this, as we've seen that overheard words can blot out what we inwardly store in these buffers. The brain, moreover, is flexible in its use of inner "sketch pad" buffers; they do not have be speech-based. Some people use visual images and so have visual inner speech.[94] Some lose their speech buffers following strokes, yet they still can think. One individual, after a stroke destroyed his ability to think and speak verbally, redeployed his visual buffers to create a new mental landscape made up of visual imagery and symbols.[95]

We are not born using our sensory and motor buffers in this way. Instead, children have to learn how to generate them and efficiently put words into them through inner rehearsal. Only when this skill is learned can "thought" be linked with them, thus turning them into mental sketch pads.[96] Abacus experts likewise have to develop special visual-spatial buffers in which to hold numbers visually.[97] The same is no doubt true for our reuse of other sensory and motor modalities. Here we can see a fundamental point about us: Our minds depend on our reusing innate brain skills—but we are not born to do this. Instead, as we grow up we learn how to adapt our sensory and motor skills to new ends.

A connection between physical strength and imagination hints at how this happens. Your physical strength is partially determined, as you might expect, by your muscle bulk. Surprisingly, however, physical srength also depends upon the neurons in your motor cortex. We know this because whereas muscle bulk correlates with added strength only after several weeks, exercise increases strength immediately. If such an increase in strength were to come somehow from the brain, then one should be able to become stronger without even using the muscles. Remarkably, this is the case. The initial increase in strength correlated with exercise occurs in merely imagining exercise. The imagination is real. Indeed, in one study, a group of people imagining physical exercises increased their strength by 22 percent, while those doing the real thing gained only slightly more, by 30 percent.[98] There could hardly be a more striking example of the plastic powers of the brain. And if you can gain physical strength just by thinking about it, you can also avoid its loss by imagining exercise. This is a fact of no little clinical importance. Those with their arms, hands, and fingers in casts lose less strength if they do imaginary exercises.[99]

And if mental exercises can be of such benefit for muscles, imagine what they can do for neurons. After all, is it not evident that education strengthens what we do with the mind, our thinking and problem-solving skills?

Unlike physical exercise, mental exercise is unlimited. Imagined physical exercise can strengthen only the muscles that evolution has given your body. Pure thought, however, can develop "mental muscles" that you did not previously possess. When we learn mathematics or music or otherwise use our imaginations, our minds are extending themselves in the mental equivalent of new limbs. Ideas, notations, and symbols—the tools of our thinking abilities—let us not only think but rework the brain so that it can do work in unique, never-before-seen ways.

Through mental exercise and neural plasticity, we conclude, the brain can remake itself. It gives us, as Carl Sagan put it, "our capacity to learn." But by itself, neural plasticity goes nowhere. It is just potential. Something must first free it to do new things.

4

Superbrain

Beautiful music can be played, but it also can be dreamed. Carl Sagan once reported that on marijuana he was able, for the first time, to hear each individual member of an orchestra play. The brain is similar to an orchestra: Its many processes occur in parallel, somehow buttressing and supporting rather than interfering with one another. In this chapter we look at how the prefrontal, human part of our brain conducts and orchestrates the rest.

In 1936, a depressed and anxious woman in Portugal worried that the surgery she was about to have would damage her attractive curls. Dr. Egas Moniz, her ambitious physician, assured her it would not. When she awoke after the operation she had no worries about her hair—nor did Egas Moniz. As the Nobel prize–giving committee noted in its citation for the 1949 Nobel Prize for Medicine, Moniz had "made one of the most important discoveries ever made in psychiatric therapy."[1] Inventor of the prefrontal lobotomy, Egas Moniz had made the discovery that much of what lay beneath our skull could be cut out with no noticeable effect.

One of the oddest facts about our brain is that its most important part can be removed and yet no one might know. As a psychologist noted of one woman lobotomized in 1949, "people meeting her for the first time usually do not detect any abnormality and consider her a proper, affable, well dressed, and somewhat taciturn lady."[2] Unlike injuries to the visual or the motor cortex, which make people blind or paralyzed, those to a person's prefrontal cortex seem hardly to alter them—at least at first. Without a prefrontal cortex, a person can do nearly all the tasks scientists have devised to check the workings of people's minds. Prefrontal lobotomies, for instance, only minimally lower a person's IQ (as measured by standardized intelli-

gence tests).[3] Four years after one individual had a large part of his left prefrontal cortex removed (for a genuine medical reason), his IQ was still over
150. He could have joined MENSA.[4] But these tests are limited. A prefrontal lobotomy does alter people dramatically, but in ways that are not
readily apparent. Important clues suggest that the prefrontal cortex plays an
essential role in humans.

For one thing, our prefrontal cortex is large. The frontal lobes make up
roughly 36 percent of what most of us take to be our brain—its large cerebral
hemispheres. The prefrontal cortex is slightly smaller, as it shares the frontal
lobe with several other smaller brain areas: the anterior cingulate gyrus and
the motor, premotor, and supplementary motor cortices. The prefrontal cortex takes up 70 to 80 percent of the frontal lobe, and so roughly 30 percent
of your gray matter.[5] This large size suggests it does something critical.

The prefrontal cortex grew to this size only recently in animal evolution. It takes up only 3.5 percent of a cat's cortex; dogs do a little better with
7 percent. It is only with the monkeys and apes that it started to greatly
expand: 11.5 percent in the gibbon and macaque and 16.9 percent in the
chimpanzee. But it enlarged most in our evolution. Between us and the
chimpanzee there was a leap in size of 70 percent.[6] But that figure is misleading. Chimpanzees have brains between a third and a quarter of the size
of our own, which means the human species has roughly six to eight times
more prefrontal cortex relative to total brain size than our nearest ape relative. No other part of the human brain enlarged so much in our evolution.
Thus, our prefrontal cortex is the part of our brains most responsible for
making us the "unique animal."

The prefrontal cortex is always working during consciousness. While
the visual and motor cortex areas may go into the brain's equivalent of
standby, this part of our brain remains steadily alert. Even when our ears are
plugged and our eyes covered with a blindfold, it is active.[7] Indeed, the only
times when it seems to rest are during meditation[8] and in the few moments
prior to going to sleep.[9] This would not happen unless it were doing something vital.

Looking at its links with the rest of the brain, we find that the prefrontal
cortex is a kind of brain for the brain—its inputs and outputs wire it solely
to other parts of the brain. Indeed, it is superwired for this, with its neurons
having up to 16 times as many synapses as elsewhere in the cerebral cortex.[10]
In effect, it maps not the body or the external world but what happens
inside the brain itself. But what exactly is it doing?

Part of the problem is that, in the prefrontal cortex, scientists face a real
case of the six blind professors feeling different parts of the same prover-

bial elephant. Some scientists say it enables us to make cross-temporal associations, others that it is for working memory or just memory, and still others suggest its governs attention or the detection of novelty. Hidden in the scientific literature are hints at functions linked to our emotions and pleasures; our right prefrontal cortex lights up in the bucolic joy of drink[11] and the bliss of orgasm.[12] In an attempt to describe its role in cognition and emotion, researchers use words like "executive," "supervisor," and "planner."[13] However, these describe only part of what it does. Like the blind professors, each coming to a different conclusion, they miss the whole.

To understand the prefrontal cortex, we need to understand the organization problem faced by the brain. The brain must integrate[14] many things at once and at the same time perform multitasking.[15] The problem—and its solution—is similar to that faced by orchestras: All the parts must function in concert.

Like our brains, an orchestra is made up of many elements, its various instruments. From Mozart to Stravinsky, composers have used the multitude of sounds instruments make to create a near-infinite variety of music. But all these separate instruments make for a potential nightmare. It is one thing for a composer to write a score and another for the members of an orchestra to perform it. In Mozart's or Stravinsky's mind, the various instruments are played as one. An orchestra may be made up of 90 or more players, so somehow all those minds and bodies must work in unison. Following the same score in front of them is not enough. Listen to any school orchestra. The score tells them only what notes to play, not how to do so together.

The brain has a similar problem: Attention, vision, hearing, touch, emotion, memory, language, movement, and so on must be coordinated. Over 200 identified skill areas exist on the surface of each hemisphere, with many more below. Like a musical orchestra, these talents let our brains do a nearly infinite number of things—sequences of actions, emotions, and thoughts. But like the notes made individually in an orchestra, they could easily not come together; not musical cacophony but cacophony of the mind: dementia and confusion.

Maestro! Maestro!

Members of a professional orchestra play their scores with an awareness of one another. But they also have conductors whose baton, hand, and body movements—a cocking of the ear here, a look of the eye there—bring the group to play in unison to the same beat and tempo. The conductor ensures that the musicians play together (horns, say, coming in after a soft string passage).

The brain does something similar. Its many talents must be coordinated into a seamless, effortless whole. A remarkable number of things can be done automatically, without attention. Driving a car requires intense sensory-motor coordination. But not only can most adults learn to do it, they can learn to do it and forget about it, turning their minds instead to the radio or a passenger in the next seat. Conversing does not stop the mental concert of road-watching eyes connected to fine motor adjustments of the steering wheel via our hands. Only when we find ourselves in a new situation, such as driving on the other side of the road or behind the wheel of a new car are we forced to readjust, and so appreciate the skillful depths of the normally unconscious mental "music" holding our actions and reactions together.

But our evolution has continuously exposed us to the turbulent, the ever changing. Like musicians who need something more than a score, early humans and our nonhuman ancestors required something that went beyond automatic behavior. They needed ongoing control. We still walk on terra incognita; much, if not most of what we do can never be anticipated and therefore practiced in advance. Most of the sentences we say that are more than a few words in length are unlikely to be exactly repeated in the course of any lifetime. New stuff happens around us, some of it potentially lethal, and we cannot always rely on our automatic responses to carry us through. If our mental lives are indeed like an orchestra, what we play most is new music.

All these abilities need an ongoing and inner source of cues to guide them. Our prefrontal cortex makes and organizes these cues.[16] The part of our brains under our foreheads is therefore a "maestro"—instead of being called *prefrontal*, it perhaps should be known as the *conductor cortex*.

Don't Shoot the Conductor!

Palm court style assemblies of players and chamber groups do not need conductors. Their organization problems are usually so simple that conductors do not aid their playing. And though it is rarely tried, a full orchestra can get by without a conductor. In a moment of radical political thinking following the 1917 revolution, there existed in Leningrad (now back to its original name, St. Petersburg) such an "orchestra without conductor."[17] The czar was gone, and they thought the conductor should go next. This orchestra relied on extensive rehearsing. (According to American conductor Bruno

Walter, the orchestra does not seem to have been too successful.) Large orchestras, whatever political theory suggests, need conductors.

The need for conducting depends on the composition. Music by Beethoven and composers after him contains many more beat and tempo changes than before.[18] Mozart and Haydn wrote smaller-scale, less dramatic works. In their time an orchestra could be led by a musician, usually on the harpsichord playing continuo. Conducting as a profession developed largely in the wake of Beethoven; though a baton had been lifted as early as 1780 in the Berlin Opera, in Britain it was not until April 10, 1820, at a Philharmonic Society concert, that an orchestra was conducted, by Louis Spohr.[19] The Leningrad "orchestra without conductor" managed by extensively rehearsing together. But even here things are made easier by the conductor, who oversees the rehearsals that bring the playing of musicians together. After this an orchestra can shoot its conductor—and get away with it.

The same goes for the adult brain. One can take a bullet to the prefrontal cortex and survive; indeed a few do this in attempted suicides. Gorily blowing one's frontal lobes out from the side, Hollywood-style, does not lead to instant death but rather to a crude prefrontal lobotomy. Adults without a prefrontal cortex, like conductorless orchestras, can get along fine (at least for a time) since mental organization has already been long rehearsed. If they live simply, the professionalism of the brain can carry such people through life. The small routines of everyday existence change slowly, allowing someone with a damaged prefrontal cortex to get by in most situations. If you do not observe them too carefully, you might mistake them, as many doctors once did, for normal.

Even without a conductor, the Leningrad orchestra had musicians who gave signals to start them off and bring their various sections and solo instruments in on cue. Similarly, something in our brains has to give the inner cues that start us doing things, keep us going, and, if need be, change what we are doing.[20] Usually that executive function belongs to our prefrontal cortex.[21] When it is injured, people tend to lose initiative. They may be able to do things, but don't get around to it. They drift. And even when they do things, small distractions drive them off course. They cannot keep mental focus.[22] They get trapped in the first approach they make; they persist in the already established. In short, they get in mental ruts. Paradoxically, for this reason, the removal of the prefrontal cortex was once a "treatment" for illness; patients acting routinely are generally less difficult and so more "manageable" by institutions and relatives.

In summary, then, the brain can organize itself without its prefrontal cortex, but it first needs practice. The brain can do this more easily for simple things, since our need for the prefrontal cortex increases when what we face gets complex. The prefrontal cortex's brain control is not the deterministic kind a puppet master has over his or her puppets. Kill the puppet master and the puppets become lifeless, but the brain can persist, especially in routines, without its "organizer." The prefrontal cortex adds supervision and direction. In computer jargon, it is an add-on or upgrade. Beethoven could never have explored the expressive "romantic" power of music if nineteenth-century orchestras had been led (like those of the eighteenth) by someone playing continuo on a harpsichord. Without a conductor, the players would get lost, unable to keep together as a single music-making entity in the face of rapidly and dramatically changing beat and tempo.

Orchestras play sequences of notes that were previously written down by composers. We survive a world that is hostile and constantly offers the unexpected by purposefully making plans to satisfy inner needs and thus creating a kind of living score, a set of inner scripts according to which the brain can improvise. Since it is fixed only in broad outline, fine details need to be extemporized when we face opportunities and difficulties. Composers are never totally explicit as to how a piece of music is to be played, and conductors and directors thus must interpret. Likewise, our inner scripts must be filled in while they are being enacted, to fit changing circumstances. And that is done by our prefrontal cortex.

The brain also holds a related inner score, not scripts of what we do but scripts of how we normally expect events to happen. These let us feel a sense of familiarity and normality when things happen as foreseen and amazement when they do not. The inner scores needed to do this are read and kept up to date by our prefrontal cortex.[23] Imagine your reaction to the following violations:

> A woman with flowing hair shaves her soapy face before a mirror.
>
> A bearded man wears a dress and paints his toes red.
>
> A doctor wears a lion mask.

A person with a damaged prefrontal cortex will let such incidents pass without reaction. When these events were staged, as described, before a woman with such a brain injury, she "glanced up at the actor[s] with an air of total indifference, and made not the slightest comment."[24] She did not detect their implausibility—without her prefrontal cortex, she had lost the ability to read the inner scripts which let us recognize what is and what is not familiar. With the loss of her prefrontal cortex, she had lost what we so easily take for granted—our capacity for surprise.

The Inner Orchestra

We have seen that the prefrontal cortex magically organizes the rest of the brain. How it does so, how it conducts mental processes, is another question. To understand it, we need to turn again to conductors.

Conductors create within themselves a "working memory" of the orchestra and its music. It starts when they study the piece they will play. All conductors can hear written music in their heads. The auditory cortex has the ability (highly developed in conductors) needed to imagine music.[25] A conductor uses this to work on the problems of interpreting a piece. Using his or her inner orchestra, the conductor can spot the difficulties that will lie in a particular passage, such as tempo, beat, timing, and the loudness of different instruments. An imaginary inner orchestra is far better than a real one for doing this, since it is played by only one person—the conductor. This enables the conductor to formulate how the real orchestra should sound later. Thus, by the time conductors start conducting real musicians, they have already rehearsed what they are to play.

This inner orchestra provides the conductor with an unrestricted realm in which to explore the music's esthetics. Buried in the written music a great composer leaves emotional, spiritual, and other subtleties that can be brought out by interpretation. But first conductors have to hear the music; only then can they find ways of challenging an orchestra to bring them out in a live performance. By the time they stand up on the podium, they have already done half their work.

But at this point they have not finished with the orchestra in their heads. Having heard the music, they now have an outer performance with which to compare with the inner one. This inner orchestra thus lets them judge how well the outer orchestra plays, and spot where musicians need help.

There is, in our prefrontal cortex, something like an inner orchestra. Yale neuroscientist Patricia Goldman-Rakic calls it "working memory."[26] A slightly more descriptive title might be "sketchpad" or "neural notebook." As we have stressed, the brain not only senses and does things but also imagines what we might see, hear, and do. This enables us in our sketchpad to model the outer world. Like a conductor with an inner orchestra, we work out in this imagined world how to handle the problems before tackling them in the more difficult real one.

Waiting for Marshmallows

Our working memory provides an inner source of cues to prompt behavior. The primate brain can thus trigger itself, rather than (as in many other animals) the external world triggering it. By giving us more options, the prefrontal cortex increases our freedom of action and so aids our survival.

Consider a chimpanzee that can see a nice, sweet, ripe banana that has been missed by others. It pays for it to wait and control its excitement until the others have moved on, and only then grab it. But to do that its brain must control the chimp's excitement, delay the urge to grab the banana (without forgetting it is there), and wait until the chimp's companions have moved off. If not, the chimp's excitement will alert the other chimps to the banana, and it will not get it. The prefrontal cortex enables the chimpanzee to grab onto an inner representation of a banana and so keep the action of grabbing on hold.

Evidence of the role of the prefrontal cortex in delay comes from many types of research. Cut out the prefrontal cortex of apes and monkeys, and they cannot do these tasks or organize their behavior across time.[27] Eavesdrop with electrodes on neurons in their prefrontal cortices, and you can spot them in cue-, delay-, or response-related activation as the brain puts actions on hold. Similarly, look at the brain activity of humans doing tasks, and you can spot the prefrontal cortex lighting up as it prepares, waits, and evaluates.[28] Working with bursts of magnetic stimulation applied over the forehead, scientists can now even temporarily stop the prefrontal cortex.[29]

The most revealing research on the importance of delay and waiting for the right time has been done on children. Psychologists gave 4-year-old toddlers a choice between two treats, say one rather than two marshmallows. The researchers told them they could have one marshmallow immediately if they rang a bell. But if they waited for someone to return, they would get two (which they would much prefer, of course). The researchers were interested in how long children would delay the urge for immediate gratification. The children differed in how long they waited, but the average was 11 minutes (usually the experimenter returned after 15 minutes). If the sweets were in the same room, they could resist only for an average of 6 minutes. However, if they learned the trick of imagining the shape of the sweets, the children would wait up to 18 minutes, even if they could see them. This inner image acted as an internal cue with which they managed their responses.

Not that any inner cue would work. The children had to focus on a cool, abstract aspect of sweets (such as their shape or color) rather than a hot, arousing one (such as their taste), in which case they could not resist. An interesting subgroup of children had spontaneously picked up such tricks (and other ones, such as distracting themselves or covering their eyes if the sweets were present). Ten years later, when they were nearly 16, the children who had figured out how to resist were found to be more socially competent and more successful at school; they had higher scores on their Scholastic Aptitude Tests (SATs).[30] What these researchers could not study directly was the prefrontal cortex, since their experiment was done long before scientists could pop someone into a brain scanner. However, brain scans on delay tasks with adults now show that the prefrontal cortex is activated when people delay responses. Their visual cortex lights up as well, suggesting that their prefrontal cortex organizes the delay by using visual mental imagery.[31] It thus seems not only that our brain uses our prefrontal cortex to free it from the here and now by making and orchestrating inner cues but that some people are better at doing this than others, giving them an advantage in life.

Without our prefrontal cortices, we become slaves to automatic responses. The French neurologist Dr. François Lhermitte tested the power of external cues on people with injured prefrontal cortices using a series of odd and funny gestures.[32] Normal people and children ignored them, but not his subjects with prefrontal cortical injuries. Instead, they copied his actions. "On being told that they had not been told to imitate the gestures," he noted, "their answer was that obviously since the gestures had been made, they must be imitated." Why did they feel this way? Normal people ignore such gestures, as we are usually guided in what we do by the cues created by our prefrontal working memory, not those made in front of us. Without an intact prefrontal cortex, people may have no choice but to let the external world rule their behavior.

Physical objects also triggered behaviors in Dr. Lhermitte's patients. He took home a woman with damaged frontal lobes. When she "spotted sewing needles, spools of thread, and pieces of fabric, she put on her glasses and began sewing in a precise manner. In the kitchen, after spotting the broom, she swept the floor; when she saw dishes in the sink, she washed them." When she saw a bed, "she tucked in the covers on both sides." The doctor walked toward her wearing a stethoscope. She lay down. The world to her was a series of prompts that switched on her automatic habits and routines. Normal people can use their prefrontal cortex to resist the cues surrounding them and thus free themselves to react to things in the way they wish. But people with injuries to the prefrontal cortex lack this ability to be guided from within.

Think Machine

Our prefrontal cortex has another function: It uses its conducting skills to invent new behavior. Again conductors show us how. They use their inner orchestra to reorchestrate old music. Leopold Stokowski is famous for doing this to Bach's organ music. (The *Toccata and Fugue in D* under his baton starts Walt Disney's *Fantasia*.) And conductors often use their inner orchestra to write new music. Mahler, for instance, was a conductor as well as a composer. If they can hear a piece of music in their heads, composers can go beyond interpreting a performance to making an entirely new one. They tune in to their inner orchestra and turn it into the most expressive organ of peace, beauty, emotion, and feeling known to our species. Beethoven, after all, had no alternative: He wrote much of his music when external deafness kept him from hearing any music other than that played in his head.

The prefrontal cortex likewise invents new behavioral performances by using its "second brain" as a sketchpad to explore various options. It can imagine.[33] Using an idea of what it wants, it can work out, play, or "pan" through possible solutions for a means of obtaining its goals. This lets it be creative and do entirely novel things.

But this is only the start of what the prefrontal cortex can do. Our imagination is not limited to sensory experiences or what we might do physically. Our brains can play with the representations and abstractions of things—concepts. We can conduct them in various mental operations and manipulations, thus letting them be objects of reflection, thought, and reason.

Try imagining what it would be like to think without a prefrontal cortex. Ask yourself, how long is a car drive from New York to Los Angeles? One woman with a damaged prefrontal cortex guessed 6 hours.[34] Asked what percentage of the U.S. population was male, she guessed 90 percent. Although we might not know the height of the tallest building in London, we would not guess that it was between 18,000 and 20,000 feet, and certainly not higher than our estimate for the highest mountain in Britain. But these were the answers given by another person with a prefrontal cortex injury.[35] Without a tape measure handy, you guess. But how do you do this? You make rough comparisons using inner cues to direct and organize ideas. You may not know how long it takes to drive to Los Angeles from New York, but you can call up in your mind other drives that you have made. You can then play around with this information, imaging a mental map of the United States that lets you compare these distances. You orchestrate what you know to extract clues as to what you do not.

Something similar occurs when we recall experiences. Recall is not the same as remembering. If we visit places or meet an individual, we can readily remember that person, since the present situation triggers past memories. But when no immediate cue exists, we must recall things. We need to invent prompts for our memory, contexts that act like a present experience.

But how do we think of them? We need to search our long-term storage of experience, and that can be an effort. Have you ever met a chain-smoking nun? The answer is somewhere in your memories, but where? To search for it, you need to think of times when you might have met nuns. Maybe when you visited a convent or a cathedral? When would you have done that? Perhaps on holiday. Or maybe visiting a dying aunt or uncle in a hospice run by a holy order? The question expands itself as to what mental records need to be checked through. As one scientific paper put it, we engage in a "conceptually driven, flexible searching activity [set] in motion by means of a restricted number of preliminary generated tentative contexts (or event-structures)." In other words, you recall by a kind of contextual 20 questions. Here we make internal cues to organize the mental activity that searches our memory. It is prefrontal cortex work: Brain scanners find that the prefrontal cortex lights up when people recall things from or encode them to memory.[36] People with prefrontal cortex injuries have severe problems with recall.

We need, moreover, to check our recollection for errors. Is a triggered memory plausible? Does it fit in with our general knowledge of the world? Does it startle us? Does it fit with our defaults? Would any nun chain-smoke? It does not seem to fit with most people's concept of a nun. We create out of mental scripts, used to judge normality, models to test whether our memories fit reality.[37] Perhaps the image we can vaguely sense of a chain-smoking nun comes not from life but from a comedy film— Whoopi Goldberg in Sister Act, perhaps? The prefrontal cortex can use its inner cues to run such checks. People with prefrontal lobe injuries fail to check their reasoning or their memories. They confabulate; whatever they say seems true to them. They lack any sense that they may have invented it or made it up.[38]

Our prefrontal cortex can also be used to check ideas. Scientists have a task they use for checking such hypothesis-testing abilities (the Wisconsin Card Sorting Test). Imagine a pack of cards, each with one, two, three, or four shapes—triangles, stars, crosses, and circles. The shapes come in four colors. Four cards are placed in front of you. You must sort the pack of cards by shape, or by color, or by the number of shapes on them. At any one time only one sorting rule will operate. You are only told when you make a mistake, and the rule can change at any time without warning. The task forces you to make hypotheses as to which sorting rule is operating. Working out which is not that difficult, unless your prefrontal cortex is damaged.[39] As might be expected, this part of the brain lights up during such tasks.[40] Further, we need our prefrontal cortex to orchestrate our mental space; it lets us mix and match and maximize our manipulation and grouping of internal ideas.[41]

Prefrontal cortex skills are also needed for learning.[42] Because it holds an image of what we should do, the prefrontal cortex can tell us where we have made a mistake. But we need to do more. As important as checking for mistakes is learning from them. Why spot an error unless we do something about it? If an error made by a person with prefrontal cortex damage is pointed out, he or she may see it but still may be unable to keep from making it again. Such people lack the skill to learn from themselves or others.

Mental planning also requires inner cues and a notepad. Effective planning is about thinking out new ways of doing things.[43] There is always more than one way of doing anything, taking advantage of a situation, using a tool, a piece of information, or an idea. We are often hindered from doing what we want. Overcoming this requires guessing and learning abilities. With our prefrontal cortex, we can play around with strategies and so change what we do.

We also need an ability to overcome habitual mental and behavioral ruts.[44] If our usual behavior and actions are not up to what we want them to do, then we improvise. It is the adapting of well-learned habits that people with prefrontal cortex injuries find hardest. We are always reusing our well-learned skills. Rehearsed, they allow us to do routine things without much effort or thought. But the world changes, so they need to be brought up to date for new situations. While it is useful to have habits, they must also be adaptable—or they can trap us. We need our prefrontal cortex to supervise our routines and attend to them so that we respond to novel situations with planned rather than impulsive or routine reactions.[45]

From modifying habits, it is a short step for the prefrontal cortex to program them. How do you write software? Not from scratch; most people use bits of previously written programs and patch them together so they fulfill new needs and specifications. The prefrontal cortex is that kind of small-step programmer. As it adapts routines, it changes them so that they grow more elaborate. In this way, adaptation becomes a kind of programming. The human brain doesn't have to wait for evolution to bring beneficial changes on a generational time scale. It can reprogram itself in incremental bits every day. The prefrontal cortex constantly refines and organizes what the rest of the brain does, so that as it builds on past abilities, new skills and competencies evolve.

We also have to organize the loading of our inner thinking cues from long-term memory onto our mental sketchpad. We do not hold many ideas in this sketchpad: Psychologists figure that in most people it is about seven at a time. And how they are held may vary depending upon the buffer used. Most people use verbal scripts, but as we've noted, a few people can hold ideas better visually.[46] Deaf people who use sign language store information in signs.[47] In any case, we need the prefrontal cortex to manage the juggling act of switching ideas from long-term memory into and out of the buffers holding them. For instance, most people can say the numbers between 1 and 10, at random, without repeating them. To do that, they need a memory manager to track the numbers that have been named and those that have not. Brain scans show us that memory management lies in our prefrontal cortex, while the temporary buffers used, as noted, for the most part lie elsewhere.[48] So the prefrontal cortex not only helps us search our long-term memories but aids their use in thought.

Our prefrontal cortex does more than just manage this sketchpad; it is also its main user. For all the skills mentioned—guessing, recollection, learning, flexibility, categorization, and abstraction—we need a mental space where we can play around, mix, and try out ideas. Neuroscientists may call this sketchpad "working memory," but we can feel it as our very selves. As its contents are mostly verbal (though perhaps abbreviated), we experience the stream of ideas and associations propelled by our prefrontal cortex as our inner voice—our identity.

The inner voice is not necessarily a vocal one, however. Deaf signers live in an inner world of sign speech or their finger alphabet.[49] Thinking in signs activates the same brain areas as does thinking in verbal language.[50] Indeed, evidence exists that as much as speakers activate their vocal organs during inner thought, the deaf do so for their arms.[51] Helen Keller from just before the age of 2 was both deaf and blind. As she could not see her hands when she thought or memorized with them, she lacked the self-consciousness to stop their movement. As her teacher, Anne Mansfield Sullivan, noted, "When a passage interests her, or she needs to remember it for some future use, she flutters it off swiftly on the fingers of the right hand. Sometimes this finger-play is unconscious. Miss Keller talks to herself absent-mindedly in the manual alphabet. When she is walking up or down the hall or along the veranda, her hands go flying along beside her like a confusion of birds' wings."[52]

Whether in vocal or sign language, the prefrontal cortex is a powerful extender of the brain, but this extension is something we need to learn. The Russian psychologist Lev S. Vygotsky noted in the 1920s that children speak as much to themselves as to other people. Fifty percent of a preschool child's speech is like this. Vygotsky and other Russian psychologists have shown that they use it to organize themselves.[53] As children grow older, this self-regulatory speech goes "underground" and becomes silent and private.

Inner speech and nonverbal cues are thus liberators. By using them as a source of inner prompts, the brain can free itself from the "here and now," and so react in its own terms to the external world. It can wait. With these cues it can imagine new actions and can orchestrate the brain to think and recall. It can even go on a podium and conduct. While the brains of conductors are as yet unstudied, researchers have found that the monitoring, manipulation, and recall of musical tones involve various parts of the prefrontal cortex.[54] Here inner cues give rise to a wide range of mental skills, such as the ability to estimate, do reality checks, abstract, be flexible, and learn.

5

Mind-Engine

Conductors do not hide in little boxes merely imagining music. No, they go out, stand in spotlights on podiums, and direct. At least in rehearsals, they may shout and stop the music-making. They may even walk out to show a player how to play a phrase. This, of course, is limited to rehearsals; not even Herbert von Karajan would interrupt his orchestra during a public performance. Then they communicate with their hands, eyes, body, and that silent instrument, the baton. Which raises the following question: If the prefrontal cortex is the brain's conductor, where is its "baton"? Somehow it must be there, since except during performance of routine skills, brain scans and electrodes implanted in the brain show it actively "tuning up" the cortex.[1] For instance, when looking for a bold letter in a word (as in the previous word "bold"), part of the visual cortex is activated twice: first when the word is identified, and then 100 ms later, after prefrontal activation organizes it to attend to the thickness of letters.[2] Another example is when you think of uses to go with nouns—say, *hammer* and *hit*. Initially, at 200 ms, this lights up the left prefrontal cortex and the nearby anterior cingulate gyrus; only half a second later, under the control of the prefrontal cortex, does the Wernicke's speech area (where much of linguistic processing takes place) light up.[3]

How does the prefrontal cortex do this? A few hints are emerging. Just as you cannot understand the role of a baton without knowing, for example, that players have eyes that follow its movements, so you cannot understand the prefrontal cortex's baton without knowing something about the players of its neural "orchestra." First, something about the anatomy of sight. The conventional story goes something like this: The retina in the eye passes an image through the optic nerve to two bulges on the thalamus, the lateral geniculate nuclei—literally, the side-jointed nuts. After this, what the eye sees goes on to the primary visual cortex. This part of the visual cortex then feeds what it sees to further, "secondary" parts of the visual cortex and then to other parts of the brain. The idea that information gets passed on sequentially along the visual system in this way is what we might call "the old story."

Things are actually more complex: The secondary parts of the visual cortex not only receive input but also feed information back to the primary cortex. The lateral geniculate nuclei used to be thought of as mere gateways that channeled to the cortex things seen by our eyes; these gateways, however, receive more axon wiring from the visual cortex than from our eyes through the optic nerves.[4] Nor does the lateral geniculate nucleus go quiet once it has passed visual information to the cortex. On average, its neurons keep active for 320 ms after retinal input—as long as the rest of the primary visual cortex.[5] Yet the neurons in the cortex are already recognizing shapes within 20 ms[6]—even the prefrontal cortex is activated after only 100 ms. None of this fits with the story of the jointed nuts' being a mere gateway to the brain. Something more complex is going on. Scientists now find that the cortex and thalamus (of which the lateral geniculate nuclei are part) reverberate impulses back and forth.

Before we look closer, however, a warning is in order: Studying the neuroelectronics happening under our skulls is hard, and what we know is very incomplete. Scientists are like military intelligence listening into enemy signal communications and picking up only a fraction of what is going on. In enemy territory, there are land lines whose communications cannot be overheard. Not all radio traffic is intercepted, and weather can degrade much of what is. Likewise, much of the activity in the brain does not reach the EEG "listening stations" scientists put on the scalp. (By the way, neurology is as full of initials as the military—EEG stands for electroencephalography, the science and technique of amplifying the tiny voltages induced by the brain upon the scalp.)

And the importance of what we do "hear" is not always clear. Is it an irrelevant hum or the very broadcasts of thought? Even if we are half correct, what remains unknown? There is a temptation to take what is found to be more of the story than it is and extend the little we know more than we should. But that is typical for science in an early stage; it has happened in the past with astronomy, chemistry, and the workings of our bodies. In a hundred years we will know much, much more. Like military intelligence, we have to go on the fragments that are available. We are beginning to get a grip on complex phenomena. Until recently the data were so fragmentary that they did not seem to make much sense, but now we can start to put together a story of some of its intricacy.

In brief, electronic oscillations appear to be a "mind engine" that lets us think, see, and know. They are to our experience what circulation and metabolism are to our bodies, the very processes of life. Moreover, the prefrontal cortex can lead and pull together these oscillations.

Neuroelectronics

Many different oscillations have been detected in the brain, some happening between the thalamus and cortex, others not. Mysteriously slow, some oscillations last 40 to 100 s,[7] while others have very fast frequencies, around 1000/s, or, to use the scientific term, 1000 hertz (Hz).[8] These differing frequencies appear to correlate with states of consciousness. Among the slowest of these brainwaves is *delta*, in which the brain is in a kind of "off" state. These oscillations happen at between one and four times a second. These are the oscillations of deep sleep. (To complicate things, there are also two other oscillations during our deep sleep, happening at slightly lower and higher frequencies.[9])

There is a problem with this off state. As anyone with an alarm clock knows, it takes time to regain full consciousness. This means the brain cannot go to sleep in its odd moments while it is doing nothing; instead it has to find a way to "idle." Alpha rhythms,[10] happening roughly between 8 and 13 times per second (8–13 Hz), represent this standby mode. They are most easily observed using EEG by asking people to shut their eyes. With the eyes closed, the visual cortex has nothing to do and so temporarily turns off and generates alpha waves. As soon as the eyes open, it is reactivated out of standby and alpha rhythm stops.

Even with our eyes shut, however, imagining objects or scenes can activate the visual cortex. When people do this, the visual cortex is no longer in standby and alpha waves disappear. Forward in the brain, the standby activity over the cortex involved in movement and touch is usually called *mu*, or *motor alpha*.[11] The standby rhythm for hearing has only recently been discovered and is called *tau*, or *hearing alpha*.[12] It is found over our temporal cortex. Hearing and especially motor alpha rhythms can occur at frequencies higher than 8–13 Hz, in which case they are known as *beta* rhythms. However, this still functions as a standby, appearing during rest and disappearing during activity.[13] Alpha is not usually found in the front of the brain, however, not because the front part of the brain cannot make alpha, but because it rarely goes into standby—even with the eyes closed and ears plugged, brain scans show it active.[14] (The brain scans also show the visual cortex to go into action when the eyes are open and turn off when they are shut.) But the prefrontal cortex can show alpha, as when people drift off to sleep or meditate. When we are falling asleep, this frontal alpha activity is only temporary, lasting 10 s or less.[15] But in meditation, such as transcendental meditation (TM), zen, yoga, and qi gong (a Chinese form of meditation), it lasts longer and changes, increasing slightly in amplitude and becoming more coordinated and slightly lower in frequency than normal.[16] Some highly practiced meditators can control their brain waves so that the alpha state remains over their visual cortex even after they open their eyes during or following meditation.[17] As with other evolved aspects of the mind, people vary in their alpha activity. Some people have a lot, others a little. How much we have seems to tie in to how we think. People with little alpha activity are high visualizers; they do not turn off their visual cortex because they use it constantly while in thought. In contrast, those with much alpha activity tend to be more verbal in their thoughts and so more likely to keep their visual cortex on standby.[18]

An advantage of alpha is that the brain can flick quickly from it into work mode—*gamma* rhythm.[19] Gamma waves are oscillations in the 20 to 80 Hz range. For short, neurologists call them 40-Hz oscillations.[20] They are triggered when the reticular formation in the brainstem puts the brain into a state of arousal.[21] Gamma bursts occur in time frames of a tenth to a third of a second.

The new discoveries surrounding gamma are exciting—they could be a kind of Holy Grail for the sciences of the mind—but reducing general psychological processes to the behavior of neurons is a problem. Although alpha, the standby state of the cortex, is fairly easily detected using electrodes put on the scalp, gamma is elusive. Gamma does not readily pass through the skull for a variety of reasons. The amplitude of its high-frequency oscillations is low. It is also, compared to alpha, more localized, fragmented, and transient. Worse for those trying to study it, muscles make electrical noise in the 10 to 70 Hz range that interferes with gamma's detection. To top it all off, where and when it can be detected, it oscillates too quickly to show up on the paper traditionally used for EEG recordings.[22] Thus, most work on the brain's electrical activity has focused on the more convenient alpha. Not being able easily to detect the faster gamma activity, scientists thought that the brain stopped oscillating when it was working and produced desynchronization, or noise.

The Binding Problem

What does gamma do? As with many things in the mind, the reason for its existence is hidden by its success. Consider your ability to see—it is extraordinary. Shut your eyes. Open them. Now look at your fingers while moving them in front of each other and this book. Consider what *doesn't* happen. Your fingers do not appear to take on the color of this page. Nor, as they move about, do they seem to jump from hand to hand. Nor do their movements transfer to these words.

Your brain has done two incredible things. First, it has segmented the visual scene before your eyes into separate "perceptual groups"—hands, book, and what is beyond them. In doing this, it has efficiently divided what you see into figures and backgrounds. Such segmentation is important because it lets you see your fingers as belonging to two hands and separate from the book, even while all are occupying the same visual space.

Second, your brain causes you to see each of these perceptual groups— objects and "gestalts"—as having different visual attributes: color, shape, and movement. That is truly remarkable. The brain processes such attributes in different cortical areas. In fact, making sense of what is before our eyes involves processes spread over two dozen different areas in each hemisphere. But in spite of this, vision is a wonderfully unified experience. Somehow, the part of the visual cortex responsible for color must be able to process the colors of our skin, this book, and what lies beyond it at the same time without mixing them up. The same is true for the other areas of the brain processing shape and motion. And just as remarkably, the processing of an object's color, motion, and shape happening in these different areas must come to be bound together as one experience. Even more amazingly, this has to happen for all the many objects seen and processed in the near instant of opening our eyes or changing our gaze. Yet the brain has no problem with this. Somehow it binds what we see with such incredible ease that we do not notice it, even though it is created by the activity of many brain areas. Your sight is performing a minor miracle. We should never cease to be amazed that our fingers do not turn the color of this book when we see both together. Or that their movement does not transfer to these words. Or get visually mixed with the world beyond. We do not hallucinate. We see.

How?

Scientists call this puzzle the *binding problem.*[23] It occurs for all our senses. The feel of our bodies, for instance, is spread over seven brain regions. It is perceived across all our senses, and yet what we see and hear happen as one experience, not as separate ones. Moreover, the problem is not limited to perception. Movement also requires the coordination of many brain areas—perceptual and motor—working as one. Thought and memory also require the processing of several areas coming together.[24] Further, we must be able to do such things as bind emotions to what we see, think, and recall. In short, *the brain has to have a trick that lets it process things across the brain as if they were all being processed in the same place.*

How does the brain manage this amazing feat of mental magic? At first scientists were puzzled. Those modeling the brain in the 1980s suggested some theories. They argued that it might involve the existence of some kind of synchronization among neurons. Though neural processes may be happening in different parts of the brain, they could work together if they did things at the same moments in time. The need to bind different objects would not cause problems, since the different bindings could fire at slightly different times. Thus, the neurons processing the color of your fingers would lock their processing with those tackling the shape of your fingers, rather than with those processing the shape of the book. Such synchronization could happen just once, but it would be easier to arrange in the brain if it happened repeatedly in oscillations. This would allow neurons to build up processing together and share information. Such synchronization need not last long—only a few tens or hundreds of milliseconds would do. Separate processing would be thus kept apart through slightly different rates of firing. This prediction was made, and scientists looked and found it to be the case. There is now direct evidence that part of the answer exists in those gamma oscillations going on between the cortex and the thalamus.

Gamma Activity

By the late 1980s, scientists had the technology to detect gamma bursts. Work using implanted electrodes—limited to research on animals—showed gamma activity binding and segmenting what the animals saw. Gamma creates faint magnetic as well as electrical fields in the brain. These two kinds of fields occur at right angles to each other, so each reveals something hidden by the other. Some research uses EEG; more recently, magnetoencephalography (MEG) has been used. Scientists are finding that hard-to-detect gamma may be responsible for even more neural synchronization than we thought. In one new technique used on animal brains, cameras peer directly at the coherence of large groups of neurons that, tagged with voltage dyes, change color as they spike in unison. This technique allows us to visualize neurons firing at the same time even at distances up to 6 mm apart.[25] MEG already lets us see gamma moving in patches around human brains. The gamma activity shifts up and down between the cortex and thalamus and shoots across the brain.[26]

As noted, the visual part of the thalamus, the lateral geniculate nuclei, receives as many axons from the primary visual cortex as it does from the retina. There is feedback, a synchronization among neurons in the thalamus and those of the cortex. Within the top layers of the cortex exist special neurons called *chattering cells* that generate gamma waves when excited. These neurons lead, the rest of the cortex follows, and gamma activity synchronizes.[27] In brained animals, including humans, this synchronization is correlated with sharpened perception: Cells link to respond to a common stimulus, such as a moving image.[28] Research on cats suggests they use gamma to segment scenes in discrete ways and time periods depending on what they see.[29]

Visual coherence depends on gamma. When sight fails to develop normally, so does gamma activity. This is the case with strabismic amblyopia, in which the two eyes are out of alignment, something that could cause double vision. One way the brain avoids this is to see using only one eye. As a result, vision in the other eye and its visual cortex is not normal. It does not bind percepts properly: Fragmented contours may come together, while continuous ones may appear disrupted. Such distortions correlate with impaired gamma. Compared to the eye that is fully used, neurons in the ignored eye's visual cortex do not synchronize very well when binding visual percepts.[30]

Gamma oscillations are reset by new stimuli.[31] Indeed, this resetting limits our ability to distinguish two rapidly repeated tones. We hear them as one sound. When we hear two tones, there are two bursts of gamma waves; when we hear them fused as a single sound, there is only a single burst of oscillations.[32] Such misperceptions occur not only when we disregard a real entity but also when we bind fragments together to see unreal optical illusions.[33]

Perception by one of your senses can evoke other responses, partly via gamma. If you hear, say, a sudden click, in seconds this will evoke a train of reactions in the electrical and magnetic potentials on your scalp. Tiny voltages in its electrical potentials switch from negative to positive in a series of peaks and troughs. Or so it was thought. New MEG research suggests that electrical and magnetic fields lock in phase but only for the earliest 20 to 130 ms of evoked responses. MEG can see what cannot be easily seen with EEG: four or more gamma cycles in 20 to 130 ms after a click is heard.[34]

Gamma, like alpha, does not apply only to vision. Indeed, most gamma work on humans has been done on hearing.[35] Gamma also operates when smelling odors and when feeling touches on one's body. And it does not happen just in one hemisphere; gamma binds visual processes across the two sides of the brain,[36] connecting areas of the human cortex as much as 9 cm apart.[37] In this rapidly expanding area of science, evidence of these oscillations and synchronizations is turning up outside our sensory cortices. Research suggests that they help the brain prepare before it does things[38] and synchronize neural activity among sensory, motor, and prefrontal cortices.[39] People react to stimuli differently; fast reactors react before fully perceiving the stimulus, and others, slower, react afterward. Such differences show up in gamma activity.[40] Gamma also links processing that takes place in the prefrontal and parietal cortices for selective attention.[41]

Gamma may also have a role in memory. While theta waves' slow frequency of 3–7 Hz is the main brain frequency used for memory storage in our hippocampus, gamma has also been found there. Such hippocampal and cortical gamma may hook up to link memory and sensation. This is the proposal of a research group headed by György Buzsáki at Rutgers University: The "coherent oscillation of neocortical and hippocampal neurons may reflect the fusion of currently perceived and stored attributes of objects and events."[42] Tantalizing research done by Hellmuth Petsche and colleagues in Vienna suggests synchronization and gamma might underlie the experience and pleasures of music.[43]

Further work suggests gamma may underlie the binding that is required to enable us to reason.[44] It has been linked to solving verbal, mathematical, and shape puzzles.[45] When people detected real words, such as "moon," among letter strings that only look like words, such as "noom," scientists caught gamma connecting neural assemblies in the left hemisphere.[46] Moreover, learning-disabled children fail to show the increased gamma normally seen during problem solving.[47] And the gamma-challenged may show dementia or disordered thought: Reduced gamma correlates with Alzheimer's disease[48] and schizophrenia.[49]

The link between gamma and synchrony may tempt us to think of it as a kind of computer clock in the brain.[50] But although both computers and brains use time to synchronize, brains are much, much slower. A 10-million-fold rate difference separates silicon and biocomputers: Computer chips clock at thousands and hundreds of millions of hertz, while the brain runs at 20–80 Hz. Yet whereas computers use a clock running at only one frequency, our brains combine many frequencies. Computers work in serial, while our brains are true parallel processors. So whereas the cycles of computers are used to order the execution of instructions, gamma oscillations temporarily link the brain's many separate areas of processing in a kind of bioelectromagnetic dance.

Not all synchronization happens in oscillations. Electrodes implanted in brains find some neurons firing in synchrony that do not appear to be doing so in cycles. One possibility is that they are part of hidden oscillations; a neuron's participation in an oscillation can only be spotted if its fellows are found firing along with it. Or neurons might be joining in without firing electrical impulses to one another; they can electrically oscillate across their cell membranes at a subthreshold level, insufficiently strong to cause them to fire. A collection of neurons may thus oscillate electrically but silently together and as a result show only "irregular" activity. Thus, even the irregular firing of neurons may not be so irregular—*desynchronized*—as was thought.[51]

Nothing is simple when it comes to the brain. A recent development is that synchronization has been found to link with oscillations, but only over distances of 2 mm or more, not shorter.[52] Other subtleties no doubt exist. What we know of the role and existence of gamma at present still depends on fragmented evidence. Oscillations may exist for reasons other than binding; binding may involve other, as yet undiscovered, processes. Oscillations occur subcortically as well as cortically. The main lesson in any area of science is that with better knowledge, things originally seen as simple are eventually revealed as much more complex.

Steering: Hands on the Gamma Wheel

Norbert Wiener first used the word *cybernetics*—the study of control processes in living beings and machines. The prefix *cyber-* is taken from the Greek word for "steer." How are the manifold processes of the brain steered? What can we say of the cybernetic processes underlying thought, which we can control even as they control us?

There is more to gamma and brain activity in general than passively binding the processing done by neurons. Brain activity can be focused. Gamma thus not only enables experience but offers a means by which the prefrontal cortex can organize the brain. To explore how this might happen, we must return to the anatomy of the thalamus.

Although we described the thalamus as if it were a single entity with two bulges—the lateral geniculate nuclei—it is more complex, consisting of many smaller parts, including two more bulges, the medial (middle) geniculate nuclei. The lateral and medial geniculate nuclei are "specific" thalamic nuclei; they are strongly linked to specific areas of the cortex: the lateral geniculate to the visual cortex and the medial geniculate to the auditory cortex. Other thalamic nuclei—the *intraliminar*, or nonspecific, nuclei—are linked widely. Their deadpan name belies their exciting function: They help shape which cortex areas bind together.[53] They seem to bring together the sensory and other information that the intraliminar nuclei give the cortex. One specific nucleus (actually a group of nuclei), the *pulvinar* (meaning "cushion" or "pillow"), may be responsible for something else: It projects to the visual cortex beyond the main area serviced by the thalamic projection from the lateral geniculate nuclei. Instead of relaying eye input to the brain, the pulvinar seems to let us mentally focus on what we see—it governs our visual attention.[54]

Other parts of the pulvinar deal with other senses, governing our focus on what we touch and hear. Steering attention is an important function. The pulvinar is the largest part of the human thalamus, and, not surprisingly, it displays gamma oscillations.[55] So gamma and other neural activity not only let the brain bind processes but also provide a powerful steering mechanism.

Yet this raises the question: Which parts of the brain steer? The story is only partially in. There are probably many processes involved, but it seems that one specialized area of the thalamus oversees how information passes between the thalamus and the cortex. In between the thalamus and the cortex is the so-called reticular thalamus, Latin for "bedroom net." The reticular thalamus sends axons into the thalamus and receives impulses from the axons, navigating information flow in both directions between the thalamus and the cortex. Neuroscientists believe it has a role (along with the cortex and the thalamus) in generating gamma activity.[56] More important, it enjoys the perfect position to act as a kind of gatekeeper, controlling what happens above in the cortex.[57] But if the reticular thalamus acts as a kind of guide or lens on cortex activity, what controls it? Who controls the controller?

The brain lets several parts of itself (including parts of the cortex, the thalamus, and even the basal ganglia,[58] to which we will come later) shape what the reticular thalamus does. Another part consists of the brainstem neurons in another "net" called the reticular formation.[59] These manage the brain's general state of arousal. For instance, they have a role in shifting the brain from sleep into wakefulness and back. Were this neural "on" switch to break, the cortex would still be technically alive but effectively "brain dead."

Other possible controllers include the parts of the cortex whose axons extend through the thalamic net. Though the processes are far from understood, these areas probably give the cortex some control over its own processing. However, we do know that one part of the cortex has special power over the thalamic net: our prefrontal cortex, which may also control the thalamic net in a secondary, indirect fashion by exerting control over the reticular formation.[60] The prefrontal cortex, through the reticular thalamus, can organize what the rest of the cortex does. In effect, the reticular thalamus serves as the baton with which the prefrontal cortex can modulate and organize the rest of the brain,[61] a baton that can adjust, area by area, the activity of the rest of the cerebral cortex.

Instead of being shifted by external cues or automatic associations, what is bound by gamma can be manipulated from within. The brain can focus on a task. Attention can be sustained, reactions suppressed. Gamma, in short, can be willed.

In the 1970s many attempts were made to train people to control their alpha rhythms through biofeedback. A good idea, perhaps, but the outcome is generally admitted to be a failure.[62] In striking contrast, the few experiments that tried to teach control over gamma were successful.[63] Unfortunately, however, the researchers did not persist. As we noted, using EEG to detect gamma is difficult. So far no one has taken up their work using MEG. But it seems like an obvious candidate for a research program.

Prefrontal cortex gamma control seems to go through the reticular thalamus. Russian scientists using implants in humans have linked activity in this part of the brain to starting voluntary acts.[64] That is one clue.

Another is that the reticular thalamus, in its position between the thalamus and cortex, controls sensory evoked potentials. As noted, some of these are associated with gamma oscillations. Indeed, one direct link has been found between gamma and the prefrontal cortex. Some subjects had their fingers lightly stimulated, and depending on which finger it was, they were to move their big toe. It is a task that makes people focus sharply on their fingers. Sensitive EEG was used to look at the gamma that appeared over their prefrontal cortex and the part of their parietal lobe that attends to finger skin touches (an area distant from the one activated by movements of the big toe). It found that when the prefrontal cortex was alerted to a touch, its gamma synchronized for five oscillations—125 ms—with the gamma happening over the touch-sensitive parietal cortex, 9 cm away.[65] How they connected we are not sure, but it could be through either the reticular thalamus or the axons linking the prefrontal and parietal cortices, or even both.

All the above lines of evidence amount to hints—circumstantial evidence—that the prefrontal cortex has a key role in controlling and shifting the brain's awake gamma activity. While we lack proof positive, the prefrontal cortex is also known to control some other brain activity, called *slow potentials*.[66] These reflect the buildup of activity in cortical neurons used in attention.[67] PET scans show parallel increases in blood flow in areas over which slow potentials appear.[68] Linked to cognition, slow potentials predict people's accuracy in doing things.[69] A stimulus presented is learned and focused on better when a slow potential is present.[70] Slow potentials appear to be controlled by the reticular thalamus.[71]

Although the relationship between slow potentials and gamma is unknown, there is some evidence that slow potentials are ultimately controlled by the prefrontal cortex. German researchers have explored whether people can consciously control their slow potentials. Wiring them up with electrodes, they showed them their slow potentials in the form of the position of a rocket shape moving across a screen. The object of a game was then to get the rocket through a "goal" at the top of the screen. Lots of slow potentials and the rocket moved to the top of the screen; not enough and it stayed near the bottom. People found it rather easy to use the sight of the rocket as a visual cue to help them control the slow potentials appearing in their brain. With experience, people found they could move the rocket up and down blindly, without looking at the screen. In effect, they could will their slow potentials.

But not all people. Those with frontal lobe injuries[72] could control their slow potentials only if aided by the external cue given by the rocket. Lacking an intact prefrontal cortex, they could not cue, or will, their slow potentials internally.

The prefrontal cortex shapes how we process some evoked potentials in the first 25–50 and 100 ms of sensory experience.[73] This is quick—40 ms (Europe) or 33 ms (America and Japan) is the time it takes a television set to refresh one picture. An interesting case occurs when we attend to clicks: A brain wave appears over the auditory cortex at 51 ms, another roughly 14 ms later over the prefrontal cortex, and then another 14 ms later over the auditory cortex again. It takes roughly 14 ms for impulses to travel the 14-cm distance between auditory and prefrontal cortex at 1 cm per ms.[74]

It has been found that when the auditory cortex is responding to such sounds, the very earliest of its evoked responses—those in the 25- to 35-ms range—seem to escape the control of the prefrontal cortex. They are not modulated, or gated, by those with prefrontal cortex injuries. Later responses are, however, providing further evidence that the prefrontal cortex is sculpting brain activity.[75] The Swedish neuroscientist Per Roland, drawing on evidence from work with PET scans, argues that the fields of brain activity are "tuned" and "recruited" by processes involving the prefrontal cortex.[76] Cooling the prefrontal cortex reversibly brings to a halt the functioning of the rest of the brain. Based on such work, researchers suggest the prefrontal cortex modulates processing in other brain areas.[77] Likewise, injuries to the prefrontal cortex on one side of the brain can bring to an almost total standstill the brainwaves on that side of the brain.[78]

The growing list of hints from the "signal traffic" picked up on the scalp is that the prefrontal cortex is in subtle and powerful control, conducting much of the rest of the brain. In terms of the military intelligence metaphor, we have an advantage. Enemies learn with time to hide their communications traffic better. But not the brain—the brain doesn't care if we listen in. In the future, our listening posts on the brain will become sharper and more sensitive in interception. MEG is a fresh technology that has hardly been exploited as a research tool yet. Increasingly, researchers are combining EEG and MEG with functional brain scans such as PET and fMRI (functional magnetic resonance imaging) to more deeply probe and understand the brain. A foreign agent's intelligence report on the brain, if given today, would finger the prefrontal cortex as being in control. Soon, perhaps, we will know details of the structure of its staff command.

The importance of the prefrontal cortex does not end here. This chapter and its antecedent do not follow on our discussion of neural plasticity by chance. Although the prefrontal cortex interests us in its own right, it may have played an even bigger role in the past. Its protean and plastic powers, we argue, had a key role in increasing brain size and in launching the evolutionary journey that led to humanity.

6

Neural Revolution

Most of the reasons given for why our brains are special are wrong. It is not brain size—the brains of whales and elephants are much bigger. Ours weigh only a puny 3 lb, compared to the 10 lb of a 3-ton elephant's or the 20 lb of a 46-foot-long sperm whale's. (The brain size of the truly enormous 100-ton blue whale is unknown. They are so vast that by the time we can get their skulls open their brains have decayed!) Moreover, humans with big brains are not always bright. The largest two human brains on record—6 lb 4 oz and 5 lb 5 oz—belonged not to geniuses but to mentally retarded people with megacephaly.[1] Nor does the ratio of our relatively large brains to our relatively small bodies explain our uniqueness. Many small monkeys and even rodents have higher or equal brain–body size ratios. Whereas your ratio is the same as a field mouse's at around 2 percent, the brain of a small monkey may be over 4 percent of its body weight.[2] Nor do we possess any truly new intracranial hardware: The evolution of our brains has only altered relative size among parts already existing in the brains of other mammals and primates. For example, our cerebral cortex as a percentage of our brain is about twice that of a rat, while a rat's basal ganglia are roughly twice the proportional size of ours.[3]

Rather, our uniqueness lies in our brain's large size compared to what would be expected for a primate of our body *weight;* our brains are roughly three times the expected mass.[4] If chimps had brains the size of humans', they would have to grow to over a third of a ton—812 lb. And monkeys would weigh two-thirds of a ton—1339 lb[5] Moreover, the relative increases in mass in our brains were selective: Our primary sensory and motor cortices have only the surface area that would be expected for an ape of our body size.[6] What ballooned up our brains was vastly expanded temporal, parietal, and especially prefrontal areas.

Those increases, particularly in the prefrontal cortex, increased our brain's potential. And not just in the ways discussed. Not only does the enlarged prefrontal cortex better organize skills already possessed by the brain, but it also changes the brain's functionality, allowing it to rewrite itself, continually inventing, as it were, new software—especially for the expanded supersoft parietal and temporal areas. This neural revolution was the ticket that enabled our species to embark on an odyssey beginning some 100,000 years ago, at the time of our hunter-gatherer ancestors.

To understand this prefrontal-led revolution, we need to look at one of the big revolutions in brain science of the 1980s. First came the discovery of neural plasticity, that the brain had the potential to do more than it had evolved to do. The second revolution was the discovery of how it might manage this. Scientists, inspired by diagrams of the brain's circuitry, tried to get "assemblies" of computer-simulated neurons—neural networks—to mimic its information processing.

The basic idea of the problem they tackled is simple: We know how transistors in their vast numbers make silicon chips work—the computer logic to do that is designed into them. But what "logic" enables the millions of neurons of our natural neural networks to work together collectively? Again, what may seem distant from the problem of the human odyssey from hunter-gatherer to you sheds light upon how it happened. In modeling the brain, scientists discovered that it, like the body, needs something to thrive. Working neural networks do not arise spontaneously; they require a "nutrient" to get them to grow. The history of our species has been that of our ability to feed more and more of that nutrient into our brains.

Neural Networks

Before we can understand what this nutrient is, we must look more closely at neural networks. In the early 1980s, tempted by the pictures and diagrams that anatomists had drawn, researchers started using computers to simulate parts of the brain. Under their microscopes the anatomists had found the brain to be made of many nodes (neurons) connected by lots of links (axons and synapses) working in parallel. Computer scientists wanted to know if they could mimic informationally what had been seen in biology. What was the logic of neurons? It was an ambitious idea. The brain is made of more than 100 billion neurons, and it has trillions of synaptic connections. Even today, the best supercomputers remain less complex than the human brain.

Worse, while computers work by running sequentially through instructions, brains do many things at once. They are parallel processors. Computer scientists thus had to find ways to make computers work in this "uncomputerlike" manner. They attempted to do so by programming distributed and parallel activity over nodes and their links. Although extremely complex,[7] for our purposes these artificial networks can be treated as black boxes that successfully simulate parts of the brain.

Briefly and very roughly, these networks are made up of an input layer of nodes, a middle (or hidden) layer, and an output layer. These nodes are linked so that a pattern of activation on the input layer spreads to the middle layer and finally to the output one. Several types of neural networks exist, but we shall discuss the most common kind.

How might such a "neural network" mimic the brain? Consider an input layer activated by visual patterns and an output layer that then says whether the input pattern looks like a cat, rat, horse, or dog. A working network takes the visual pattern of a dog on its input units and lets this activation spread from these input units to the output ones. If properly trained, one output unit will be activated more than the others—recognizing a dog. In this way, the visual input pattern of a dog gets recognized as "a dog" by the network. It should be stressed that the activation of network computing does not spread sequentially through the net but passes across all of its nodes. A neural network thus "thinks" by spreading and interacting activation.

How this all happens depends upon what are known as "weights" on the links to each node. Weights add the activations and multiply their sum by a fraction between zero and one. The resulting activation is then passed on from a node to its links with other nodes. The important thing to note is that the thinking power of a network comes from its having the right weights. With the wrong ones it will get the wrong answer, or none. Thus, getting the weights right is the key to getting the network to "think"—and the source of a problem.

The weights have to be learned, and this is done by training. Given inputs, the network spreads activations through its nodes, leading to output responses. It then uses these responses to train itself. After errors are made, the various weights get changed so that, in effect, next time, given the same input, they will make the right response. This process of adjusting weights is repeated many times. Eventually, the weights evolve to the values that let activation spread so that for any input the network comes up with the right output. For instance, in the above illustration, at the end of the training process the network will settle on weights that let it recognize a visual image of a dog as a dog rather than a cat or something else.

Scientists cannot model everything, so they fudged some of the details needed to make their networks function like those in real brains. They slipped into the habit of talking about the above training as learning. But in most cases, learning was the one thing they did *not* get their artificial networks to do.

Informationally, they gave their neural networks a free lunch.[8] And as the saying goes, there is no such thing. The free meal lay in the error correction given during training.[9] This contains a catch-22: How can a network (or the processes training it) know when an error has been made unless it knows the correct answer already? Either the network knows, in which case it does not need to learn, or it does not, in which case it is not in a position to train itself. Scientists turned a blind eye to this informational double bind at the heart of neural network training. It might have been fine if they had acknowledged this major glitch, but they did not. Instead, their research papers talked about having modeled the very learning process taking place inside the brain—just the thing they had not done.

Bad science? Not exactly. Let's be positive and note the irony that from this mistake *we* can learn! That the computer scientists had to fudge is a profound insight. You may simulate neurons and the way they pass information among themselves, but you are still left with the business of how information arrives there in the first place. That is what cannot be simulated and must be glossed over. And that is a discovery: It means neurons must be sensitive—highly sensitive—to the existence of learning feedback. What functions they can acquire depends upon the original arrival of information needed to learn new functions. This is the information paradox, and it suggests a new picture of our brains. As our bodies require nutrients and energy to grow, so our brains need information to learn. They cannot develop new functions if denied the information needed to learn. They do not just manipulate information, they acquire it.

This insight changes how we understand neural plasticity, and how it allows the brain to work in an unlimited number of ways. Neural networks and neural plasticity are two sides of the same coin. Though not emphasized by those working on them, neural networks can take on nearly any computational task—functionally, they are "neurally plastic." But first they must be trained.

Human brains, especially modern ones, exploit their neural plasticity. Our brains have discovered "workarounds"—ways of getting around the information paradox—that enables them to tutor their potentials into complex functions. This is a deep observation that changes our appreciation of ourselves, and it is what distances us from the earliest humans. On the surface of it, we might seem to be discussing some obscure aspect of the computer simulation of the brain. But below the technical surface a new story is emerging. As a way to create our minds, neural plasticity can let our brains learn a near-infinite variety of skills—but not easily. The brain needs to provide itself with something that will create the workarounds required to free that potential. That something was present in our brains in addition to neural plasticity—and only in our brains. We have already met it; it is the prefrontal cortex. Evolved for one thing, it came to do quite another—provide the workarounds needed to exploit neural plasticity. In doing so it created the "missing link" between ape brains and our own.

The Cortical Catalyst

Let us look at the prefrontal cortex again. In Chapter 4, we note that the prefrontal cortex did not determine IQ. Damage it or cut it out, and IQ is not affected as measured by standard IQ tests. Indeed, lacking a large part of his left prefrontal cortex, one person managed an IQ of 150. However, there does seem to be a link. John Duncan of the Medical Research Council Cognitive and Brain Sciences Unit in Cambridge has argued that the "conventional belief in neuropsychology that frontal functions are rather unrelated to psychometric 'intelligence'" is, in fact, wrong.[10] IQ measures two kinds of intelligence, called *fluid* (also known as *g*) and *crystallized* intelligence. Crystallized intelligence is the intelligence we gain from what we know. This can be unaffected in prefrontal cortex–impaired people because it is based on knowledge stored in parts of the brain that remain intact.

Fluid intelligence is much more interesting. It concerns, as Duncan puts it, the ability of the brain to mentally program "an effective task plan by activation of appropriate goals or action requirements." Now, normally, when you measure someone's intelligence in terms of what they know (crystallized intelligence), you also indirectly measure this fluid intelligence. That is because the amount you know depends on the amount you learn. Other things being equal, a high ability to acquire knowledge leads to high levels of knowledge. But not always: You might lose your ability to learn new knowledge while retaining what you already knew. This is the situation of high IQ in the prefrontally injured. If you specifically test them for fluid intelligence, they are not the sharpest tool in the shed. Not to put too fine a point on it, they have lost the ability to learn.

Let us look closer at what Duncan means by the ability of the brain to mentally program "an effective task plan." Lose it, and you suffer what

Duncan calls "goal neglect." To give an example, people can be instructed in how to tackle a problem, and yet when the time comes to do it, the instruction slips their minds. They cannot remember requirements that they have been explicitly told. This is not because they do not understand these requirements. People showing goal neglect can have perfect awareness of what they have been instructed to do. The problem is that they cannot incorporate the information as inner cues to organize themselves. This being the case, goal neglect is easy to overcome: You simply give people feedback—external cues—that prompts them to do the right thing. But it is not just those with prefrontal injury who show goal neglect; it happens to all of us when we try something very new or attempt to do many tasks simultaneously, overloading our prefrontal cortex.

Researchers use instructions to test goal neglect. But goal neglect applies even more to our spontaneous tackling of problems. After all, we not only follow instructions but also invent them along the way. Every time we aim to do something, we create, as Duncan puts it, "new candidate goals by working backward from currently active supergoals." Our aim—in Duncan's word, our "supergoal"—cannot be immediately met, so we break it up into small goals that bring us nearer to it. Setting achievable goals requires a facility in working out what subgoals need to be achieved and an ability to accomplish them in a planned and constructive way.

Here we are halfway to understanding how the prefrontal cortex can turn the potential offered by neural plasticity into new brain functions. The prefrontal cortex acts as a neural circuit programmer. The first step in learning a new skill is to have a supergoal—the desire to be able to do something. From this we can sketch out subgoals needed to learn it. Many of these subgoals will concern information workarounds. The information needed to learn a skill will often not be available unless we do special things, such as

- Create or seek out situations that will give us useful feedback.
- Focus on particular information that helps us achieve success in learning tasks. If need be, learn about which errors best tutor our performance.
- Persist even if our initial attempts do not seem to get us very far. We get nowhere without practice.
- Spot what skills we already possess that can be modified.
- Break down the skill we seek into subskills that we can learn individually.
- Discover supergoals that make us enjoy all the above—find how to make it fun.
- And, most effective of all, cheat by getting others to show us.

This is only part of the story of prefrontal brain programming. The brain is the supreme associator. It can pick up something happening in one place and bring it to bear elsewhere in the same brain. Remember, the prefrontal cortex can activate an imagined sensation or an imagined motor action. And it can coactivate, bringing such activations together into working memory to produce novel skills.

We can look at the prefrontal cortex as a learning catalyst. Just as chemical catalysts create new organic substances by bringing existing substances into reaction by temporal and spatial conjunction, so the prefrontal cortex creates novel information associations by activating their components elsewhere in the brain and linking them within its working memory. Indeed, the analogy works the other way as well: Catalysts can be described as the working memory of molecules, since they work by holding substances on-line that would otherwise rarely meet and so give them the opportunity to react. As chemical catalysts release the potential of carbon, oxygen, hydrogen, nitrogen, and other atoms to create the complex organic molecules of life, so the prefrontal cortex brings elementary brain processes together to create the complex thoughts, experiences, and feelings of minds.

In sum, the prefrontal cortex can activate neural networks in parallel in situations that give feedback to the resulting information processing. And seek out the special error feedback that can tutor new emerging skills. And practice and persist at this conjunction. And spot what already existing skills can be creatively modified. And break these skills down into subtasks. And make all this fun, and relatively easy, by drawing on others who have already learned.

Take, for example, how you came to know how to sound out these letters. That is a skill that required your brain to orchestrate two skills together, those processing visual image identification and those devoted to speech sounds. By themselves, they would not normally link. But the prefrontal cortex can bring them together into association. The best way for it to do this is with the help of those who already know the link. Parents and teachers point to letters and say sounds. They present to us the association we need to hold in our working memory, thus entraining this association in our brains. Teaching aids help us by showing us letters together with interesting pictures of things whose names begin with the first sound we must learn—A and apple, B and boat, C and cat, and so on.

The prefrontal cortex also helps us teach. If the prefrontal cortex can help us seek and create learning situations, we can do this for others. Likewise for persistence, practice, breakdown into subskills, motivation, and even finding others better able to help.

The prefrontal cortex is itself the most plastic part of the brain, the most adept at picking up superskills. This is so for two reasons. First, it lacks the quantity of links between the brain and the external world that might put the brakes on its own neural plasticity. Second, the links it does enjoy with the rest of the brain mature later, much later: Some aspects of the development of our prefrontal cortex are not complete until we enter our second decade of life.[11] Synaptic density does not stabilize at adult levels until about age 16.[12] The prefrontal cortex is still refining the white matter links within itself well into a person's late teenage years.[13] Around puberty is a particularly active time of refinement of its circuitry.[14] As one neurologist, Ivica Kostovic, comments, "The final maturation may be a very late event in frontal association areas."[15] Also (as discussed more fully later on), compared to those of other apes, the human brain is developmentally delayed. A chimp brain does much of its growth inside the womb, something we are forced to do following birth. Such developmental delay—*neoteny*—is important because it defers the period of the brain's greatest neural plasticity until it is in an environment of much greater potential learning, the complicated world *outside* the womb. A consequence of this delay is that the prefrontal cortex's competence to change the rest of the brain stretches long into childhood and even early adulthood.

Moreover, research on adolescent monkeys suggests that their prefrontal cortices respond better than other parts of their brains to an enriched learning environment; a month after exposure to enriched environments, their prefrontal cortices had increased their activity by some 35 percent, while those of animals not exposed to an enriched environment had slightly decreased their activity. These "enriched" monkeys outperformed the others, both cognitively and socially.[16] As the most neurally plastic species, we can choose to put ourselves in stimulus-rich environments that will increase our intelligence. But child psychologists find that the development of prefrontal working memory—the key agent in reshaping the brain—takes time. In 4-year-olds, it is limited to basic sensory and motor functions; by age 8 it expands to include complex problem solving. Even then, it has hardly begun to develop the ability to "integrate complex processing demands" that it will possess as an adult.[17] Likewise, the ability to inhibit actions and focus attention develops in a way that is linked to the maturation of the prefrontal cortex.[18] Thus, the development of the prefrontal cortex's power over the rest of the brain can itself be shaped and expanded by what it learns, and so even bigger changes to the working of our brains can be effected.[19] The brain's inner teacher can be taught to be a more effective teacher.

Curiously, what is special here about human brains is also common to the rest of biology. Nature is a great reuser. Over hundreds of millions of years, limbs have been changed into flippers and wings—and even into the hands with which you hold this book. The most basic processes by which neurons work, such as receptor-mediated voltage changes across their surfaces, evolved from processes that exist in cells outside the nervous system. The revolutionary innovation of the human brain was to put neural plasticity and the prefrontal cortex together to do something new. Neural plasticity had first arisen as part of brain development, but it became increasingly important in its own right as it unlocked computational powers in neural circuits taught to do new functions. Human evolution expanded brain size. As a result, lots of free cortex and neural circuits were available for new tasks. Particularly enlarged was the prefrontal cortex. This expansion enabled enhanced delaying, better attention control, and goal organization—and then increasingly took on the new role of tutoring novel skills in this new brain.

A New Approach to Our Origins

Neural plasticity made the human odyssey possible; the prefrontal cortex made it happen. This can be summarized in a short formula in which NP stands for neural plasticity and PC for prefrontal cortex.

The formula is NP + PC = human mind. Neural plasticity plus the prefrontal cortex led to our minds.

But did neural plasticity and the frontal part of our brains go further and change what it is to be human? Or were hunter-gatherers, and all those people between their time and now, like us? Do we have, for want of a better word, different minds (though living in the same biological bodies and with the same genes)?

There have been, after all, many changes since the dawn of our evolution. Anthropologists and archeologists have shown how the lives of people in the past were different from our own. Indeed, the varieties of ways in which people have lived would on the surface seem incompatible with our being the same species—though of course we are.

Before we can probe these questions, we need to explore further what the brain does. The brain faces the orchestra problem—it has lots of skills that need to be organized. Previously, we focused on the prefrontal cortex as a conductor. But to carry on the analogy, what of the orchestra? What abilities can the prefrontal cortex—and so the mind—call upon? The brain is certainly not entirely plastic, as our genes shape its connections. Some 200 areas in each of our cerebral hemispheres are wired in. The brain uses this vast orchestra of connections differently in different skills. Maturation decreases somewhat the opportunities for neural plasticity, but considerable flexibility remains. So brain function as a whole remains open: While its wiring has been laid down, function can still be modified. To understand what might have been open to change and how it made us what we are, we first need to understand the skills evolution has genetically bequeathed to us.

Here we face a problem. If our brains are not what we thought they were, then perhaps neither are its skills. Indeed, our ideas of what our skills are may be even more misguided than our notions of what our brains are. Although we suggested above that the prefrontal cortex organizes such things as attention, vision, hearing, touch, emotions, memory, language, and movement, we must admit now that we did so only for convenience. It was not a suitable place to raise questions about the inadequacy of these notions in regard to the complexity of the processes they describe. In fact, none of these everyday words fully expresses the subtlety and richness of what the brain does.

To illustrate this as simply as possible, consider how ill-prepared even the common idea of vision leaves us for understanding the brain. "Vision," by common usage, suggests a process; but it is now known that it is built out of many processes, subdivided into at least 32 areas. Before the eye's input gets to the cortex it goes through two vision systems. One sees motion, and the other color. They largely come together in the primary visual cortex. Nonetheless, the overall visual system continues to stream this division into higher areas of the cortex. Vision goes into a *what* stream that identifies things and into a *where* stream that locates their positions.[20] And even this description is a gross simplification.

Neither is hearing a single process. We can use our ears to sense where we are (echolocation and spatial hearing) and to identify things, such as a cat from its purr. We can also hear speech, but even this is not what we take it to be. Hearing speech also involves seeing lips; it is easier to hear what someone is saying when we can see their mouth movements than when we cannot.[21] It is also easier to learn to speak: Blind children are slightly handicapped in learning speech compared to those who can see.[22] And what we see affects what we hear: A person seen to be pronouncing the sound *ga-ga* but dubbed with the sound *ba-ba* is heard to say *da-da*.[23] So part of our auditory cortex is also a "visual cortex" having access to what our eyes see.[24] Incredibly, the brain sends signals to our ears that cause them to emit sounds, otoacoustic emissions that counteract the effects of background noise; indeed, speech is more intelligible in ears that can make such sounds than in those that cannot.[25]

Nor can we neglect emotions, attention, and memory. Like vision and hearing, they are not what we take them to be. As the prefrontal cortex upgrades our brain, it alters our skills. To understand the nature of our intellectual, emotional, and sensory processes, therefore, we must look to what they used to be in our ape ancestors.

With this in mind, we can recast our formula in a new form. If we define "ape mind" as that suite of sensory, motor, and other skills belonging to prehuman apes, we can say:

NP + PC + ape mind = human mind

To see how that happened, we first need to look beyond human brains and see what shaped the evolution of ape brains.

7

Machiavellian Neurons

His proper name has become an adjective: *Machiavellian*. Niccolò Machiavelli (1469–1527), Italian statesman and philosopher, was the ultimate ruthless political manager, or at least that is suggested by what his name now implies. Ironically, the negative associations of his name may have less to do with his advocacy of cold-hearted strategy—trickery and fear mongering to gain power—than with his honesty about life as it actually plays out in politics. He may have been the first—at least in the West—to systematically observe the ruthlessness of power accumulation, but the subtle crafts he noted are older than our species.[1]

More than the technical toils of science and civilization, it was our social interactions that catapulted the ape brain into the human one. Modeling others—unpredictable, hierarchical, erotic, plotting, guarding, loving, hating creatures that we are—makes banging flint into arrowheads look like child's play. Our social—Machiavellian—prowess, both accelerated and was itself enhanced by changes in the prefrontal cortex.

Ever since it first created brains, evolution has pushed them into the most extreme places on Earth. The brains of some sea birds control wings that fly across ocean skies for 3 to 10 years without stopping. The brains of sperm whales take them 10,000 ft deep. Though next to each other, the worlds in which these brains live might as well be in different universes.

The biggest difference between the worlds in which brains live, however, is not physical but social. While for some brains life is a lonely struggle against a cold and predator-filled world, for many life is shared with tens or even thousands of their kind. In some ways this "social" world is an easier one, as it can increase an animal's chances of surviving predators, meeting a mate, and finding food. But it can also be a harder one, since living with others demands brains. The number of individuals linked in a civilization has a direct effect on the brains of its members—society is the kiln in which the clay of intelligence is fired.

Surviving with others was a key to the expansion of our prefrontal lobes and their subsequent ability to upgrade old skills for new purposes. More than any other animal, we live in a world of purposeful interactions. Other animals, such as ants and bees, might be more social than we, but their sociability is automated and instinctive, while ours is brain powered.

Grooming, Suckling, and Paying Attention

Sociability arose in stages. For the earliest groups, collective life happened among strangers. Fish in a shoal are part of a mass of individuals; they do not live in a social group per se. One fish in a shoal is much the same as any other. Dwelling in anonymous togetherness, their brains have only the dimmest awareness of each other. They live in a world of blind reactions that hold them physically to their group.

Evolution removed this anonymity, and allowed individuals to react to those with whom they lived. The simplest form of social recognition is ranking. Higher-ups dominate others in access to mates and food. Such high-ranking individuals tend to stand out in size and appearance; large antlers, bright coloration, and roaring sound mark high-status red deer, for example. Individual animals in a group have different looks in order to show social status, not identity. In many species, members of the group react only to each other's perceived status; they do not treat each other as true individuals. Eventually, many variations arose in the animal brain by which anonymity was lost and individuality was discovered. For some it was found with one individual. Swans and other birds that pair for life are familiar with their visually distinct partners. But this individuality still is not social at the "Machiavellian" level; it makes for pair bonding, not organized flocks.

Other animal species evolved beyond mate familiarity. Such sociability arose several times among mammals—in whales and dolphins, in carnivores such as dogs (and one variety of cat, lions), and in nearly all primates, such as monkeys, apes, and ourselves. These organisms had the brains that let them spot their fellows on an individual level. Dogs in a pack, dolphins in a school, and monkeys in a troop are not strangers to each other; their brains create interpersonal skills and social emotions. Brains processed who was who. Groups morphed from chance gatherings to complex, intertwined relationships. Status hierarchies tagged along in a new, more interconnected world of individuals working out their lives in a web of mutual attachments and acquaintanceships. Warm greetings, friendly get-togethers, and individual personality entered into evolution. For the first time an advantage lay in the social skills needed to survive friend and foe alike.

In monkeys and apes, especially, this led to brains that intensely needed each other. Instead of the company of strangers, they wanted to be with those they liked. Monkeys and apes search each other's fur for skin parasites as a way to be social no less than for hygiene. It is a way to say "I like you" while doing a favor. Although we may squirm at the idea, people also probably groomed each other before the existence of daily bathing and flea powders. Back massages and hair tousling may be social "echoes" of ancestral grooming. Chimps, once they find a parasite, dispose of it with their teeth (otherwise it might escape and hop onto them). This makes a characteristic chattering sound. (We did the same—it is one of the first meanings of the word *chat*.)

Such contact is used to manage emotions. We may feel we are the only species that misses absent friends, but apes do as well, perhaps even more intensely. Lacking tranquilizers, they have each other. Grooming helps keep their arousal under control by inducing their brains to secrete tranquilizing morphinelike substances.[2] This is particularly needed after separation. Chimp friends—or brothers, or mothers and their children—often become highly excited upon meeting after a few days separation. For a chimp, a few days is a distressingly long time. Missing each other deeply, they hug and calm each other down by grooming. Grooming is the social balm of apes and is useful in many situations. Jane Goodall records seeing an example of this in the band of chimpanzees she studied in Gombe National Park in Tanzania. A male she called Figan, after being hurt in a fight, was crying for his mother, Flo. She ran a quarter of mile and groomed him until she could quiet his whimpering. Figan, however, was no baby—he was at least 23 years old![3]

This illustrates something else. With ape brains, mammals had come to need the care and presence of their mothers, not only when young but for the rest of their lives. These bonds changed social existence. Now, unlike other animals, an ape mother and her children could know each other as adults, paving the way for new kin relationships. A mother, knowing the children of her children, could become a grandmother. The experience of life-long kin attachments also extended to brothers and sisters. Relatives started helping each other. Primate blood was becoming thicker than water: Monkey mothers helped their adult children, and monkey grandmothers took an interest in and promoted the welfare of their grandchildren.[4] In chimps, such relationships tend to be rare since females usually leave their mother's group. But where chimp daughters do stay in their mother's band, chimp grandmothers can take a keen interest in helping their grandchildren. As Jane Goodall notes of one such grandmother, "Melissa is a very attentive grandmother and has spent relatively more time playing with little Getty than she did with any of her own offspring."[5] Primates were a package deal: With them came the extended family, whose brains had the software to follow individuals through intricate social bonds.

Political Cortex

Not all the bonds were nice. When the connections among individuals are increased, so are their powers to destroy competing individuals. With the rise of sociability, simple social ranking led to complex power struggles. We may think of politics as a human affair, but those studying the lives of monkeys and apes find "politics" the best word to describe conflict in the primate world.[6] Apes and monkeys engage in power games in which they play off each other. They form short-lived alliances, then gang up on others to wrestle for social dominance for themselves. (Other carnivores, dolphins, and perhaps whales also gang up like this.) Such alliances are needed to win better social ranking.

At the top of the group is the alpha, the top male. Using alliances, those lower on the social echelon can gang up on the alpha and let one of them take his place. (The occupier of the number one rank can likewise make alliances to keep would-be usurpers in their place.) Jane Goodall saw the above-mentioned chimp, Figan, when he later became alpha chimp, help increase the rank of a youthful admirer and imitator, Goblin.[7] But once helped, Goblin played politics. With Machiavellian intrigue against Figan, he took over the alpha rank his mentor had held for 10 years. Figan, after this betrayal, made sure he had allies not only to regain his lost position but also to keep Goblin away.[8] Within a year, however, Goblin returned and got rid of Figan. Humans are thus not the only animals to play power politics; it is everyday life for chimps.

Such politics requires skills—brain skills. With whom do we share food, with whom not?[9] With whom do we cooperate? When do we compromise? Half the art of politics is knowing when to back down, accept a temporary defeat, reconcile, or await another day. It requires fluid "Machiavellian" intelligence to plan and project, anticipate social reactions, and act in accordance with presumed outcomes. Goblin knew when to play it cool and when to pour it on.

Goodall records Satan (so named because he stole part of her manuscript) meeting Goblin one day in the forest. Satan, traveling alone, had stumbled upon him peacefully chewing palm nuts. Goblin immediately sat up and stared at Satan with his hair erect. Satan's hair, in response, stood up, but he carefully avoided challenging Goblin by looking away from his eyes. For nearly a minute there was tense silence. Would there be a fight or not? Then Satan broke the silence by walking up to Goblin. He started to groom Goblin's rear. After a short while they turned to face each other and Goblin started to groom Satan in return. At first they were tense with loud clacking of teeth. But gradually they relaxed. After 20 minutes they went off feeding together calmly.[10] Ready to fight at first, they settled, for a time, for a peaceful accord.

A Mother's Love

Although the question rarely gets asked, it is in need of an answer: How did the skills for sociability arise? In evolution, nature reuses old abilities to new ends. The social skills primates use in their Machiavellian political machinations, we argue, owe a hefty debt to those that enable parents, especially mothers, to care for their young. Sociability evolved in tandem with parenting. Developments that helped the brain do one tended to help it do the other.

Not all animals look after their young. Most cold-blooded animals (with a few exceptions, such as crocodiles), such as fish, lizards, turtles, and snakes are "cold blooded" to their children. When they hatch, young reptiles often hide from their biological mother, lest they become her next meal.[11] Evolution has led to smarter parents caring more for fewer offspring. Mammals and most birds care for their young. And, as any 10-year-old naturalist will tell you, so did the warm-blooded reptiles from which birds evolved—dinosaurs.

Consider birds: They evolved "advanced" features of parenting. In most birds the care of chicks is shared between both parents. In contrast, most mammalian fathers, after mating, leave the resulting mothers-to-be. Instead of looking after young, they tend to invest their energies in looking for new females with whom to mate. In consequence, most mammals tend to grow up under the care of a single parent, though there are a few exceptions. But though bird parents share the caring for their young, it is much more limited than the care of mammalian mothers. Mammal mothers adore their children, and the roots of our sociability lie in the nature of their special love.

Although devoted, birds are stupid, which is another way of saying they tend to be trapped by instinct—habitual behaviors. A robin, for example, was seen to feed one of its young a piece of meat too large for it to swallow.[12] Then it chucked the young chick and the half-ingested food from its nest. Nice housecleaning but lousy parenting; the bird's instinct to clean the nest made it remove what it took to be a foreign object. Unfortunately that object was a piece of meat with its own young chick dangling from it—a real-life case of "throwing the baby out with the bathwater"! Chicks are also blind when viewed from the vantage point of our plastic brains. Konrad Lorenz discovered that ducklings imprinted in their brains the first image that they saw after hatching as their mother. When that image was not the actual duck mother, they followed Konrad himself around!

Birds have a more instinctive type of brain, partly because they fly and thus can't afford to have big, heavy brains. They may have something akin to a cerebral cortex—the *wulst*—but it is not as complex as a mammalian cortex. It lacks layers and a reticular thalamus to modulate processing with the thalamus below. Thus, birds cannot be flexible with their instincts, and so they get tricked; mammals, by contrast, evolved smarter, less rigid, more Machiavellian brains.

Mammals are better mothers because they give birth, though a few exceptions, such as marsupials (like kangaroos) and monotremes (like duck-billed platypuses), exist. And all mammals (kangaroos and platypuses included) suckle their young. After all, biologists call us "mammals" for a reason: female mammals have mammae—milk glands or breasts to feed their offspring with. The word *mamma* is tied phonologically to suckling.

Milk and suckling create extra opportunities for ties between a mother and her young. Milk not only gives nutrients and energy but also other useful substances like antibodies and even interleukin-10, an anti-inflammatory regulator for the digestive system.[13] Human breast milk even contains benzodiazepinelike substances. Benzodiazepines (such as Valium and Librium) bind to receptors in the brain, which means that breast-feeding babies imbibe a natural "tranquilizer."[14] Mammalian mothers, through their care, also regulate the sleep, activity level, oxygen consumption, neurochemical levels, and heart rate of their young.[15] And, equally important, they regulate their growth; touch, whether by a human or rat mother, is needed for young to grow.[16] In rats, touch when young shapes later adult behavior.[17] More than instinct-programmed food providers, mammal mothers care for, protect, and actively nurture their young.

It is a two-way process. When the mother goes away, her children can call her back. We can see here, in auditory cues, the origins of spoken language. Crying arose in part because mammals, in general, have better hearing (especially at high pitches) than reptiles or birds. Indeed, biologists identify the rise of the inner ear bones (which act as bioamplifiers) as the point at which our reptilian ancestors turned into the first mammals. Their infant cries, from the high-pitched squeals of squirrel monkeys to the soprano bawling of the human child, repetitively undulate up and down.[18] Each infant's cry tends to be distinctive, identifying it to its mother. And as any human mother will tell you, unlike the distress squawk of the bird, different cries of her children carry different meanings. Mammal mothers respond to the vocally announced needs of their young.

The heightened care that mammal mothers give their young was crucial in mammalian evolution, because it prepared the way for communication skills that could be reused in true adult sociability. Mothering created brains that attached and cared not only for young but also for friends and kin. Not only do monkeys and apes get a fix of morphinelike substances when they groom, but so do mothers when caring for their offspring.[19] Primate societies, including the dark flower of human civilization, did not appear from nowhere but sprouted from the seeds of the mother-child bond.

The roots of sociability can be seen in the attachments littermates share with each other. The less instinctive brains of mammals have much to learn, and they do so by playing. All young mammals play.[20] Usually, such play is with parents and other young. A few birds—ravens, jays, and parrots—also play, but play in birds appears to be limited to those species with the highest brain–body size ratios. And many mammals never stop playing. Chimps, monkeys, hyenas, dogs, and female lions as well as humans continue to play as adults. Even sea lion females visiting tide pools in the evening chase and mock-fight with each other. Many mammals play before mating. Birds engage in elaborate instinctive courtship rituals that show well their heritage from reptilian ancestors and their ritualized, hard-wired, cold-blooded neural processing. Before mammals mate, by contrast, they play in a way that, it is claimed, "except for the size of the animals, is virtually indistinguishable from the play of juveniles."[21] It is not only humans that engage in "foreplay."

Although in most mammals such early family emotions are gradually forgotten, they nonetheless provide a ground on which the skills needed for adult sociability can be built. In other social mammals the emotions of childhood remain to form an important part of adult life. When in pain, young mammals cry—they whine, bleat, bay, howl, wail, or whimper. Such "pretalk" brings, if not their mother, then other members of the pack, herd, or troop to the infant's aid. That response to another's pain starts early. Even day-old human babies are distressed by another baby's cries.[22]

As adults, many mammals keep this need for someone else. In monkeys, separation cries link not only child and mother but the individuals in a group. In us, the chirp and the bleat of the pager or cell phone are technological updates of such separation cries. When whalers harpooned a whale, others of their pod (conveniently for their killers) tried to help their distressed companion. Chimpanzees, when stressed and in fear, clasp each other the way they did their mothers as babies. So do spider monkeys—and, if we are honest, we must admit that we do too.[23]

Humans have pushed the mammalian penchant for play to unprecedented heights. Think of baby talk and clowning around, the sweet nothings whispered and goofy vulnerable displays demonstrated by lovers, or the playfulness of poets, inventors, and artists. The skills the brain evolved for childhood allowed some mammals to experience vocal and playful attachments as adults. In short, the evolution of mothering ushered in not only caring but adult personal lives and problems in monkeys and apes— with the associated Machiavellian manipulations, social problems, and opportunities.

Social Tensions

Don't take life personally, they say; but still we do. Group life, while it can be loving, can also become quite nasty. Living with others is not so easy. Tensions flare up when things get stressful. Different members of a group compete, sometimes beneath a surface of compliance and partnership. Primates in the wild challenge each other for food, status, and reproductive opportunities; yet they tend to hunt together, collect food together, and scare off predators and individuals from other groups together. Where studied, mammals with complex social lives experience social dramas as alliances form and break down or when ranks change. But under the tensions, they must find a way to live together. Life in a group may be harsh, but the outside world is harsher.[24]

Beyond refashioning maternal relations into social attachment, mammals had to evolve means of reducing social tensions. From Charles Darwin to Franz de Waal, scientists have claimed animals have a kind of morality.[25] Although its foundations are subject to philosophical debate, there is no question that codes of conduct ward off tensions within a social group, greasing the wheels of social interaction. This is true not just for us but also for other social animals.

Thus, animals in a social group find themselves living in a world governed by "oughts," a social "morality" of "right" and "wrong" that shapes what is and what is not allowable. Codes of conduct are especially well developed in chimpanzee cultures.[26] Consider Goblin and Satan again. Goblin stares at his competitor Satan. Satan tensely averts his eyes and approaches Goblin's rear and grooms it. He could take a bite at it. When Goblin turns around and they look at each other, Goblin could do likewise. But they do not. Instead, they groom each other and go off feeding, relaxed in each other's company. Chimpanzee life is full of such tension-reducing techniques, socially sanctified and learned modes of averting aggression.

Chimps, so interdependent in a band that they can rarely risk hurting each other physically, have developed bloodless ways to challenge status. For instance, instead of a true attack, a charging display using branches will be made. Not that tensions always disappear with such antics; sometimes blood is spilled or deaths even ensue. After 8 years at the top, on the morning of September 19, 1989, Goblin was deposed in a brutal and bloody attack. If a veterinarian had not by chance been visiting Gombe, Goblin's life would have been over.[27] But such violence within a group is rare; it runs against the grain of finding ways to cooperate. This ability to hold back in order to preserve the group had to happen if we were to evolve personalized social lives. Somehow chimps' brains had to find within themselves, most likely in the prefrontal cortex, a means to hold off "hot" emotional reactions.

Self-control helps the group, but it can also serve the individual's self-interest and aid his or her ability to play those games of life. Jane Goodall recalls trying to feed the adolescent Figan bananas after more powerful males had left the feeding area. But, excited, Figan made food calls that drew bigger males, who grabbed the bananas. A few days later, however, when Figan was given bananas he attempted to silence his food calls. He was successful, but as Jane Goodall notes, "calls could be heard deep in his throat, almost causing him to gag."[28] Perhaps in such restraint lay his future abilities to rise to become number one. Holding back like this is, in a sense, the opposite of pride: It is the mother of all social skills. In social situations we are confronted with a host of conflicting options. Should we get quick and easy relief of tension or bide our time? Using alliances for one's own ends requires restraint, stemming of the tide, suppression of the urge—in a word, self-control. One must not take out tensions upon partners until it can be done to advantage. Politics is about selecting the right time; it requires self-discipline as well as intelligence. Such restraint requires extra brain power. It must be learned. Play is one of the ways we learn it. In playing, young animals use their brains to learn the skills they will later use to compete and cooperate.[29] Through play-fighting, for example, they learn the limits of what they can do and how to stop themselves from exceeding them. Behavioral taboos, like social skills, are learned, usually by pretending and play.

With the rise of greater sociability, an explosion occurred in the depth and number of relationships a group could hold. Indeed, social complexity snowballed as monkeys evolved into apes and us. We know a larger number of other members of our species and in greater depth than do other animals. But the more individuals an individual knows, the smarter he or she must be to keep on friendly terms, and this requires good social know-how, with abilities in remembering, recognizing, and dealing with others.[30] Individuals, moveover, needed to learn how to manipulate their relationships and alliances to their own benefit.[31] Brains needed the ability to understand the social politics of their band and know how and when to juggle cooperation with competition. That required assessment of social situations. Who was best to pair up with in alliance? Whom should they avoid? These pressures led to even bigger and brighter brains.[32]

After all, not every brain is born equal to ponder and answer such questions. Consider once more Goodall's chimps Goblin and Satan. Although Satan is older and nearly three times the weight of Goblin, he never made it to the top of the chimps' "who's who" list as Goblin did, for the simple reason that he lacked the social skills needed for making alliances. Not only was Goblin able to get Figan—after 10 years of alpha rank—to help *him*,[33] but he also had the skill to know when to break with the relationship and go for the top himself. Figan likewise was socially skilled and able, however briefly, to make a comeback.

High rank helps chimps reproduce their genes—politics has its rewards. Female chimps in heat will mate with many troop members while ovulating; such promiscuity, in which the identity of the father is not known, is rather typical for them. But the top dog, head honcho, leader of the pack, alpha chimp does not have all the females to himself. Male chimps also father offspring by being more normal, in human terms (and more "deviant" in chimp terms!): by persuading females to go on romantic little getaways, isolated "consortships." This requires advanced social skills, since a female in heat will not disappear into the woods with just any male. Successful consortship in males demands an ability to get along with females and an ability to spot an opportunity and take advantage of it.

Social competition gave evolution a sharp cutting edge to select the genes that build brains better able to compromise, cooperate, and win in the social games of life. Playing social politics may be hard, but it rewards skilled brains. Goblin probably fathered more of the next generation of chimps than did Satan. Over many hundreds and thousands of generations, smarter, bigger brains more advanced at social skills evolved. These brains, in turn, led to still more intricate social situations. A feedback loop thus evolved, linking brighter brains to more complex societies.

Not surprisingly, the part of our brains able to assess a situation and to delay, organize, and activate our behavior according to worked goals—our prefrontal cortex—exploded in size during ape evolution. Remember the story of the children whose ability to resist the temptation of marshmallows suggested that they used their prefrontal cortices? Whether in apes or in Machiavelli's Renaissance Italy, planning future benefit requires brains. And the same parts of the brain that are used in abstract thought are used in delaying gratification and plotting social intrigue. By and by, over the evolutionary years, selection for sharper and larger prefrontal cortices led to our species.

Fission-Fusion Bands

The dispersal of food also shaped our brains. Leaves are everywhere, but edible fruit is scattered and separated in small concentrations on trees and bushes. That poses different problems to brains: Herbivores, who do not require wits to find their food, can live in herds. Fruit eaters, in contrast, must trek off in small numbers while finding a way to stay together in a group.

Some primates live both in small feeding parties and as members of a much larger group. Members of such a group have a territory and know each other well. But the group of which they are a part rarely meets up as one; they are a band, not a troop. Instead of living together, its members go off in small parties or on their own roaming for food. It is an adaptation to gathering widely dispersed foods in small parties without losing the benefits of belonging to a large group. A band or large group, in spite of fragmentation, is nonetheless a social entity.

Chimps and bonobos—a kind of chimp previously called pygmy chimpanzees—live this kind of life. Both sexes search for food in small parties (usually of three to four but sometimes many more and of mixed or single sex) that come together daily.[34] But these parties are not the true social groups of chimp life; come the next day, the same chimps may or may not be together. Different parties form fluidly on a day-by-day basis. In this way, the chimps casually meet their friends and allies, and their sense of belonging to a larger group is renewed. The larger group or band has a territory, and parties of male chimps will go off to patrol its boundaries against foreign chimps from other bands (at least in common chimpanzees). Chimps outside the band, considered aliens, are attacked furiously. As Jane Goodall remarked, "If they had firearms and had been taught to use them, I suspect they would have used them to kill."[35]

Primatologists call such groupings *atomistic* and the process of breaking apart and coming together *fission-fusion*.[36] Their lives are "fused" together; they know who is a member and who is not. If males on patrol meet adults who do not belong to the group, they attack. *Fission* means that the groups rarely meet as a whole, as members are usually fissioned off or fragmented into short-lived subgroups. A chimp may feed with two members of its group one day and another two the next.

We too are members of a group with whom we spend little time. Though belonging to social groups, we spend most of our lives on our own or with only a few others. At work, for example, we usually spend our days with a few people, with some of whom we are friendly. That is true if you work at a word processor, with a hoe in a field, gathering wild foods, or setting traps and hunting (as did the earliest humans). Only rarely, such as during a wedding, do the members of an individual's wider social group meet face to face. Though chimps may separate, they will meet the rest of the members of their band over a few days. In this lies a huge difference between us and them: We can remain apart from our groups for much longer periods. We have a "superfission-fusion" social life.

Fission-fusion existence selected for better communication skills so that apes could find a way to retain identity as a group. Greater communication amplified alliances, leading to more sophisticated social lives. Primates became more complex, competitive, and cooperative, with increasingly subtle plans and means of bringing to fruition their goals and supergoals. Not surprisingly, then, chimps, who are more social, have bigger brains for their size than gorillas.[37] These factors—communication, social interaction, and increasing braininess—were in full swing some 6 to 8 million years ago in the social lives of the apes that branched off from the common ancestor of chimps and gorillas, the ancestors that evolved to become us.

8

The Troop within Our Heads

Unlike other animals, we have free will, or so some have claimed. Much religious doctrine has asserted that Man , being God-made, was given a will to choose. Even if we are not religious, free will is something we feel in our lives and our emotions. (And if it is not true, we have the *illusion* of free will, which is even stranger!) We may not always be rational—we may be stupid, ignorant, selfish, and even callous—but we imagine ourselves to be the masters of our actions and beliefs. We seek to learn our own lessons and to be wrong in our own way. Above all, we want to do our own thing, not what others want us to do. We have, after all, a brain packed with a large prefrontal cortex that can organize itself by inner cues.

But does that make us free? Some have questioned whether, in fact, we are. "A freeman?—there is no such thing! All men are slaves; some, slaves of money; others, of chance; others are forced, either by mass opinion, or threatening law, to act against their nature."[1] Such were the dark words of the ancient Greek playwright Euripides. Was he right?

When looking at freedom, we need to distinguish between physical and psychological freedom. Roughly, the first is that made by the nature of our society and technology, the second that made by our minds. When you have no money, someone has a knife to your throat, technology is not advanced enough, or the law forbids something, you lack physical freedom. But often we do not lack money, no one is threatening us, technology lets us do just what we want, and the law gives free rein to our actions, and yet we still cannot carry them out. Somehow our freedom is blocked at the source: our minds.

The problem is social deference. Sociability takes precedence over our own inner cues. We defer to superiors or the group, even when we have a right to assert ourselves. Social pressures and situations arise that trap us so that we find it hard to say no or take a stand. How often do you go into a restaurant and get poor-quality food or service but remain silent—even when asked, "How is everything?" Some of us are skilled at complaining, but all of us defer in one way or another to the "right thing." We might be able to organize our behavior by inner cues, but these cues themselves can be overruled. Although we may be loath to admit it, sometimes we become like puppets of the wants and expectations of others.

Such deference can be surprisingly strong. Consider your ability to ignore orders from a stranger. It's easy. If someone politely tells us to do something we do not want to do, we can walk off. But perhaps we are overestimating our freedom. Imagine that you answer a newspaper advertisement asking for volunteers to take part in a short experiment. The advertisement says you will be paid for your help. A scientist in gray lab overalls greets you and explains that you will take part in a study of learning. He pairs you with someone else who has answered the same advertisement. Following the toss of a coin, you take the role of "teacher" and the other person takes the role of "learner." The learner goes into another room and is wired with electrodes. When the learner makes a mistake on a learning task, you are to administer an electric shock. The scientist in gray overalls asks you to give increasingly strong electric shocks when the learner fails to "learn." What is the chance that a few minutes later you will obey the scientist and, as the learner fails to learn anything, give increasingly stronger shocks—so strong that they torture and kill the learner?

A young academic at Yale asked psychiatrists and psychologists whether people would do this.[2] Universally, he found that such experts on human nature thought that they would not.[3] In the early 1960s, Stanley Milgram, in one of the most famous (or infamous) experiments in the history of psychology, then went and showed that people would. Of course, Milgram did not let people give real shocks, but through an ingenious system of acting and fake equipment he led them to believe that they were. No doubts exist that his subjects believed that the shocks they were giving were real. Most of them did not enjoy electrocuting the other person. Indeed, they fought a stressful battle between the urge to stop and the orders given to them to continue. They felt deeply that what they were doing was wrong and kept asking whether they had to continue. But when pressed, they still obeyed. The man in the gray overalls did not force them; he merely told them to keep giving stronger shocks: "It is absolutely essential that you continue" and, "You have no other choice, you must go on." Nothing stopped them from telling the man in the gray overalls where to put his "*******" machine and walking the hell out—they were not locked in. They were just told to stay and give killing shocks. It was not bars, but words—and something in their minds—that imprisoned and "forced" them.

Milgram's results have been a *cause célèbre* ever since. After all, showing that ordinary Americans would kill if ordered is not a small discovery. It cuts to the heart of our deeply held feelings about humans in a free society. We like to believe that given freedom, we will use it for good, not for what we know is wrong. But here was a scientist showing that we can all be easily tricked into giving up our free will and electrocuting another human being.

Milgram's experiment offers a mirror in which to see history and ourselves. Before, when people looked at the horrors in Nazi death camps, they could tell themselves that they would never do such things—"Oh no, not I." But Milgram took away such comfort. Here were people like us giving up responsibility for their actions and, however much they may have disliked doing it, obeying orders to kill. We can no longer tell ourselves that we are so different. Somehow our free will can disappear, and with it our ability to do what we know is right. It is not a pleasant thing to discover.

Nor is it pleasant to find that animals can act far better. *Monkeys* trained to pull a cord for a food reward stop when they see that their action causes visible and auditory distress to a monkey in the next cage! Some refuse, even if they have to go without food for days.[4] Many theologians have denied that animals have a soul, but if they don't, then perhaps it is an evil appendage they are better off without. Here we have a very dark finding about ourselves, a "bestiality" unique to the human animal.

A Scientific Taboo

We do not want to look. Before we press further into our social natures and what underlies them in the brain, we need to ask why so little is known. Surely one of the most fascinating aspects of ourselves is how we affect each other. It is both part of the richness of our lives and one of the ways in which we lose freedom (an understanding of the loss of which is needed for understanding our evolutionary origins). It is an area as important as it is understudied.

The problem can be traced to our individualism. We see ourselves as masters of our actions and prefer to be ignorant of the ways in which we are not. One of the most striking findings of Milgram's experiment was the failure of psychiatrists and psychologists to predict its result. If anyone could have, it should have been them. But they did not, and part of the reason is

that our philosophy of individualism makes us underplay how far we influence (and are influenced by) each other. Our independence is overrated. We
take it for granted that people are free from others if we observe no visible
restraint. This is not an unreasonable belief, but often it is just wrong. We are
more than reluctant to accept this, for we find the subject distasteful. We
would rather avoid finding that people give up their freedom than do the
research that could show us how to keep our independence.

No one today would be allowed to repeat Milgram's research. To put it
mildly, strong voices in psychology complained that Milgram had gone too
far and treated his subjects badly. To show obedience, he had had to mislead his subjects. He could not tell them what he was really studying or ask
them to give real electric shocks, let alone fatal ones. So he lied and paid
someone to fake the screams and pleadings of a person being electrocuted.
In other words, he deceived people. Was he wrong to do so? There was a
strong reaction to his use of deception. Today, no ethics committee would
permit such research.

But was *deception* really the reason people objected? Some doubted this
and did their own experiments. They asked people to evaluate the research
done by Milgram, but they deceived them by varying the description of
what he actually did and found. Half of the subjects were told the true
nature of the experiment and that the person's screams were faked. The
other subjects were not. Each of these groups was then split again, with one
half being told the true outcome of the experiment—that people obeyed—
and the other half what was predicted but not found—that people didn't.

The presence of deception in a description of an experiment did not
alter people's judgment about its morality. People judged the use of deception to be acceptable, but only if the experiments did not find that people
obeyed.[5] Thus, what people really objected to in Milgram's research was its
finding, that people easily give up their freedom. Any research that shows
that we are born unfree, and suppress our individuality with regard to others or our group, touches a raw nerve. We would rather not know.

Individualism has thus made our social natures one of the last taboos,
at least in science. If psychologists study it, they usually do so from an individualistic point of view. Indeed, one social psychologist, C. Daniel Batson,[6]
has complained that it has "become a taboo topic—like sex for the
Victorians—that we social psychologists politely avoid, especially in public."
Research, he observes, "has been to stay close to the surface . . . assuming
perhaps that the embarrassing problem of our social nature will disappear."
As Batson further notes, "Perhaps the reason that social psychologists have
spent little time on the question of our social nature is because they already

know the answer. . . . Our behavior may be highly social; our thoughts may be highly social; but in our hearts, we live alone. . . . We are social egoists." As another psychologist put it, "From Freud to sociobiologists, from Skinner to social cognitionists, from Goffman to game theorists, the assumption in Western psychology has been that humans are by nature asocial individualists . . . social relationships are instrumental means to nonsocial ends, or constraints on the satisfaction of individual desires."[7] A Chinese anthropologist has called it a Western "hang-up."[8] Scientists would prefer it if we were machines that used others to satisfy our individual wants and drives. That is how they mostly study us. It poses problems for them that they (with the rest of us) are, most of the time, warm, feeling, and caring social creatures.

But of course that is what we are. Being a social creature does not normally limit our freedom, but in some circumstances it can. If we want to be free, we need to understand why we so often give our freedom up. It is a question we need to ask to understand what skills lie behind our sociability, our mind and its odyssey. But we cannot start to do so until we acknowledge that we have social natures.

The Unfree Mind

Where can freedom be found? Perhaps in a flock of estuary birds? Flying together at high speeds, thousands of birds maneuver with precise coordination. The flock flies this way and then that. It turns as if a wave has passed through it. These "maneuver waves" start slowly in a few individuals but quickly spread. Unless the individual birds reacted together, the flock would disperse, exposing isolated birds to predators. Sometimes it is "smart," in a survival sense, to give up your freedom and fit in with the group.

Once started, a wave travels through a flock at about 70 birds a second.[9] Surprisingly, this is much faster than a single bird's reaction time. Thus, individual birds cannot have seen their neighbors and said to themselves, "Hey, they've changed direction—I'd better copy them." Something else besides copying is synchronizing the birds. Somehow they see themselves, even if only for a short time, as part of a whole. They see the wave maneuver and time their own change of flight with it.

Individuals cease to be individuals in many ways—not just when flying together. Humans can react physically as a group; a wave of legs passes down a chorus line at roughly 10 dancers every second. As with birds taking off, this is too fast for movements made in reaction to neighbors. A similar thing, no doubt at a deeper level, organizes a jazz jam or a basketball team. This suggests that people are good—surprisingly good—at synthesizing their actions into a larger whole. Soldiers marching in step with each other are not doing so as individuals.

We all have a sense of "we" that identifies us with "our" group and favors "us" against outsiders. We have our fraternities, sororities, and other old boy and girl networks. We seek out people who share the same club, school tie, or accent.[10] Much of this activity is harmless, but our loyalties also have their darker side. When loyal group members are found to be doing wrong—committing sexual or physical abuse, faking data, or taking bribes—other group members protect them. The bonds among group members may make them treat the whistle-blower, not the wrongdoer, as the criminal. They do this especially if the whistle-blower is a member of their in-group—one does not squeal, tell tales, or inform on one's comrades.

Social psychologists find that we easily become prejudiced. It takes the smallest hint that you belong to one group and other people to another for you to favor "your own" group. The reason you belong to one group rather than another may be no more than a preference for abstract artists, Paul Klee rather than Wassily Kandinsky. You need not even meet and interact with the members of your own group, but prejudice will nonetheless rear its ugly head.[11] It may be our football team, school, town, or nation or the color of our skin. Once fully identified with that "we," people become sensitive to the needs of their group and callous toward other groups. Outsiders cease to matter. The stronger our identification with the "we," the blinder we become to the humanity we share with the "them." Out of this psychology comes the nasty side of history and the human race: the world of "ethnic cleansing," genocide, racial prejudice, and global terrorism. Thus, we may be born alone, but we quickly learn to identify ourselves with a group, leading, in some cases, to barbaric consequences.[12] (In passing, we should note that the march of science and ethics—at least in the West—has led in large degree to a rational rejection of this social psychological tendency to favor the groups to which we belong.)

But we say more than "we" with our social group. In each of us is a silent "we," shaping our lives with those to whom we are close. A hidden harmony exists in our minds that shapes our ties with each other. Social animals are social not only in what they do but also deep in their brains.

Many neurochemicals govern neurons. Hidden from consciousness, they regulate our emotions and our lives. We may think we are free, but it is a freedom lived within a neuropharmacological cage. Science is just starting to understand this neural underpinning of sociability. Brain chemicals rise and fall together among people who are physically or emotionally close. For example, the menstrual cycles of female friends and roommates living in student halls become synchronized[13]—a process regulated by these chemicals. In most cases women are not aware that this is happening. Menstruation is regulated by the *hypothalamus* ("below the thalamus") through the pituitary gland. Through smell, a woman's hypothalamus picks up and harmonizes with the menstrual cycle in others. Housed together, men also harmonize the chemistry of their brains, but in their case it is their daily cortisol rhythm.[14] This relates to another observed synchronization, this time in our basic rest-activity cycle, which paces our lives at intervals of roughly 1 to 1 1/2 hours. This "biological clock" accounts for periods of dreaming and dreamlessness and also, during the day, for a shifting between our tendency to rest or be active. The clock is easily overridden in the modern world, but in hunter-gatherer societies it can be seen shaping the activity patterns of everyone living together in a village.[15] Our brains, it seems, seek to be orchestrated as part of a larger social whole.[16]

Probably brain-based, the subtle interactions between us and those with whom we are emotionally or physically close are "tuned" by external events. When we delight in someone's company, we synchronize our movements with theirs.[17] Our faces and hands are anything but waxworks when we chat. We are active together in duets of eye contact, tones of voice, facial movements and body gestures, and physical contact—handshakes, pats, and hugs.[18] And we do this unconsciously and spontaneously as part of the flow of being with someone. A smile is the brain's way of saying it's happy to another brain. Fake smiles do not fool. When we "put on" a smile, we do not move all the muscles that light up a genuinely spontaneous smile.[19] Spontaneous smiles sport not only turned-up lips but a gathering up of wrinkles around our eyes (at least if we have not had a botox facelift). Our faces communicate more than we realize or control.

These expressions of ourselves affect our neurophysiology. We might stonewall or "stiff-upper-lip" the emotions we see in others, but if we open up, they affect us—and affect us deeply. Even the visual clues seen when looking at a video are affecting. People looking at and passively empathizing with a stranger on a video undergo significant changes in their own physiology. Provided they can empathize with negative emotions, they synchronize with what is happening physiologically in the stranger.[20] And if we can be so measurably affected by the weak contact of a video, the neurophysiological transfer among physically and emotionally close people— laughing, crying, fighting, loving—must be far greater. Our brains probably tune in far more deeply to what is going on in those close to us than we realize (or want to—many a relationship has faltered because the depression of one gets picked up and depresses the other).

Two important shapers of close relationships are the peptides *oxytocin* and *vasopressin*. These two brain chemicals regulate social, and especially sexual and maternal, bonding. They are secreted by the hypothalamus into both our bodies and our brains.

Their "nonsocial" secretions into our body must be briefly noted to avoid confusion with their "social" secretions into our brain. Oxytocin is secreted into the bloodstream after the sex act, following the birth of children, and when mothers nurse. Oxytocin also stimulates labor. Indeed, some women in labor are given artificial forms of oxytocin to induce contractions, though in this case the natural oxytocin being mimicked is made not by the brain but by the uterus. (The hypothalamus does make it during birth, but the uterus outdoes it, producing 70 times as much.) Notice that all these situations are linked to bonding events. Vasopressin's nonsocial functions concern limiting water secretion. Most of us have experienced dehydration after drinking alcohol. One of the less publicized effects of alcohol is that it inhibits vasopressin production and so causes our kidneys to pass water. Medically, failure of the hypothalamus to make vasopressin can lead to diabetes insipidus, which can cause people to drink and pass up to 20 qt of water a day. The body and brain secretions of oxytocin and vasopressin, as well as other chemicals, are kept apart by the blood-brain barrier.

In the brain, vasopressin and oxytocin help make us faithful lovers. The prairie vole is rare among mammals for being monogamous. In these voles, both vasopressin and oxytocin are secreted after extended bouts of love-making, lasting for hours.[21] They also help parents form loving bonds with their young. But these brain chemicals can act differently in females and in males. In females, oxytocin acts to induce bonding; it has a somewhat more complex role linked to erection and ejaculation in males. Vasopressin secreted during sex makes prairie voles affectionate to their partners and hostile to territorial intruders. Does this happen only in rodents? One study has found that receptors for oxytocin in a monogamous primate, the common marmoset, are distributed in their brains as are those of monogamous voles.[22]

Vasopressin and oxytocin may have similar effects in us. They may "drug" us into being more or less monogamous and attentive, not only to each other but to our children: Oxytocin in particular is linked with infant attachment.[23] Perhaps because of this, oxytocin induces relaxation, lower blood pressure, and a general antistress pattern.[24] It has been suggested that oxytocin is responsible for the antidepressant effects of Prozac and similar drugs.[25] We speculate that the rare phenomenon of a remarkably pleasurable, even tingling sleep—you will recognize the description if you've experienced it—that appears to occur only between more attached, longer-term lovers (rather than new loves) is oxytocin-induced.

Another important group of brain chemicals is our natural opiates, endorphins. As noted in the previous chapter, social primates release them in their brains when being groomed, grooming, or caring for young. This is also likely for humans. Indeed, evidence exists that people who commit suicide have changed levels of opiate receptors, suggesting that their experience of social isolation may originate in a lack of natural opiates.[26]

Again, this is just the tip of an iceberg; such processes could shape our lives, our happiness, and our mental health. Is social isolation the other side of being in love? Do people differ in their secretion of and sensitivity to such brain chemicals? We do not know. In spite of its fundamental importance to all of us, we know little about how and when our bodies tune in to each other. One of the richest facets of our lives may be passing us by unobserved and unacknowledged.

Other People

All this makes us extraordinarily social animals. It is not only chimps who intensely need each other's company. We would guess that some two-thirds of our lives—our cares, anxieties, and actions—are spent with other people. Even when we're alone, most of our concerns are about other people—what did I say? Did I hurt him? Do they still love us? Will she forgive me? We rarely talk purely to pass on information. We talk to be friends—to chat, gossip, and bond. Not for a few minutes but for hours. Seventy percent of the talk when professors and their students eat together is socializing, talk about people.[27] Our brains predispose us to love and befriend but also to hate if we are rejected. Most activities are more enjoyable when done with others than when done alone.[28] Solitary confinement is a torture we might imagine we could bear, but few have brains that can.

The founder of American psychology, William James, put it well:

"A man's social me is the recognition which he gets from his mates. We are not only gregarious animals, liking to be in sight of our fellows, but we have an innate propensity to get ourselves noticed, and noticed favorably, by our kind. No more fiendish punishment could be devised, were such a thing physically possible, than that one should be turned loose in society and remain absolutely unnoticed by all the members thereof. If no one turned round when we entered, answered when we spoke or minded what we did, but if every person we met 'cut us dead,' and acted as if we were non-existing things, a kind of rage and impotent despair would ere long well up in us, from which the cruellest bodily tortures would be a relief; for these would make us feel that, however bad might be our plight, we had not sunk to such a depth as to be unworthy of attention at all."[29]

Such need drives even lone sailors, thousands of miles from other people, to make a kind of social life. Lacking another person, they "interact" with their boats. They report that they personify them, attuning themselves to their craft's every movement. As the psychologist Myron Hofer of Columbia University notes of a boat, it is "as if her sensorimotor stimulation had replaced that of social interactions."[30] We can also imagine the presence of other people. Shipwrecked sailors, Antarctic explorers, and prisoners marched between concentration camps all report that they have felt someone else with them; in their moment of need they experienced someone watching over them.[31] Children have imaginary companions—more of them than perhaps adults suspect. Even as adults in the isolation of the modern world, we make loving bonds with our dogs, cats, and other pets. Their companionship gives us the contact we seek, but sometimes fail to find, with other humans.

Social contact is the unacknowledged "supervitamin." Not having enough of it damages your health as much as eating the wrong diet, smoking, or not exercising—perhaps more so. Studies of people who live alone show that a lack of daily contact carries about the same risk of shortening lifespan as smoking a pack of cigarettes a day![32] Unlike the case of cigarette smoking, however, no health warnings are issued about living alone. Here is one: A paper published in 1988 in *Science* concluded that social loners are *twice* as likely to die as those sharing their lives with another. Have a heart attack, for instance, if you live alone, and 6 months later you have a one in six chance of another; if you live with someone, the odds are reduced to 1 in just over 11.[33] Not surprisingly, therefore, social contact matters when we are ill. Those recovering from cancer have longer survival times and are less likely to relapse the more they meet other people socially.[34] Perhaps it is because people under social stress and lacking social support have weaker immune systems, whereas social support gives them a boost.[35] People need human contact so deeply that it is ingrained into our body's mechanism for fighting off disease.

The brain's need for others and their love is necessary for us to grow up. In 1915, nearly 100 percent of babies in New York orphanages died. New York orphans are luckier now. An American doctor visiting an orphanage in Germany spotted on a sterile ward a fat old woman holding a baby. "Who's that?" he asked. "Oh, that," he was told, "is Old Anna. When we have done everything we can medically for a baby, and it is still not doing well, we turn it over to Old Anna, and she is always successful." Touched and massaged, premature babies thrive better, increasing their weight by up to 47 percent more than if given only standard care.[36] Rat pups deprived of touch stop making proteins needed for growth.[37] Tender loving care, so unscientific to scientists, is vital if we are to live, especially as babies.[38]

In the 1950s, a team of scientists headed by Harry F. Harlow took baby rhesus monkeys from their mothers and forced them to grow up in social isolation. As adults, they were disturbed. They wanted a mother, or if not that, any cuddly contact. Given the choice between a soft, artificial "mother"—a cloth figure with something like a face—and a wire frame with an artificial nipple, they chose to spend most of their time with softness. Hugging something soft was more important to them than food. When scared, they ran to the soft "mother." Not surprisingly, they grew into odd and "psychotic" monkeys.

Psychologists have found children also need a "mother" and to bond with "her." In the young, this bonding is not always with the biological mother (the father or foster parents can be as good) or with just one person (it can be father, mother, and/or another caregiver). And children differ in their needs. Some show strong attachment, others weaker. Children actively shape the kind of bond that is returned: A few children do not like being cuddled; others need it like life. And, of course, a child's need for attachment changes as he or she grows older. But those who once ran orphanages and children's hospitals did not want to know. They thought hygiene, not contact, was more important to a child's well-being.

But the effects of not having early attachments go deeper. Postmortem examinations carried out on the brains of the isolated monkeys mentioned earlier now show us that the lack of social contact when they were young had affected their brains. The structure of their brains was normal, but their neurotransmitter systems were not.[39]

This finding in primates of neurotransmitter impairment as a result of lack of early attachment is a worrying finding for us human beings. The way our human brains grow makes us even more sensitive to such damage. The impairments found in rhesus monkeys affected the last parts of their brains to mature. Yet rhesus monkey brains are nearly finished at birth. Not so human ones. Our brains continue developing for a year to complete what monkey brains have long finished by the time they are born. Indeed, rhesus monkey brains grow only 70 percent further in size after birth while humans' grow a massive 340 percent—nearly five times as much. Our brains are thus probably much more vulnerable to neurotransmitter malfunction as a result of lack of evolutionarily normal attachment. When we consider the spate of street, prescription, and designer drugs taken by people of all kinds to battle depression and search for euphoria, for "something missing," in their lives—along with the fact that drugged parents tend to be less atten- tive—we can glean the structure of a vicious circle.

Compared to monkeys, we are born mentally only half made. Our men- tal vulnerability, the flip side of our neural plasticity and intelligence poten- tial, makes it easy for us to be brain damaged. This is not mere theory. In the early 1990s, American families adopted infants from Romanian orphan- ages. In the United States, doctors found their early neglect had left its mark on their brains. For a start, their brains were often physically flattened from lying with their heads in one position in their cribs. That would not in itself matter, because brain shape does not affect brain workings; but there were significant changes in development going on inside. Harry Chugani of Wayne State University Children's Hospital in Detroit did PET scans that found that such children have 50 percent higher brain activity at rest than normal, but not in all parts of their brains. Isolation had impaired their pre- frontal cortices so that these had lower activity than other brain parts.[40] The behavior of such babies is consistent with a prefrontal defect: They control their emotions poorly, have a tendency to stare blankly into space, and form only superficial attachments.

There are more worrisome findings. In 1994, Adrian Raine, Patricia Brennam, and Sarnoff A. Mednick of University of Southern California found that violence in adults was linked to the effects of the combination of birth complications and early maternal rejection. Birth complications eas- ily damage the brain, since difficult labor can restrict oxygen to it, but what was important here was that while neither birth complications nor maternal rejection by itself raised the risk of being violent as an adult, together they did.[41] Of 4269 consecutive males born, only 4.5 percent were in this group, yet they committed 18 percent of its murders, armed robberies, attempted murders, rapes, and assaults. Maternal rejection had somehow added an extra load on brains already burdened by a difficult birth.[42]

Why does the brain need attachment? The answer is that every social mammal from its earliest hours needs its mother (or another caregiver) in order to physiologically function properly. Its mother is a kind of network server to which it allocates important brain functions via contact and attach- ment. Young brains and their bodies, for instance, need someone to regulate their sleep-wake cycles, activity levels, temperatures, immune systems, hunger and toilet functions, oxygen consumption, certain neurochemicals, and heart rates.[43] Such regulation through attachment must have specific advantages, because it puts the young at risk when separated from their mother. Without her, their bodily regulation goes haywire.[44] These bodily tasks are the physiological equivalent of house temperature regulation done by thermostat-based central heating or an air-conditioning system. It might seem that our brains should be able to perform them itself, like a stand-alone PC. Why would evolution allow all hell to break loose when a caretaker is not there? It says something profound about how our brain works that we transfer such tasks to our mothers. Indeed, it is this delegation that creates what it means to have an emotional attachment. We are deeply social beings. Our brains do not develop or work by themselves. They are part of the human network.

Parental dependence early on helps human brains develop bonds needed later in life. Being tied to a mother provides a child with emotional "working models" of how to attach with others.[45] Natural selection has come up with this trick to make sure we are adequately social. Indeed, one of the major discoveries of modern empirical psychology is the extent to which the quality of infant attachments provides a template for later romantic and parental relationships.[46] A poor attachment between child and mother leads, later in life, to poor attachments with romantic partners. Moreover, it strongly affects one's attachments as a parent to one's own children.[47] Other network bonds that underlie our friendships and even a faith in God have been linked to these early regulatory roots.[48]

Secure later relationships can do much to undo unsatisfactory early bonds. Children have a strong ability to find parent substitutes, when available, to compensate for parents who fail them; many a romantic partner and therapist have done likewise. But early attachments are so powerful that they usually leave their emotional mark.

Emotional attachments at all ages affect the working of our bodies and our overall health. Disruptive relationships literally (physiologically) disturb our self-control, our homeostasis, putting our well-being at risk. For instance, people without relationships—compared to those with them—have disordered immune systems and consequently many more colds.[49] Nearly 9 in 10 university students who described their parents as unloving and uncaring suffered ill health by their mid-50s (heart disease, high blood pressure, stomach ulcers, and alcoholism); in contrast, only one in four students who had two caring and loving parents suffered such things.[50]

These attachments, moreover, shape how we learn. When a mother is scared, her young look at her and what is scaring her and pick up her anxiety. Psychologists call this *maternal*, or *social, referencing*.[51] Such emotions guide children, especially when they feel uncertain. Young children will venture across a piece of glass covering a foot-wide gap if they can see their mother is happy that they cross it—and will not if she isn't. Children (and also young monkeys) learn positive emotions this way as well. They see whom their mother is friendly with and they develop a similar liking for them. When playing with toys and meeting strangers, they use their mothers as a kind of reference and learn what to fear and what to like from her. We probably never cease to use our attachments in this way to guide what we feel (although the way we do it may change). Children follow their parents' religion and often their politics and way of life. Psychologists studying groups find that we use the norms of our group as our own.[52] After our mothers, we make social reference to our peers, social class, and society. We conform.[53] While other animals may bond through physical presence and contact, we learn to use all manner of rituals, ideas, memories, and recollections to keep "in touch." As Myron Hofer puts it: "When regulatory interactions are repeated in association with the same pattern of maternal olfactory, auditory and visual cues, the two become linked in memory, so that cues denoting mother elicit expectancy. . . .With experience and maturity these expectancies become hopes and memories, resulting eventually in the formation of the complex set of predispositions, acquired responses, and internal states referred to as attachment."[54] Here are processes shaping us as we grow from childhood to adulthood that are as important as they are unstudied.

Nonimmediate Sociability: The Troop Within

So far we have discussed only the contact emotions of our sociability—but another dimension underlies it. As we hinted at the end of the last chapter, two kinds of sociability exist, an immediate and a nonimmediate kind.

Many monkeys never leave the sight, smell, and sound of their troop. They are physically with friends and their group. Their presence is sensed around them in the troop's vocalizations, faces, and smells. But there is another kind of social existence: life in a fission-fusion chimp band, where social life is fragmented. The group is never physically present as a unit because it splits every day into smaller parties. As a result, the members of a chimp band roam far away from each other each day. This presents problems. It stops an individual chimp from looking around and locating a given friend or relative who might have gone off with others for the day. With group, kin, friends, and allies not immediately present to the senses, members of fission-fusion groups developed a secondary sociability—they internally mimicked group attachments no longer present. In effect, they needed a way to carry their physically dispersed "troop" with them when it wasn't there and did so—in their heads.

Humans live in social groups and have relationships that are not immediate. Our brains therefore have to recreate internally what is not physically
around us. This gives our sociability a new dimension: We are with others
in our minds and emotions even when we are alone. This hidden dimension
is so much part of us that we fail to see it. But it is perceptible if we can
step outside our familiarity with ourselves. Try imagining being from Mars.
What would seem odd about our species? Perhaps that few of us feel that
we are totally isolated. Even when physically alone, we are filled with concerns that presume the existence of others.

That we internalize social emotions is not always so obvious. Pride, for
instance, does not seem at first sight to involve us with others. But it could
not exist unless we cared whether other people thought well or badly of us.
Our self-esteem is an index of the extent to which we gauge whether other
people would want us in their group. Being wanted by others pumps up our
self-esteem; that which makes others reject us deflates it. Self-esteem is thus
a kind of "sociometer" making us feel good or bad about whether groups
would like us as members.[55] It is the same with self-consciousness, shame,
and guilt. We feel them when we are alone, but they are social in their concerns. As a result, things in our social world matter—and matter a lot. Our
stubborn prides, private shames, sickening guilts, silly vanities, and angry
resentments answer to an evolutionary purpose.

No doubt aliens would find it odd that we spend so much of our lives
dressing up and looking after our appearance. After all, it is not a little
bizarre when we consider that the one person in the world who cannot normally see what we look like is ourselves. Nonetheless, self-consciousness is
not a small thing. Whole industries—fashion, cosmetics, and toiletries—
would not exist without it. And aliens might find it odd that when we claim
we are expressing our "individuality," we look much the same as everyone
else, especially to anyone outside our immediate social group.

But something is odd about these emotions: They make us sensitive to
people and social groups that do not exist except inside us. We feel a loss of
pride, dignity, or self-esteem often when no one but ourselves can see our
hurt. It can sometimes be so sensitive an issue that, paradoxically, we cannot
even discuss it in public! Or we might feel these emotions even if the other
people toward whom we feel them are not alive. Many a person has felt
pride, guilt, remorse, or shame in regard to what a dead parent or friend
might have felt about them. These emotions, we suggest, attest to what
might be called "the troop within our heads," our internalized social world.
In it, humans experience a sociability not tied directly our senses but carried
around in us. In our brains are created aspects of physically absent people.

This hidden sociability is linked to the prefrontal cortex. Primates living physically in their groups, such as gibbons and macaques, have smaller prefrontal cortices than those that do not, such as chimps and us. While we devote 29 percent and chimps 17 percent of their cortices to prefrontal cortex, gibbons and macaques devote only 11 percent to it. We should not be surprised that the prefrontal cortex is somehow involved in creating this troop in our heads. The creation of inner cues is, after all, what our prefrontal cortex does for us when we act, recall, and think. The prefrontal cortex also allows us to extend our minds by letting us carry around the sociability needed to live in fission-fusion bands. Inducing emotional inner social cues when we are alone lets us be in a group even in its absence.

Could inner sociability trap us? Pride and self-esteem tie us to the opinions and judgments of others. Needing to be thought well of limits us to doing what presents us in a good light. We seem not so much concerned about our appearance as ruled by it. We are social even in the absence of society; we crave acceptance and sicken without it. Then there are the shared social rules and feelings—morality and the common kindness that springs from empathy with others. These should, and usually do, restrain us. When we talk about freedom, we rarely mean the freedom to pass someone by when they have had an accident and need our assistance, nor the freedom to murder. But then, these restrictions on what we might do are not the ones that we fear; indeed, they are the ones we want. The troop in our head, however, may shackle us in other, less desired ways that underlie the issues raised at this chapter's start about deference and not making a "scene."

9

Our Living Concern

As you read these words, your heart pumps between two-thirds of a pint and two whole pints of blood around your brain each minute.[1] Under your skull, the blood pulses in through arteries at speeds up to 80 cm per second.[2] It keeps your brain alive with firing neurons. Somehow all this works to make our sense of selfhood and existence, the "me" feeling, including those private yet social emotions of pride, vanity, and self-esteem. Here the physical emerges as the mental—as personal consciousness. Our brain's mass of neurons, by a strange and mysterious story, is also "I."

How all this happens is one of the best-kept secrets of nature. Science in the last few hundred years has found out how the other organs of our bodies, such as our hearts and livers, work. But it has only begun to pick the lock of thought, understand the secret of brains. Most of what has been found are boring details for the specialist. As a result, we have only the faintest idea of how the brain creates this sensation we each have of "me-ness."

But we do have clues. Scientists are finding within our brains strange, silent powers. Let us call them *mindmakers*, these processes that weave this sense of being a "me." This chapter looks at the mindmakers that theoretically give existence its animated feel, the feeling of being alive. They are clues to understanding such things as our freedom and the links between the prefrontal cortex's inner cues and our hidden sociability. Such links have been open to neural plasticity and so underlie the rise of our intelligence and our capacity to pick up Carl Sagan's "extrasomatic knowledge." Here ape minds were reworked into human ones. But before we ask how and what happened, we must first understand what they changed.

Imagine that you can put your hands inside your skull. Place your two hands together on the top of your head and press down. Let your fingers slide down between your two cerebral hemispheres until they stop at the bundle linking them, the *corpus callosum* ("body of hard skin"). Your fingers touch the two middle facing sides of your cerebral hemispheres. Slide your fingers forward. You now can touch the inner (medial) sides of your frontal lobe and several areas of interest. At the top of the frontal lobe the supplementary motor cortex folds inward. Part of the prefrontal cortex in front also edges around. But the main thing you touch is your *cingulate cortex*. (It covers a bundle of axons called the *cingulum*, meaning "belt" or "girdle"— that's what it looked like to early brain anatomists.)[3] You have two of them, facing opposite each other in the left and right cerebral hemispheres. They are linked across the corpus callosum. Another part of the cingulate, the *posterior cingulate*, extending backward to the parietal lobe, will be encountered briefly later on. But it is the forward end—the *anterior cingulate*—that appears to be the key player in making us who we are.

A Hidden Observer

First, what do minds do? It should be an easy question. Indeed, for thousands of years philosophers have argued that consciousness gives us the self-knowledge to answer questions like this. But learning about our minds is hard: Consciousness does not give us privileged access to its inner workings.[4] Like fish unaware that they live in water, we are unconscious of many of our own mental processes. A computer may come with a manual listing the basic things it does, but we are not born with any such knowledge. Simple self-reflection certainly does not give it.

The problem is akin to that science met regarding the workings of our bodies. Only in the centuries since the Renaissance have doctors discovered even roughly correct ideas (earlier ones were hopelessly wrong). Not until 1628, for example, did William Harvey discover that blood circulated due to the heart's pumping. People knew that our blood and hearts were needed to keep us alive. But what was unknown was their function. The same is true now of our minds and brains. We know they are linked—that you cannot be conscious without a working brain—but we do not know how. That is changing.

Smart social brains needed skills for living together. To do this, the brain expanded in places. Key to the evolution of this expansion was expansion in the medial area of the frontal lobe, particularly the anterior cingulate, the part of the brain we had you imagine touching. This part of the brain underlies the skills of restraint Figan and Goblin showed to each other. In the anterior cingulate lurks what might be called a "hidden observer."

PET scans let scientists take pictures of the brain as it works. The medial areas of the frontal lobe, especially the anterior cingulate, the part of the frontal lobe next to the corpus callosum, shows up on these scans in many tasks. These tasks require a common ability called *"attention-to-action."* It is something we never knew we did. Put simply, before we react to a cue or do something, our brain focuses and looks ahead, checking whether or not it is the right thing.[5] It is brain work of which we are not immediately conscious. We are familiar with our ability to focus on the world outside; here, we discover, is another attention hidden inside us that focuses on what we do or think.

Don't Murder Me

When faced with a stimulus, we have a range of responses, and they can conflict. Therefore, we need to focus carefully on which of several responses is best cued by the stimulus. Scientists can see this attention-to-action process at work by using colored written words. Consider saying the color name of the ink with which these words are printed. Easy—it's black. But suppose the words were printed in different colored inks, such as yellow, red, and blue, and the printed words were the color names BLUE, YEL-LOW, and RED, so that the words did not match the inks they were printed in. Now you have to focus on naming the ink color while not saying the color name you see in that color. You respond slowly, as you try to avoid making errors. Somehow you have to focus on giving only one of the responses cued and suppress the other. Psychologists call this the Stroop Effect, after the American psychologist John Ridley Stroop, who first studied it.[6] (The term has since been widened to include other situations in which a stimulus offers two competing responses, such as naming words printed on pictures.[7]) Similar attention to our responses is needed when we focus on saying words beginning with one letter at random or making a fresh sequence of random movements of our first and second fingers,[8] or if we prepare to make a quick reaction to one cue but not to a different one—a go–no-go task.[9] Indeed, this happens in thought whenever we face conflicting mental options.

Attention-to-action may seem a strange skill, limited to weird tasks that scientists ask people to do in laboratories. But everyday control of our bodies and our potential actions needs constant attention. The world is cueing us to react all the time, but we control ourselves. Nothing alerts us to the presence of this skill, because it works so well. We do not realize how the fragile sanity of our lives depends on this hidden attention.

Some things never happen with intact brains: For instance, when we are putting on our pajamas, our right hand does not grab and tear them and need to be wrestled to the ground. Yet, nightmarishly, it could happen if we lost attention to our actions. One person, after a stroke, described such a fight with his right hand. According to him, this hand had "a mind of its own" and was "always trying to get into the act." Neurologists call his problem "alien hand syndrome." But his right hand is not an alien. What has happened is that he has lost the attention that normally stops his hand from being externally cued to do things. Part of the left side of his brain sees his pajamas, treats them as a cue, and makes an automatic reaction to undress. At other times this hand may for the same reason impulsively grasp bed-clothes and sheets. If he sees a tool, his hand may seek to use it. This odd experience is common in those with injuries to their left anterior cingulate, the nearby supplementary motor area, and the corpus callosum that links the two hemispheres.[10] The damage to the left supplementary motor area has had the effect of releasing urges elsewhere in the motor cortex, and these urges trigger the right hand's doing things. This man should still be able to control his right hand, but with his left anterior cingulate damaged, he has also lost the capacity for attention on the side of his brain that could spot and inhibit it. If this were all, then his intact anterior cingulate on the other side could intervene. But it cannot, because the links between it and the wayward hand have been cut. Thus his right hand is no longer monitored, so it can react and do things. Eerily, it has started to have a "will" of its own.

Such brain injuries reveal how dependent our lives are on such conflict monitoring. It seems obvious that our hands work together—until our brain fails. Then we witness what we otherwise would never grasp: that although we may feel we are a coherent whole, feeling is in fact not so spontaneous. It involves brainwork, attention to our actions.

We need this mental checker—what Freud would have called the *über ich*,[11] or *over-self*—that helps us deal with our options and reactions, especially in our social worlds. The smoothly operating cingulate over-self monitors us and stops us from breaking both behavioral taboos and self-imposed restraints. It is an inner focus that lets us select how we react.

This attention-to-action restraint seems to have played a big role in enabling social primates to live together in troops and fission-fusion bands. When apes learned to carry a troop in their heads, it was done partly through upgrading of their cingulate over-self, their inner self-observer. We can see it more clearly in its extreme operation, in socially inhibited people. Such people are constantly thinking about what others think of them.

Compared to most of us, their anterior cingulates are hyperactive, checking their every move.[12] They are constantly attention-anxious over themselves.

The next time you board a crowded commuter train and all the seats are filled, ask someone to get up for you. It is a request you can get away with. Research suggests most people, if asked, will give up their seats. But will you ask? Not if you are like those studied by researchers looking at people's obedience to such requests (not surprisingly, Milgram was behind this work).[13] The students used as researchers boarded trains but often could not bring themselves to ask for a seat. Even when they did, they often found themselves acting as if they were ill to justify it. Their asking for and getting a seat went deeply against their sense of fairness; it is taboo to ask for someone's seat in a train without a good reason, such as age or illness. We—or rather our brains—have a subtle sense of what we can and cannot do. This puts boundaries on what we do when we are with others. It makes us self-conscious about our motivations. We are, however, rarely aware of these boundaries, as our attention invariably stops us from even approaching the point of breaking them. We just sense it as an error or wrong and do something else.

The anterior cingulate has a role in managing our freedom. That might sound like an odd function for our brains, but circumstances exist in which we feel lucky *not* to be *totally* free. We can see this problem in the cockpit of the Airbus A320. It is a highly automated passenger aircraft with state-of-the-art aids to help pilots. One of them, the Flight Envelope Protection System, limits their freedom of action. It is a safety system in the aircraft's computers that attends to the pilots' actions and keeps them from going beyond the plane's design limits. This is not something a pilot needs to worry about in normal flight, but it is crucial in moments of crisis. For instance, when looking out the window, a pilot spots a light aircraft about to collide with the plane. There is no time to think, only to react. The Flight Envelope Protection System, by restricting pilots, lets them pull the aircraft sharply up to 2.5 g (Earth's gravity,) the design limit for forces acting on the airframe. Due to the Flight Envelope Protection System, they do not have to ponder whether the aircraft will go beyond this limit or even tumble out of control—the Flight Envelope Protection System will do that for them. So, paradoxically, freedom is sometimes made easier by smart attention. We live in many such protective envelopes.

Attention to our actions helps us delay our immediate response to annoyances. For instance, our boss may criticize us. We may want to tell our boss to go to hell, but we do not. We may not want to kiss the boss's ass, but to put it in the crude terms of ape politics and corporate survival, we cover our own. Satan no doubt would have liked to take a bite out of Goblin, but his brain stopped him. We have the ability to halt our hot-tempered and rash reactions. In this way, we gain a useful safety valve. We daydream away our frustration rather than acting on it. Something inside attends to the difference. Knowing the difference, we release our tension internally. Fantasizing about killing our boss does not make us murderers. When we cool down later, we can find better ways of managing our frustrations and work out a way to explain our problems to him or her—or look for another job.

In fact, it is remarkable that so few of us do go on and commit murder. And, if we are to believe criminologists, it is not because we fear punishment.[14] In a world where we could all commit the perfect murder (published accounts of serial murders give all too clear do-it-yourself instructions), few of us would—even if we were paid. People worry about the lack of morality among us, but most people find it hard *not* to be moral. Serial killers may engage in recreational murder, but they number only dozens in populations of hundreds of millions. They are the exceptions. Indeed, most crime is done by a tiny percentage of the population. We may have weak wills, but most of us have strong "won'ts."

As we will not hurt others, so we will not hurt ourselves. You wouldn't just stick a needle into your thumb or bite your lip until it bleeds. Though some people are masochistic, most of us are not. It doesn't occur to us to hurt ourselves.

This protection lets us anticipate harm. Sometimes it is only a "hunch" or "vibe," sometimes it is true grueling anxiety. If we know we are going to have an electric shock, we can hardly attend to anything else; our heart races and our hands sweat. This anticipation is created by our anterior cingulate.[15] Not only does the anterior cingulate attend to our actions, but also it links such attention to our autonomic nervous system (ANS), to our sweating and increased heart rate.[16] These physiological responses help protect us. When was the last time you had to stop yourself from putting yourself in the way of dire harm?

Remember those people in Milgram's experiment. They found themselves giving electric shocks—something they would never have done unless they had been told to. Stop-emotions have one defect: They do not work well against other people's orders. Somehow people are less likely to attend to their actions when they originate from outside. People who would never think of murdering another human readily obey an order to kill when in uniform. Whether it is right or wrong that this happens (and there is a case for killing people in wartime), the ease with which this happens is striking—and unexplained (but see the next chapter).

If you still doubt that we live cocooned in protective attentive envelopes made by our brains, consider pain. It is, and usually acts as, a benevolent dictator. (We do not mean to imply it is always to our good. This dictator can also be malevolent.) When we are hurt, it is pain and its "don't move" signs, not thoughts of our well-being, that stop us from moving an injured or burned limb and so causing our bodies more harm. The pain makes us attend to nothing else.[17] When we are ill and injured, our sensitivity to pain increases.[18] In this way, it makes sure we care for our bodies. Not with an argument but with an unchallengeable experience, it simply does not give us a choice. "No," it scrawls stridently across our flesh, "you can*not* move your arm; no, you can*not* get out of bed; no, you *will* stay still and focus on nothing else except on how *not* to move your arm." And we obey.

Pain is a specialized and usually short-lived protective system for the body. We should be thankful for it. If we were born without the ability to feel pain, we would be constantly injuring our bodies, especially our joints. We would not rest injured body parts and so would be susceptible to infections, which could lead to death. Pain is a wake-up call, a sort of spatial version of an alarm clock telling us what we can and cannot do. Of course, it is imperfect, reflecting its trial-and-error origin in evolution rather than in a divinely crafted design. We need pain, even if terminal disease and injury later in life make its protection ominous and pointless.

Tags

Pain illustrates that attention to our actions is only half the story. If we have options, we need "tags" to help us make choices. There is no point in being able to focus on what we might do if all the options seem equal. Thus, we need to be able to mark some out as "do not do this" and others as "OK, you can do this." Our options are tagged with emotional colors so that some attract us and others stop us.

The color of pain comes from various pain nuclei deep in our brains. The nucleus sending pain tags to the anterior cingulate has the hideous Latin name of *ventral posterior inferior thalamic nucleus*.[19] This is the neural source for "do not do" tags that create pain. Although brain scans cannot spot this or other pain nuclei, they can see the anterior cingulate (particularly on the right side) at work when we experience pain, whether in disease or pain research studies.[20]

Other parts of the cortex also light up during pain, but this is a sensation of location, not an experience of suffering. The anguish of pain comes from its rule over attention-to-action. Damage the anterior cingulate, and "pain" ceases to focus attention; patients report they still sense their pain, but it no longer distresses them.[21] It no longer *tags* their attention. A similar effect helps explain how morphine works. Only the anterior cingulate contains opiate receptors.[22] So morphine kills not the sensation of pain but its emotional distress. The anterior cingulate is one of the areas in the brain that shows reduced activity after acupuncture.[23]

Pain tags link to only part of the anterior cingulate, which is really a collection of areas with different inputs, outputs, and responsibilities. Other parts deal with emotions (whose source we discuss shortly), while yet others are cognitively organized by other parts of the cortex (especially the prefrontal cortex). And different parts of the cingulate specialize in controlling various parts of our bodies, such as our hands, our vocal cords, and our autonomic nervous system. Here, inside us, is a rich system for paying attention to what we do, a system that is involved not only in external and physical matters but also in the purely mental realm. When we think up "random" numbers between 1 and 10, our anterior cingulate lights up.[24] And it lights up when we select what verb fits with a given noun, say "hit" with "hammer."[25] So the anterior cingulate, guided by the prefrontal cortex, can attend to what we do not only in the outer world but also in the inner one of thought.

Alien Thoughts

Another instance of such internal attention occurs when we daydream. Not only can we make a movement, but we can also imagine doing it. We easily imagine doing things, often using the same parts of our brain as for the action itself. Think of an almost-dream, where you imagine getting up and using the bathroom, or answering the door, or turning off the alarm clock, or reaching for water, but don't actually do it. The brain needs to spot the difference. Did I really move my arm? Or did I only imagine doing it?

Lesions to both anterior cingulates cause people to forget, at least temporarily, how to distinguish reality from imagination. They talk about having done things that they only dreamed of doing.[26] As one such person commented: "My thoughts seem to be out of control, they go off on their own— so vivid. I am not sure half the time if I just thought it or it really happened." Fortunately this is a temporary loss of reality, lasting only a few days. Though part of the anterior cingulate may be injured, some of it still survives, as does the prefrontal cortex. These can adapt to the task previously done by the anterior cingulate, thus letting a person regain a sense of reality.

And we may need such attention to spot the difference between what we and others do. Did I say that? Or did I overhear it? The anterior cingulate is implicated in this self/other awareness. It has axon links to the auditory cortex, with which we hear speech. Work on monkey vocalizations suggests that the anterior cingulate uses these links to prevent us from wrongly hearing our own vocalizations as originating from someone else.[27] Here is another skill we are not aware of, simply because it is so good.

But it doesn't work smoothly in everyone. Schizophrenics typically hear voices. Careful observation shows that they are slightly mouthing the words that they hear as hallucinations. The voices bothering them stop if they keep their mouths open[28] or if they softly hum a single note repeatedly.[29] Unable to recognize themselves as the source of their own quiet vocalizations, they perceive them as coming from outside. Schizophrenics can have various brain malfunctions, but a key one stems from the presence of fewer than normal inhibitory neurons in their anterior cingulates.[30] And in schizophrenics this secret organizer of perceptual reality also receives reduced blood flow compared to the rest of the brain. Recently, brain scan results have directly linked these anterior cingulate deficits with auditory hallucinations. Schizophrenics hearing voices show abnormal anterior cingulate activity during their hallucinations.[31] This suggests that the anterior cingulate of such schizophrenics is not working normally to spot the self/other boundary with regard to their own voices. The impairment causes them to experience not alien hands but alien words.

A Mother's Love Revisited

How did the skills needed for paying such unconscious and conscious attention arise? Part of the answer, we suggest, is mammalian mothering. The instincts of a mother for her young need "eyes." Attention-to-action gives blood instinct that sight. A robin that throws its own young out of its nest in obedience to blind instinct is not attending to what it is doing. Mammals, we argue, would first have evolved attention-to-action to stop the blindness of a mother's instincts.

Parents need such inner eyes to battle bad instincts. The world gives organisms conflicting cues. For instance, a mother and father must be caring and tolerant of their newborn baby yet many other nonparental instincts interfere. Like robins, we have instincts to clean the nest and remove unwanted nuisances, but we can focus on our actions. Parents face a kind of Stroop situation, not words of different colored inks and names but the different feelings evoked by their young. Children are not just bundles of joy;

they are also irritants. They demand attention, awake us nightly from our sleep, and later, rebel and disobey. Our instincts are to hit as well as to care for them. We are distressed when parents fail and batter their children. But the greater surprise is that so few parents do. Mothers and fathers are tolerant, extraordinarily so. Something is keeping them that way. If anyone other than one of our children kept us awake night after night, we would quickly turn them out. Children can get away with all their "sins" because something in us watches our actions. It creates a protective envelope around our children. Indeed, it does something that is in some ways even stranger: It makes us love them. They become part of our living concern. We extend our protective cocoon of "self" to include them.

In the evolution of the brain, attention to what we do arose with our anterior cingulate. Birds do not have this valuable mammalian part of the brain. Cut it out and mammal mothers neglect their children.[32] Remove it from their young, and they cease to play and cry. Here, in this core part of the mammalian self, one of the family-bonding brain chemicals—vasopressin—has increased numbers of receptors.[33] As noted above, the anterior cingulate also has more than its fair share of opiate receptors. According to Jaak Panksepp (Distinguished Research Professor Emeritus of Psychology at the Medical College of Ohio at Toledo), the brain's natural morphinelike endorphins play a key role in a mother's attachment to her children; they are released during activities such as grooming that relieve the distress of separation.[34] Attention to the care of others, a key brain skill that evolved with mothering, was later, we suggest, taken over and adapted for doing other tasks in the rise of complex sociability. Skills that aided mothering could be turned by evolution into the restraints needed for friendships, behavioral taboos, the inhibition of aggression shown between Satan and Goblin, and no doubt in everyday life by us all.

Brain Colors

We now need to turn to the tagging and coloring processes behind many of our emotions. We must trawl deep in our brains to find what colors our attention and so makes our feeling of what it is to exist. It begins in a bit of brain no larger than a thumbnail, our hypothalamus. Found under the thalamus, this structure is the starting point for controlling our raw needs. The hypothalamus is the seat of our hungers, thirsts, sweats, shivers, and fatigues. In this part of our brain lie thermostatlike mechanisms turning our appetite, need for water, body temperature, and sensitivity to stress up and down. The hypothalamic "thumbnail" also makes us flee or fight; it holds the master key to the protective and the defensive reflexes of our autonomic nervous system.[35] And, more subtly, it aids our survival by regulating our hormones and immune system. Here the brain (through the pituitary gland) controls menstruation, pregnancy, lactation, and growth and fights disease. Thus, it makes sure our bodies survive and grow, and, by being deeply linked to sexual attraction, it makes sure we reproduce our genes. Although little is known about the role of the hypothalamus in our social lives, it may play a key part. Differences in the size of one part of the hypothalamus are linked to whether men are sexually attracted to their own or the other sex.[36] It also plays a role in the synchronization of menstruation among women; oxytocin is made here, though its bonding effects happen elsewhere in the brain.[37] The hypothalamus is also a key part of the body's stress circuitry. Stress, as we all know, is strongly dependent on whom we are with. (As noted elsewhere in this book, large gaps exist in present science about the brain. For instance, a recent report finds the hypothalamus activated when people imagine music[38]—something quite unexpected.)

The hypothalamus, however, is blind. When stimulated, it reacts to everything. Such indiscriminate reactions will not aid our survival. They need to be fitted into the world outside the brain. So the brain has other areas turn the emotional pigments of the hypothalamus into the emotional tags that color our lives.

The first is the *amygdala* (Latin for "almond"—early anatomists of the brain saw a nut). We have two of them, one in the forward part of each temporal lobe.[39] On average each one is about 3 cm³ in volume, with the right one being slightly larger than the left.[40]

The amygdala ties the emotional tags of our hypothalamus to our senses. We are not neutral to the world, as it constantly interests, moves, and affects us. At times, it can fill us with fear, horror, or sadness. Or it can be friendly, beautiful, sexy, or even joyful. When we look around, things can suddenly grab our attention—scare or lure us—depending on whether we have seen, say, a wasp, a naked person, a smile, or the glance of someone's eyes. The amygdala paints these extra emotional colors onto an otherwise neutral, "gray" world. (The left and right ones paint with different palettes of colors; for instance, the right more than the left is responsible for fear).[41]

It is a coloration we can lose. After injuries to our amygdala, we may fail to recognize fear in someone's face.[42] Such injuries can have other effects. One young man sustained a head injury from a motorcycle accident that cut the link between his vision and his amygdala. It changed him. He stopped hiking afterward, as he found the scenery dull, and "all the same." He ceased subscribing to *Playboy;* though his sexual urge remained, photos of naked women no longer excited him.[43] He still had his vision, but the emotional spectacles through which he previously saw the world had been shattered.

The amygdala also allows us to detect emotions more subtle than fear or sexual feeling. Some of its cells respond to other individuals, picking up on the subtleties of emotional expression in their eyes, faces, gaze, and "social actions."[44] It can tell us the difference between a friend and a stranger. Ultimately, the amygdala affects our behavior. Monkeys without their amygdalas lose fear and anger. Humans with injuries to theirs become "petlike," "compliant," apathetic, and emotionally blunted. People have had their amygdalas surgically cut out in an attempted treatment for incurable self-mutilation (though it is done rarely). One such person was a 27-year-old woman. According to her therapist, she was changed: "She could not describe spontaneously what the desired states of marriage or motherhood might involve; nor could she imagine more than the surface aspects of visiting a relative—'driving there and back' described the whole situation."[45] She had lost the emotions that color our lives and give marriage and motherhood meaning. It stopped her from mutilating herself, but at the cost of losing a key part of who she was—her social emotions.

The amygdala not only colors our emotional senses but smartens them. Hearing, vision, and our other senses come together in multisensory maps on our temporal lobes. Our amygdala lies under them. In this position, it acts to change how we sense things.[46] A male rat without an amygdala will mate with females, but only if he can experience them directly.[47] He will not learn visual association that might lead him to them. Similarly, monkeys without these intracranial "almonds" cannot learn associations needed to avoid hunger. The amygdala thus acts to tag what the American philosopher Charles Peirce (1839–1941) called *indexes*.[48] Afraid of fire, we can also panic at seeing smoke. Smoke is not fire, but it tells us there is one. Peirce saw this connection, noting, "Anything which focuses the attention is an index. Anything which startles us is an index. . . ." The amygdala heightens our perceptions so we feel anxious not only at the sight of wasps but also at hearing them buzz. We need to learn to fear not only in one of our senses but across all of them.[49]

The amygdala can go beyond subtle heightening to vividly coloring with Day-Glo what we sense. If somebody is raped, hit over the head, shot, or bombed, adrenaline floods the body. The amygdala takes special note, making sure the trauma is "remembered" on even chance associations—a lover's crucifix like one the rapist wore, the blade of a ceiling fan seen while being bludgeoned. It now is on constant alert to possible repeats of the original situation. So in effect, the amygdala works by changing our perceptions.[50] It makes an insect into something that causes our heart to pound with anxiety and fear, and it causes a child to smile with delight upon seeing his or her mother. Wasps and mothers' faces not only are visually different but are also emotionally sensed and responded to in quite different ways. The experience of seeing a wasp or scorpion is colored with anxieties, that of seeing one's mother with stay-and-enjoy feelings. In this way, the amygdala shapes our behavior through our senses. Imagine a wasp landing or a scorpion crawling onto this page. You would not merely note its presence; "WARNING SIGNAL" would be going off in your mind. You would quickly do something to get rid of it. These signs impose their authority over us via the hypothalamus, but their intelligence comes from the amygdala.

As we've seen, the anterior cingulate—the brain's second tagging area—seems most involved in our brain's inner attending. It watches actions for nonemotional reasons, but it also plays a key role in laughter and other expressions of emotion.[51] Here the anterior cingulate has extended from attending to our actions to attending to our interactions with others. Why cry if we believe that no one will hear, or laugh if everyone turns to you in sudden shocked silence? Emotional expression depends on how others attend to us. What are tears and sobs but means of grabbing the attention of people who might aid us? Laughter changes the attention of a group: What could have been serious becomes, at a stroke, discounted as such. Here attention-to-action goes beyond merely having a backstage role and becomes a player. The anterior cingulate is a motor area controlling vocal actions that interact with others' attention. Through these it controls tagging and the shifting of tags not in ourselves but in the experience of others. Stimulate it, and people will giggle or weep. Lack it, and other people have to prompt you to reply, because you will never spontaneously seek to say anything on your own.[52] It is in the anterior cingulate that social emotions meet our motor system.

Body Emotions

Hidden between the frontal lobe and the temporal lobe of the brain is the *insula* (Latin for "island"). It is the third and last brain area involved in processing sensations received from the body. It is here that emotions such as disgust, fear, and anxiety link with our experience of our mouths, stomachs, and chests. We do not feel sick or anxious in the abstract but inside. The nearby amygdala tags not only what we see or hear but also our body's outside and inside. Here our emotions meet our bodies.

Visceral emotions are obvious body emotions, but there are others. (Not all of them are associated with the insula; some are handled nearby, in brain areas where representation of other parts of the body is linked to the amygdala.) The insula is an area scientists have for some reason shied away from. When someone massages us or holds our hand, we feel relaxed and comforted. These are feelings happening in our bodies, like disgust, not in the abstract but felt in our muscles or skin. Similar body feelings happen when we are about to stand up in a lecture and become the focus of attention. We feel our bodies tense and become excited. Emotions also exist in our posture. You can tell a lot about people's moods by how we walk and stand. Confident and successful, we are upright and hold our heads high. Depressed and rejected, we slouch, cringe, and almost hide within ourselves. Tell off the family dog, and it will cower and put its tail between its legs. And so will children, except it will be their heads that are lowered. Our bodies carry, express, and, indeed, embody our emotions.

But this is only the beginning. Though we are not taught to think about it in this way, we actually have a special organ for creating and experiencing social emotions. Look in the mirror and you will see it. The human face is extraordinary. Consider those strange objects, our eyes. They are far more than the video cameras of the brain. We keenly attend to them in others, something human evolution made easier by exposing a little of the white sclera surrounding their darker irises and pupils. And what about our mouth? Far more than a hole for food, it is surrounded with muscles to fold our lips into a thousand and one subtle expressions. Our face is laced with muscles whose only function is to contort it into smiles and frowns. Even our skin gets in on the act: When embarrassed blood rushes to our face. While most mammals express a few emotions through their faces, no other makes as many as we do. Chimps can laugh, cry, and raise their lips to show their teeth. But our human face, much more naked of fur and more mobile in its skin, is capable of many more subtleties of expression, even some that chimps cannot make, such as disgust and surprise.

Our faces not only communicate but also experience. If we put on a sad or happy face, we can actually induce the emotions indexed by such expressions. Studies on the electrical activity of the brain when people make a spontaneous, or "Duchenne," smile or a carefully rigged imitation of one suggest that their brains react in similar ways.[53] So moving your face in an expression can not only reflect your emotions but cause them—a strange circularity.[54] Faces not only express; they feel.

We experience these bodily feelings in our insula (and perhaps in nearby somatosensory fields that process the sensations from our skin, as well as the orbital prefrontal cortex, discussed next). In these areas, the amygdala tags a representation of what our body feels—a map of our innards and mouth.[55] Here is where our emotions meet the sensations felt in our flesh and guts.

Pain is sensed here not as something ordering our attention but as a sensation (remember, those with damaged anterior cingulates could still feel pain, it just ceased to bother them). Brain scanners find the insula and nearby somatosensory fields lighting up when electrodes on a person's skin are heated up to 50°C.[56] Here is not the suffering made by our anterior cingulate but the sensation of its location.

There are also suggestions that the insula oversees sensorimotor links. The above tasks linked to the anterior cingulate are done with focused attention; they are novel and fresh to those doing them, not routine. They require the brain to attend to which response should be linked to which outer or inner cue. But we often make automatic selections after practice. What we do has ceased to need the same kind of attention and so becomes habit. We know what is right by "feel"; we learn by memory, by heart, develop a knack; it becomes second nature. Tasks that activate the anterior cingulate when novel and fresh stop activating it when its associations become practiced and are redirected to the insula.[57] So it is at least for the above-mentioned task of selecting verbs to fit with nouns—"hit" with "hammer." When these verb-noun links are novel, they activate the anterior cingulate (and prefrontal cortex). But if these pairings become practiced, eliciting them lights up the insula instead.

We mention this not just to elaborate on what the insula does but to hint at how much more we have yet to find. From what was previously known about the insula, we would not have guessed that it might perform such habit associations. When we describe our understanding of the brain as terra incognita or an uncharted ocean, we are not exaggerating. As we might learn about an unknown continent by focusing a viewing device more accurately on it, we are now learning about the brain. In decades to come, books like this will be written solely on the anterior cingulate, the amygdala, or the insula. There will have been a knowledge explosion about how such mindmakers create us. Imagine the shock people felt in 1628 when William Harvey first proposed that blood circulates. Now think about blood transfusions, heart transplants, clotting factors—or visit the cardiology section of a medical library and wonder at the rows of journals and books. Be stunned. We are at the threshold of an information explosion about the nature of our own mental existence.

Mixed Feelings

Our amygdala, anterior cingulate cortex, and insula may be smarter than our hypothalamus, but they are limited. The world to which they adapt is one of immediate external cues and consequences. They ignore the future. But we need to tag certain actions as ones to stop even when they cause no current harm. This is particularly true for apes in fission-fusion groups. Fragmented social life puts strong demands on behavior. In contrast, if your troop and relationships are consistently around you, social coexistence with them is simple from the brain's viewpoint. Your senses can use their presence as external cues to organize your social interactions. You do not need sophisticated expectations: The stimuli to guide you are, after all, physically all around you.

But for those that live in fission-fusion groups, these external stimuli need to be replaced by internal cues and scripts. Their brains must somehow internally reinvent the experience of belonging to a troop even when they are alone or in a small party. That, in turn, requires a skill for recalling being with fellow group members. But most of the emotional brain cannot go beyond the here and now. The hypothalamus, amygdala, and anterior cingulate do not enable an animal to carry the inner cues and expectations needed for fission-fusion social existence. Next to the anterior cingulate and insula, the brain evolved another part capable of doing just that.

Our Orbital Cortex

Imagine you can reach into your brain again and feel in front of your anterior cingulate cortex. You've put your fingers on the part of your prefrontal cortex called the *orbital prefrontal cortex*. It is named after your orbits, the circular hollows in your skull that hold your eyes, as this part of the brain works behind and above them. It is part of the prefrontal cortex discussed in Chapter 4, but it is only half. Above and to the side of the orbital prefrontal cortex is the *dorsolateral prefrontal cortex*. Roughly, the orbital part of your prefrontal cortex is your emotional prefrontal cortex, and the dorsolateral part is your nonemotional one.

The orbital prefrontal cortex has links with the amygdala, anterior cingulate, insula, and hypothalamus.[58] The links go both ways. Not only do these evolutionarily older parts of the brain tell the prefrontal cortex what is happening to them, but they let the prefrontal cortex manage and so upgrade what happens in them. Through links with the hypothalamus, the prefrontal cortex can change what the hypothalamus tags as necessary to our survival. Through those to the amygdala, it can control the amygdala's emotional tagging of what we see, hear, and otherwise sense. Similarly, its links with the visceral "gut feelings" of the insula let it control them and so our autonomic nervous system.[59] Through these links the orbital prefrontal cortex creates and fine-tunes what startles, terrifies, or pleases us.

It also has strong links with the anterior cingulate. Though we have discussed these areas of the brain in isolation, they of course work together. Here is an example: When the anterior cingulate focuses on something, a part of the prefrontal cortex is activated along with it. For nonemotional tasks, this is the dorsolateral prefrontal cortex; for emotional ones, the orbital part lights up.[60] For instance, in the above-mentioned selection of verbs to fit nouns, the anterior cingulate lights up along with the dorsolateral prefrontal cortex. While the anterior cingulate is able to deal with the attentive and selective aspects of the task, it needs the working memory of this part of the prefrontal cortex to direct and guide it. When the anterior cingulate activates in anticipation of our getting an electric shock, so do parts of our orbital cortex.[61] This part of the prefrontal cortex holds in working memory what we anticipate or are anxious about.

Some nonsocial examples illustrate how the orbital prefrontal cortex could control tagging done elsewhere in the brain. Present a rat with a loud noise at the same time as an electrical shock to its foot, and it will freeze later—even when it only hears the sound. Gradually, such freezing will stop if it hears the sound with no more simultaneous foot shocks. This is similar to the way a victim of a violent assault can gradually lose his or her fear of the scene of the crime by revisiting it.

But, interestingly, if the rat's equivalent of our orbital cortex is damaged, it cannot overrule the fear reactions happening in the amygdala.[62] It is the same with us. After we have been in a car crash, the amygdala and other parts of the brain may react to make us anxious about being in cars. But we do not want to learn such a fear. The prefrontal cortex gives us an eraser: By reasoning, talking about, and writing down such emotions ("rewiring" the sensations through the cortex), we can overcome and eventually extinguish them.[63] However, in some people the ability to overcome such emotions is weak, especially for very traumatic experiences. If you were a GI in Vietnam in 1968, you would be unlikely to forget the head of your best buddy on a plate, his mouth stuffed with propaganda leaflets, following his capture by the Vietcong. It may well come terrifyingly back to you for the rest of your life. Some people in particular have difficulty in recovering from such bad experiences. They have problems in using their prefrontal cortex to unlearn fears laid down elsewhere in their brains.[64] Their anterior cingulate fires up when exposed to images of war, while the parts of the orbital prefrontal cortex that are supposed to cool it are silent.[65]

A similar thing happens to people with a condition known as *atypical facial pain*. It is a chronic nonlocalized sensitivity to pain in the face. It usually affects women. Ordinarily, pain activates people's anterior cingulate cortex (and, at least in women, their orbital prefrontal cortex). In those suffering atypical facial pain, however, the anterior cingulate is more than normally active, while the orbital prefrontal cortex is inhibited—basically, turned off. One interpretation is that in ordinary people the prefrontal cortex becomes activated to reduce the pain in the anterior cingulate—the logical thought sensors turn on to "explain away" the pain; hence the activation. But in those with atypical facial pain, the prefrontal cortex fails to switch on and tell the pain in the anterior cingulate to go away.[66]

10

Doing the Right Thing

What is the greatest hurt you might face? Going blind? Deaf? Being paralyzed? Not suffering any of these impairments, we cannot dismiss how awful they would be. But at the risk of being wrong, we will suggest that there may be one thing that hurts even more: being socially rejected and not having a friend.

One 7-year-old girl had the misfortune of having a brain hemorrhage in her left orbital prefrontal lobe area. It left her in a coma for 5 days.[1] She survived and recovered all her faculties—or so it seemed. Her parents must have been relieved. She seemed normal except for being a slightly slower learner for the next 3 years. But they could not see into her future. Hidden away in her brain she had suffered an injury that was later to leave her blind and crippled, not in her eyes and her body but in her sense of good and bad and in her ability to live with others.

Her left prefrontal cortex injury had been extensive. Worse, it had stopped the right prefrontal cortex from working and damaged the white matter links under her two anterior cingulate cortices. These injuries stunted her personality. This was not noticeable at first. Like others with prefrontal injuries, she seemed to be normal. But then things changed. When she entered adolescence, she failed to develop a sense of morality, self-care, or common sense. As her body grew, her mind was left behind; she had a child's emotions in an adult's body. A year after leaving high school, she married her first boyfriend and had a baby, but she was unable to care for it and, following their divorce, lost custody. Afterward she would have up to seven boyfriends at a time.

As an adult she had the practical sense of an immature adolescent. She could never stay in a job for more than a few weeks. She kept telling her employers how they should run their businesses—not that she knew how. She replied to ads to become an airline flight attendant or store manager without the right experience and qualifications. She lacked any sense of strategy. She hoped to open a bookstore, but she made no real plans. Asked about the financial or practical steps needed to do this, she could not give any. Her friendships were short-lived. She failed to grasp another's point of view, and, as her neurologist noted, she "made hurtful statements, asked embarrassing questions, and gave impulsive responses." Today, without knowledge of her prefrontal brain hemorrhage, she would probably be classified as a psychopath. Others might call her a sociopath or say she had an antisocial disorder. In the nineteenth century she would have been described as "morally insane." But she was not born that way.

Adults with orbital prefrontal cortex injuries are slightly less affected than children.[2] Having already learned many social skills, they become psychopaths with tact. But much of what they have learned does gets lost. They cease to have the inner sense they once had of right and wrong. They lose their empathy for others, and they do not register the least shock when shown pictures of mutilations or explicit hard-core pornography.[3] They are impulsive, unable to stop themselves acting on the spur of the moment. They do not feel guilt. They find it hard to hold a job. They have social skills but do not use them. They make unwise business decisions. They usually end up socially crippled and bankrupt.

They also find it hard to keep a steady relationship. Our sexual emotions need managing. The orbital prefrontal cortex has links with the areas of our brains controlling sex.[4] Indeed, our prefrontal cortex is the only part of our cerebral cortex with such links. Brain scanners find that the right prefrontal cortex lights up in men masturbating to orgasm.[5] Through its links with subcortical sexual centers, this part of the cortex probably discourages us from making love in public. It enables us to turn sex into lovemaking and so into part of a relationship—to have one lover, not seven.

What of people who are born psychopaths? Like those with orbital prefrontal injuries, they lack the inner sense needed to make wise and moral judgments. Descriptions of their lives are much like those given above of those with early or late injury to their orbital prefrontal cortices. Some research suggests that they may have weak prefrontal cortices, some that they do not, but this research concerns only the skills of the nonemotional, dorsolateral part of their prefrontal cortex.[6] Other, still tentative work finds defects in psychopaths also found in those with orbital cortex injuries. For instance, the labeling of smells is impaired after damage to the orbital cortex, and psychopaths, in comparison to controls, have the same defect.[7] Both also have defects in their startle reactions to visually disturbing pictures.

Studies using interview-based tests of those involved in violent crime find that 73 percent have evidence of brain damage. The figure for nonviolent criminals using the same kind of tests is 28 percent.[8] What if, instead of an interview-based test, we could look directly inside the brains of psychopaths, especially those involved in violent crime? The technology to do this now exists, although it is still new and far from simple. Except for one study (discussed below), the research has not begun. Psychopaths are not easy to find except in prisons. The machines needed to look into their brains are not portable. So practical problems exist. Judges do not sentence people to a prison sentence and order PET and MRI scans (though a precedent has been set with John Wayne Gacy, the serial killer of 33 young men, whose brain was removed for examination after his execution).

In spite of these problems, a study has been done on 22 impulsive murderers, including a serial killer of 45 people. The activation of their brains was compared with that in a group of impulsive nonmurderers while they performed a task involving sustained attention that needed the use of the prefrontal cortex. In spite of the varied nature of the murderers and the low resolution of the scans, impairment in various prefrontal areas was found, especially on the left side of their brains.[9] But this study was limited; it failed to assign a task sensitive to orbital prefrontal cortex impairment, and it averaged across a group of people not all of whom were necessarily psychopaths. New brain scanning technology such as fMRI now exists that allows individual people to be studied while trying to do tasks that require a working orbital prefrontal cortex. Although you may not until now have heard of your orbital prefrontal cortex, this term will increasingly be bandied about in expert testimony and discussions of criminality.[10] What will happen to our notions of criminal responsibility when violent murderers are found to have defective brains? This question may well soon be, if it is not already, troubling judges, lawmakers, and juries.

Are you curious about what it would be like to be a psychopath? You can find out, in a limited fashion. There is a simple way to stop your prefrontal cortex from working for a few hours and come back to normal. Go 2 or 3 days without sleep. The brain turns off the parts it can do without. The blood flow around parts of your orbital prefrontal cortex would drop by a quarter, that around the dorsolateral prefrontal cortex by slightly less.[11] Not surprisingly, you would lose your ability to concentrate, plan, and make good decisions. You would be irritable and might say or do things that would lose you your job or screw up your relationships.[12] Your sense of judgment would leave you. Though not yet a psychopath, you would be well on the way to becoming one.

Moral Cortex

For those of us who are not psychopaths, it is easy to make practical and moral judgments. We know automatically that some things are best avoided while others should be sought. We work out how best to fix something, how best to live and interact with others. In our language for these judgments we use the same words—usually good and bad for practical ones and right and wrong for social and moral ones, but we easily swap such words. A repair can be "right" or "good," cheating someone "bad" or "wrong." The main thing is a concern with doing the "right" thing.

That problem applies equally to both practical and moral questions, since in both cases outcomes matter. On practical grounds, some decisions are unwise; others would have outcomes that are morally or socially undesirable. We are constantly judging where our actions will lead us morally and practically. It is so much a part our lives that we assume there is nothing to it. But the existence of people like the brain-damaged 7-year-old girl mentioned above shows that seeking and making such judgments is not spontaneous. Again, neural "sleights of hand" in the brain do much of the work without our noticing it. And the practiced performer in this case is the orbital prefrontal cortex.

Lack of judgment and concern about the future does not equal lack of thought. People with damage to the orbital prefrontal cortex can think of many things they might do, but they do not anticipate well how they will feel about them. Thus, even if they can list options facing them, they are stumped as to how to choose among them.[13] Most of us, by contrast, automatically know which outcomes we would like and which we would not, and so are concerned, often worryingly so, about which might actually happen. You can see this in games like Monopoly, in which card draws and dice throws with uncertain results offer penalties and rewards. Before we pick up a card or throw a die, we anticipate. As shown by the technology that measures skin conductance (used in so-called lie detectors), normal people playing card games with random reward show apprehension and concern as to what will happen next. But when tested, seven people with damaged orbital cortices all failed to register an anticipation response. They were insensitive to their own future.[14]

We anticipate in part because we imagine our own emotions. People with orbital prefrontal injuries do not. Here is yet another unconscious brain task done so skillfully we normally have no reason to suspect its existence. Spend a minute recalling a sad situation in your life. Avoid anything that might cause you feelings of anger or anxiety; focus on the sort of memories that make you want to cry. We can all, in this way, make ourselves feel sad. We can relive in memory the death of a relative or childhood pet. If we linger on it, tears may come to our eyes. It should come as no surprise that this internal cuing of memories requires the advanced abilities of our prefrontal cortex. When brain scanners look at people imagining sad situations, they find they used their orbital prefrontal cortex.[15] Psychopaths, however, lack the ability to imagine feelings—they are "emotionally shallow."

Why should our ability to imagine emotions be connected to our sense of the future and our judgment? Think back to what the prefrontal cortex does: It is a sketchpad. The examples we gave earlier of the skills this function enables were all nonemotional; the areas of the prefrontal cortex discussed, after all, linked mainly to the nonfeeling parts of the brain. But the orbital prefrontal cortex also links to our amygdala, insula, and hypothalamus. It thus can take bodily and facial feelings created in these areas and use them as "somatic markers"—cognitive colors for sorting out different possibilities, such as "I like it" or "Thanks, but no thanks." This lets us imagine what we will feel like given a particular outcome. We can foresee, for example, whether an action will cause us hurt or delight. The dorsolateral prefrontal cortex lets our brain work out on its sketchpad the kind of things that are likely to happen—it answers the *how* question. The orbital prefrontal cortex lets us "feel" distress or reward that we see coming, helping us decide what to do—answering the *what* question.

We often describe this feeling as telling us what we "should" or "ought to" do. We can now see why psychopaths are crippled. They may have a working nonemotional prefrontal cortex, but it is blind without an emotional one. They will do things the rest of us would avoid because we "know in our bones" or have a "gut feeling"—at least as imagined by our orbital cortex—that we will not like them.

The above judgmental "oughts" concern mainly our own welfare, but as we've seen, social animals cannot do all they want to each other because if they did, it would break up their group. As a result, they are concerned not only about the future possibility of their own distress but also about that of others. Our emotional sketchpad gives us a capacity for empathy—something essential for social coexistence.

Our orbital prefrontal cortex thus enables the emotions and sensitivities that make up friendship. We have a sense of our friends' likes and dislikes, and we know when they're hurt. We feel guilt and remorse when we fail our

friends. Friends do things for each other: Following a good turn we feel grate-
ful and have a sense that we need to return the favor. A whole palette of
emotions arises in our minds to guide how we act toward our friends. They
are felt spontaneously. Our emotional sketchpad links these social emotions
with our expectations about the nature of friendship. If we cannot make
these links and feel what binds friends together, people end up thinking that
we are taking advantage of them and stop being our friends or even cooperating with us.[16] Psychopaths lack not only empathy but also the sense of
guilt or remorse that leads to tit-for-tat reciprocation. The above-described
woman who had an orbital cortex injury at age 7 notably lacked lasting
friendships; she did not have the emotional insight to keep them alive.

Much of what we feel is determined by what we imagine others think of
us. Our parents might have told us that "sticks and stones may break your
bones, but words will never harm you," but we know that is rubbish—words
and other people's thoughts do hurt, and very badly. Why do they, and why so
powerfully? Why does someone thinking or talking about us in a certain way
make us self-conscious? Why do we so readily experience dignity, pride, dis-
honor, shame, remorse, self-consciousness, and self-respect, not just with our
friends, but with virtually anybody? Why is it that we can feel these things
even in private, when it is only a mere possibility that someone has witnessed
what we feel so badly about? Nobody, for instance, has to actually notice that
our underwear is showing or our fly is open while we are dancing at a ball for
us to be embarrassed. We can even feel shame about experiences that no one
would want us to feel shame about—after rape, for instance.

How do neurons lead to these feelings? The answer, we suggest, is that
we humans carry our social group around in our heads. As we've noted, this
is what allows us live to in fission-fusion social groups. Our brain creates an
extra dimension by turning the ties and taboos of our social existence into
emotions. We are never really solitary; we have continuous inner dialogues
about what we are feeling, plus self-conscious dramas about who we are to
others. Our brains are heightened or hampered by what social psychologists
call a *social ghost*—our self as known, if only potentially, through the eyes
and attentions of others.

How do we learn these social and moral emotions? Likely we flesh out
the details of our emotional sketchpad on the basis of how familiar, trusted
people react to others. We have already met an example: A young child or
ape observes its mother, and if she is scared, so is the youngster. We also learn
through such social referencing. What part of the brain is responsible for this
emotional tutoring? Those working on social referencing in monkeys suggest, again, that it is the orbital prefrontal cortex.[17] Supporting this sugges-

tion is evidence that the orbital prefrontal cortex processes our perception of faces. Indeed, this is one of the first things it does. Scans on 2-month-old babies show the orbital prefrontal cortex at work when they look at pictures of faces.[18] Interestingly, social referencing is linked to the development of a child's capacity for attachment, and this, Allen Schore of the Department of Psychiatry and Biobehavioral Sciences at UCLA School of Medicine argues, is linked to the orbital prefrontal cortex.[19]

It is tempting here to leave the brain for a moment and look for connections with philosophy. After all, in describing the brain's creation of social "oughts," we are discussing nothing less than that which enables us to be moral. Consider the key question in ethics: Can we turn matters of fact and logic into statements about right and wrong? This is the *is/ought problem*. The Scottish philosopher David Hume (1711–1788) proved we cannot. New facts about our brains do not change this. Our brains may find it easy to turn *ises* into *oughts*, but this leaves the basic problem of ethics untouched. If David Hume were still alive, he would tell us that new facts about the brain fail to help us know which are the right *oughts* and which the wrong ones. Different brains might come to different judgments, so we still need philosophy to work out how we might judge which of them is valid. Facts about the brain explain only how and why we make the judgments we do, not whether they are morally valid judgments. There is no neurobiological theory of morality, only a neurobiological theory of moral abilities—which is quite a different thing. So research on the orbital cortex does not dispute traditional problems in ethics, though it may suggest new directions for inquiry.

Profoundly, neurology tells us that we are born with an ability to be moral thanks to evolution. Touch your forehead, and there below it, your orbital prefrontal cortex (plus a lot of natural selection) enables you, like all of us, to live in social groups. It gives us moral abilities of which we are not normally aware but that are as basic for our lives as sight or memory.

We have not just a sense of right and wrong but, as important, a concern with ethics. Able to feel right and wrong, we attempt to seek what they are, —not, ultimately, by social referencing to our mothers, but by means of reason. The brain puts each of us on a quest. We seek answers—and raise questions. What in the distress of others should make us feel that something is wrong or right? Our actions bring consequences on others; what makes them moral or not? What kinds of entities are entitled to moral consideration: the unborn embryo brain? future generations? laboratory animals? pests? These are questions raised by our brains and our evolutionary past, but these brains and their history cannot in themselves provide an answer. Evolution only made a brain prepared and motivated to live in a world in which it is concerned with *oughts*. But evolution and the study of our brains, of course, cannot go further to give us answers—what these *oughts* should be. To understand what these *oughts* might be, our past—or that of our brain—is a poor place to find guidance or precedence. How we find the *oughts* we should follow is a difficult problem. There is still much work for philosophers.

Social Brains

We are like archaeologists finding a fragment of an ancient wall that is known to have once been part of a great city. Outside the neurosciences are vast expanses of knowledge and theory about society and social behavior. What we know about the brain rarely connects with them. Neurologists do not yet readily talk to philosophers, sociologists, anthropologists, and social psychologists, but perhaps they should. When we study the brain, we study something hidden behind every human action in society and history. In a way, it is brains, not people, that riot, rule, conform, rebel, fight, work, lead, conspire, and commit suicidal martyrdom. And these phenomena are driven by a variety of emotions made within brains, such as pride, guilt, and outrage about rights and wrongs.

Is there more to the link between the brain's emotions and society? History and the social sciences teach us one broad generalization about ourselves, and so about our brains. We—our brains—rarely act independently of what is socially expected or assumed. Once born into, or given, a role, most people live it. Our sense of right and wrong is the same as that of our society. This is especially so with people in non-Western societies who reject Western individualism as rude and unmannered. They value collective obligations, especially to their in-group, although they experience these not as obligations but rather as a means of belonging. They take themselves to be a part of their group and not individuals separate from it. To quote Hazel Markus and Shinobu Kitayama, two cross-culture psychologists: "[I]n some cultures, on certain occasions, the individual, in the sense of a set of significant inner attributes of the person, may cease to be the primary unit of consciousness, instead, the sense of belongingness to a social relation may become so strong that it makes better sense to think of the relationship as the functional unit of conscious reflection."[20]

But such experiences are only a matter of degree. Even sociologists of Western societies see people as having a *collective consciousness*. The term comes from nineteenth-century sociologist, Émile Durkheim. It binds societies and cultures by giving them a shared sense of what is normal, ordinary, expected, good, bad, right, and wrong. How correct such sociological theory is is difficult to say; we tend to be skeptical of heavy theorizing. But without question, social psychologists and anthropologists have found that we—whether Western or non-Western—rarely act independently of the social world into which we were born.

Our modern individualism should not blind us to this fact. What we do has to fit in with what our group does. We are social apes. We did not evolve from animals that were solitary and met only for mating, disappearing for the rest of the year from each other's sight. We evolved from apes that had to find a way to live together. We thus need to share socially predictable ways of doing things. This, of course, does not mean we cannot be self-oriented in our actions. We are out to get the best for ourselves; behind social cooperation is social competition. But we do this like other apes, against a background of finding ways to live and work together with our social group. We may be seeking our self-interest, but we generally do so in the terms of society, not against it. We may be selfish, egotistic, and self-centered, but we do not want to be psychopaths.

The socially acceptable emotions discussed above fit in with our individualism. No one wants to be unable to feel shame or guilt or not to know right from wrong; we do not want to be "morally insane." But what about social emotions that all but drown our individualism? Conveniently, we tend to sweep them under the carpet. What follows is a sketch for a social psychology of our lack of freedom. One of the biggest events in the human evolutionary odyssey, we propose, was the loss of freedom by hunter-gatherers to elites, an event repeated around the world after people progressed to farming. What could have let it happen so easily?

Brain Surrender?

A problem exists regarding our prefrontal cortex and our sociability. Are the cues produced by our prefrontal cortex socially independent? We suggest they may not be. After all, nothing requires that our brain make its cues blind as to what goes on in our society and our relationships. Our brains are not wired together by extra-bodily axons, but sociability among us can make them act as if they were. It is as easy—perhaps easier—for us to pick up cues from others as it is to find them autonomously within ourselves.

No law of physics suggests that our brains cannot do this, and much of history and social psychology suggests that we do, frequently. We have lots of words for it: obedience, deference, submission, and servility. We talk as if people surrender their independence in doing them, but of course, more truly, it is their brains that do. All people are obedient or deferential because their brains first make them so. If brains can link emotions and thought to invent empathy, self-consciousness, and pride, they can also invent servility and deference. All the brain needs is to hear or guess what others want and internalize that as its own inner cue. A brain that can use such cues to make itself free can also use them to make itself a slave.

Sometimes we surrender ourselves not to a person but to the "done thing." Indeed, this is more common for Western people. We are much freer than people in other societies and at other times, but we still get trapped. Our brains learn that in certain situations we should act in polite ways. If someone asks us to help them, we feel we ought to try to do so. Such expectations are powerful. They can trap us and make it difficult for us not to feel obliged to do what, in fact, we would rather not do. We can find ourselves socially cornered.

Our attachments organize us from without. When uncertain, children look to their mothers and other members of their families for cues as to what to do. We may also internalize cues from the group. Nothing in social psychology contradicts this. Where it exists, the evidence suggests that when adults are uncertain, they turn to their group or those in "authority" for a sign of what is right and wrong. Do we not defer our judgments to hear what our leaders and our peers are saying? Rather than being an isolated part of growing up, the orbital cortex's social referencing may be the first stage of what later becomes the social psychology that shapes our adult minds. Social referencing, instead of fading away, emerges as social reverence, deference, and willingness to be indoctrinated.

We might want to deny that people give up their freedom so easily to situations or to others. It is possible to rationalize obedience as apparent submission by people still acting freely in their thoughts. And in many cases this is true: People under duress have little choice but to do what they are instructed to. If you know you will be court-martialed for disobeying military orders, then you follow them. If you have a knife to your throat, your inner cues may be fully independent, but your external actions will not be. But is this true of every case in which people do what they are told? We have already met an example that suggests not.

When we brought up the people who took part in Milgram's experiment, obeying commands to electrocute another person—something they never would have done of their own will—we suggested that something overrode their moral attention. Now we will suggest what it was: They surrendered control of the inner cues organizing their behavior to the experimenter. Instead of using their own sense of caution, they looked to the man in the gray lab coats and let his external cues organize them: "It is absolutely essential that you continue," and "You have no other choice, you must go on." However much they thought inside or even verbalized aloud that what they were doing was wrong, it was the order from outside that controlled them.

But why did people so easily let themselves be externally organized? They struggled inside, laughed nervously, and protested, saying they wanted to stop. They pulled their earlobes and twisted their hands, saying, "Oh God, let's stop it." They clearly did not want to keep on giving "shocks." But the emotions behind their protests could not impel their brains to make the inner cues needed to walk out, tell the experimenter, "No, I refuse to go on," or go and apologize to the person they had started to electrocute.

One reason might be the way Milgram set up the experiment. He initially presented it as a "helping in a learning" situation. People readily go along with tasks if you ask "a small favor." As noted, we are brought up to help and be polite. But in agreeing to help another, we consent not only to their request but also to their cuing. Once given up, control is not easy to reclaim. We feel committed.

Observing the situation from the outside, it is easy to feel annoyed that Milgram's subjects didn't tell the experimenter to go to hell and walk off. But they were in it, unprepared and with their normal reactions caught off guard. Go back in your own mind to when you have been asked to do "a small favor." It is quite difficult to get out of. It is easy to protest from outside the frame, but once in it, the edge is hard to see. To break out, you must assertively face the other person and challenge his or her control of the situation. To stop the experiment, Milgram's subjects would have had to take a stand and say something like, "Yes, I will do your experiments—but only those I think are reasonable and ethical, of which this is not one." But most people do not like challenging others. We would rather not get into the situation of having to assert ourselves and say no, particularly with a person of apparent authority.

Milgram's subjects were caught in another way. It seemed to be acceptable to the person running things that they were electrocuting someone. They were not surrounded by signs saying, "THIS IS A TRICK—BEWARE!" Their brains had to internally create their own warning signs. And the entrapment developed gradually. The subjects were not asked to give "450-volt" killing shocks to start with—something that might have forced them to make a decision. The first shocks were 15 volts, slightly painful but safe. The people allowed themselves to obey (it seemed all right), and then the situation shifted.

This gradual acquisition of control is, interestingly, also the ploy of telephone con men. They start the conversation acceptably enough and then carefully maneuver the conversation. After politely asking questions, supposedly for a marketing survey, they shift gears into the hard sell.

Whereas Milgram tested obedience, con men test greed. Both utilize unfamiliarity combined with unexpectedness. Both throw people into a new situation. Americans do not expect to be asked to torture other people—certainly not in answer to a small ad in a newspaper. People likewise rarely meet people who on the surface are honest and trustworthy but who below it are working one big, calculated lie. They face a stark choice: Either think totally independently or turn to someone for guidance. And doing the latter has been made the easy option. After all, here is someone they trusted only a few minutes ago. They went to this person's laboratory to help in a scientific experiment, or let him or her into their home. Unless their brains can start being alert to their own inner cues, they are trapped.[21]

Telephone cons rely on the fact that people caught by surprise can be controlled through their politeness (and, of course, greed). People give up their life savings over the phone for worthless gems, penny stock shares, and Florida swampland. They could put the phone down but instead write checks and give credit card numbers. This is something that happens not just occasionally to the elderly and to dimwits, but every day to articulate, thinking professionals. (The best professional con men will not consider carrying out scams for less than $50,000.) It was estimated that such telephone fraud costs Americans more than $1 million an hour.[22] That is not only a fact about crime but one about human psychology. It says more about our minds than a shelf full of social psychology research papers.

Brain Ecology

Our orbital cortex makes many kinds of judgments, which do not always mesh. For example, we want to look after our own interests in both the long and the short term, and we seek to be good and moral people, if not in substance then at least in appearance. But these aims often conflict: Should we buy a new consumer luxury item or give that money to aid others less fortunate than ourselves? Studying may get us good exam results in the long run but keep us from going out with a girlfriend or boyfriend tonight. What feels good or bad often conflicts with what we sense is good or bad morally. We may make the "right" judgment, but our feelings may disagree.

These conflicts arise because the processes of the prefrontal cortex can work to different ends. A piece the area of a magazine cover such as *Time* tucked in the front of our skulls, the prefrontal cortex is really a collection of smaller prefrontal cortices, perhaps around 60 specialized conductors and composers of our behavior and thoughts (given 200 areas in each cerebral hemisphere). Twenty-two have been counted on the much smaller macaque monkey's orbital cortex alone,[23] and there must be many more in ours. Each of these has different inputs and outputs—concerns, memories, experiences—from and to the rest of the brain. There should be dozens of working memories controlling our wants and scruples and "coloring" the goods and bads, rights and wrongs we feel. There is no reason, however, that they should all agree. We tag our actions differently in different areas: morality, empathy, self-interest, duty, social situations, obedience, and greed.

Two and a half thousand years ago Plato described the mind as divided into rational and irrational parts—as he put it, the charioteer and his horses. Freud articulated our mental conflicts in terms of tension among the id, ego, and superego and differences between conscious and unconscious motivations. Transactional Analysis (TA) counselors interpret our inner conflicts by speaking of an inner Child, Adult, and Parent. Such therapists see these parts as sometimes working together but usually not doing so.

Perhaps analogously, researchers today looking directly at the brain find that a particular problem can activate several different areas simultaneously. Usually they work harmoniously, but again, sometimes they conflict, with one part grabbing control of a process that would be better controlled by another.[24] One of the ways the brain deals with a mental task is to switch certain areas off.[25] Various parts of the brain seem to compete for activation. As in ecology, a struggle between conflicting processes creates the larger whole. Perhaps it was in this internal battle that Milgram's subjects were trapped. One prefrontal loop area was tagging what to do in terms of empathy with the person they were electrocuting. But another competing area tagged their actions for compliance to authority. Obedience cuing, in most cases, won out.

The Left and Right of Freedom

Brain scanners show subtle but strong differences between the two hemispheres of the brain. The right one is strongly involved in face recognition, voice identification, emotions, visuospatial skills (e.g., turning shapes around in one's mind), and skills with a "broad" focus. The left one is strongly involved in speech, verbal reasoning, and skills with a "narrow" focus. When the left cerebral cortex is in command, people reason in a theoretical, decontextualized, and deductive way. When the right is in command, people refuse to make decontextualized deductions and approach problems in an empirical way.[26] In memory, the right prefrontal cortex retrieves memories, while the left encodes them for storage.[27] Activation of the right prefrontal cortex is linked with depressed immune responses and the left with stronger ones.[28]

The two sides are linked to different aspects of being with others.[29] The right prefrontal cortex is activated by novel situations and interruptions, the left by continuous active engagement. The right hemisphere is concerned with the nonverbal and "negative" side of emotions—withdrawal, disgust, distress. The left is concerned with positive, environmentally engaging approach emotions, such as joy and interest. Depressed adults show reduced left prefrontal activity.[30] Ten-month-old infants with greater right prefrontal activity cry more when separated from their mothers than other infants do.[31] Those who look on the positive side of themselves rather than ruminating on failures not only show high self-esteem but have more left prefrontal cortex activity.[32] But not all positive emotions are left-sided: in men (women were not studied), while the rest of the cortex shows reduced blood flow, the right prefrontal cortex lights up at sexual orgasm.[33] It is also activated by the pleasures of drink and possibly by sensory deprivation in flotation tanks.[34] There are also hints that the level of the right brain's involvement and that of the left's inhibition are linked to hypnosis and hypnotizability.[35]

At present, no one knows why these differences exist. Why are different tasks done separately by the two prefrontal cortices? Some research suggests that the two sides of the brain, while often cooperating, sometimes compete to carry out tasks either might do, each seeking to inhibit the other.[36] Could it be that the dorsolateral part of Milgram's subjects' left prefrontal lobe wanted to stop the experiment and get out and that the orbital part of their right sought to enact the suggestions of the experimenter, overrode the left, and caused them to obey?

This chapter may seem to suggest everything reworked in us lies in our frontal lobe. But what our prefrontal cortex does works with other parts of our brain. Most of our mental processes are spread around the brain in various subtask specialties. Nothing is localized; it is just described that way for the convenience of discussing it. Though we talk about skills being localized in a part, they are more like spaghetti on a plate. Try to pick up a few strands on a fork, and the whole plateful seems to come up. Everything done by the brain is highly interwoven. We will move on to look at some more mindmakers that work closely with the prefrontal cortex.

11

Where Memories Are Made

A man wakes up one day in the hospital with his head bandaged. He's been in a bad accident, and his problems have only just begun. When visiting hours roll around, he awakes to find that his wife is not his wife. She is gone. Instead there is a *doppelgänger*, an exact duplicate, in her place. So, too, with his children; they've been replaced by duplicates. At his bedside consoling him is a new wife with the same name and the same looks as the old one. She even brings along a "replica" mother-in-law. They're spitting images of the old family, yet, he "knows" they are not the originals. He wishes it were science fiction, but it is not; it is horror, because it is real.

In comes the doctor:

Doctor: How many children do you have?

Patient: This can be confusing, but in any event I am responsible for eight sons and two daughters.

Doctor: That is confusing.

Patient: Well, between the two girls they, oh, I suspect they had five each.

Doctor: You have to slow down now. Two girls? You are married twice? Let's go through this again. You are married once.

Patient: Married once.

Doctor: How many children did you have from that first marriage?

Patient: Four boys and one girl, but the second girl who I am married to now, she picked me up on December 5th a year ago.

Doctor: She picked you up?

Patient: Yes, at the hospital to drive me home for the weekend. They don't look alike, but they are similar, both brunettes and they are similar in appearance.

Doctor: So all of a sudden you came out of the hospital and you went
         back to the same house with a whole different family.
Patient: Well, that is about the size of it, yes.
Doctor: How can you account for that?
Patient: I don't know. I tried to understand it myself, and it was virtually
         impossible to understand it.[1]

The patient has Capgras syndrome. A traffic accident has severely injured his frontal lobes and the right side of his brain. And the family he knew in his head has been split into two.

The man with two wives is not alone in experiencing doubles. Someone with the initials R. K. awoke to discover that he could not be the true R. K. His appearance and circumstances had changed from what he remembered, and thus, with a merciless logic, he observed he could not be the same person. "If I were really me I'd have a place of my own. I'd be working."[2] A blind woman awoke one day to find that her faithful companion for 10 years, a cat, had been replaced by a bad duplicate that was cruel and ill intentioned toward her.[3]

The mind does many conjuring tricks that make experience seamless and whole. In spite of surface changes, things usually remain familiar. This is true of even the most basic thing, the sense each of us has of our own continuity. I am I, even if I get a suntan, move to another country, or change my name. Our sense of who we are can adjust to these changes. R. K. shows that our sense of continuity can break down if our brains cannot update new details of how we look and other aspects about ourselves. *"Cogito ergo sum,"* penned the French philosopher René Descartes (1596–1650): "I think, therefore I am." But as John Locke, the English philosopher (1632–1704), noted, identity is linked to memory. Therefore, it may be truer, if less pithy, to suggest, "I memorize the past with the present, therefore I have been and so am now." If we do not flow from someone in the past, who can we be?[4]

Memory has traditionally been seen as a kind of filing cabinet into which our mind stores experiences, but it is more complex than that. Our memory not only stores but recalls and identifies. To do so it must be able to link the present with the past. It must be able to to sense *continuity*.

For instance, when we meet people, we can remember old details about them. No one stays exactly the same, yet remarkably, this does not bother our sense of who they are. A new haircut or coat does not make us experience someone as a new person with a different past. We easily spot the person under the changes and thus can quickly recall our past experiences of them.

The continuity of our selves and others is all so obvious that we take it as if there were no skill behind it; it is simply never a problem. It therefore comes as a shock to discover that our brains must be doing something, and doing it well, to create this seamless sense of continuity behind people's identities. The man with two wives shows this by his misfortune of having lost the ability to do it.

Normally our brains update the links that let our memory retrieve past details about the people we know. Here lies the problem for the man with two wives. His present experience of his wife is not being used to update this memory link with his memories of her. Without this, his brain cannot join his experience of her as she is now with the memories he once had of her. His brain thus experiences a woman who is very like his wife but who is not, at least in his memory, linked with the woman he married. She both is and is not his wife; in a way, she is like a twin, similar but not the same. He copes with this bizarre situation by experiencing her as a duplicate.[5]

What keeps these bizarre experiences from happening to you? The answer lies in your *memory headers*.[6] They hold information about someone's current identity that links it with stored memories about them. Your memory headers are constantly updated with new cues. While stored memories can be permanently changed, or updated, with new factual information, memory headers are always changing but do not store information beyond what is needed in the moment to recognize someone's identity. By holding identities long enough to summon what you need from stored memory, they let you recognize someone, however much he or she changes. They link the past and the present for you as you move through life.

A common experience, especially in old people, is explained by this division of memory into headers and storage records. Sometimes we can think of many details about someone, such as to whom he or she is married and where he or she lives and works. But however hard we try, we cannot recall the person's name. Instead, we just keep thinking of more facts about him or her. It is as if we have gained access to the memory file storing details about the person but cannot find the "header" with which the file is labeled.

Headers are not the only way we recall our memories. Marcel Proust, on a winter's day, tasted tea given him by his mother. In it he soaked crumbs of squat, plump little cakes called *petites madeleines*. Suddenly a shudder ran through him. "I had recognized the taste of the piece of madeleine soaked in her decoction of lime-blossom, which my aunt used to give me." Up came long-forgotten memories of his youth in Combray: "the old gray house upon the street. . . ."

Smells and tastes link places and things to memories, but we cannot call them up at will. Instead, we must encounter them. However much Proust might have thought about his madeleine cake dipped in tea, the memories would not flood him. He needed the real thing.[7] The right smell or taste can be a powerful trigger, calling forth chunks of our memories that have long been out of reach. As Proust puts it: "When from a long-distant past nothing subsists, after the people are dead, after the things are broken and scattered, taste and smell alone, more fragile but more enduring, more unsubstantial, more persistent, more faithful, remain poised a long time, like souls, remembering, waiting, hoping, amid the ruins of all the rest, and bear unflinchingly, in the tiny and almost impalpable drop of their essence, the vast structure of recollection."[8] Tastes and smells can act as a kind of headers for the emotions that go with experience, but we have to chance upon them. By contrast, we can usually recall verbal headers, such as names, when we need them; but perhaps we pay a price for this accessibility: Such memories are not so strongly linked to past emotions.

The Roots of Identity

How does the brain update memory headers and so produce the sense of someone's having a continuous identity? There are two ways. First, although people may change, not everything about them does so at the same time. We identify people through a number of sensory clues that are not necessary or sufficient by themselves; this tends to happen on the right side of our brain through tone of voice and how a person looks.[9] With this side of our brain we also identify those who are emotionally near to us as opposed to those who are not; we sense them as "close."[10] If the right side of the brain is injured, the ability to recognize emotional closeness goes. This is what happened with the "man with two wives" described above. However, this does not totally explain his problem. If emotional recognition has been destroyed, he would be unable to recognize his wife. But not only does he recognize her, he says she was like his "old" wife. What he cannot do is link his current experience of his wife with the old identity by which he already knows her. He should be able to—we can update our headers in a further way: by what others tell us and what we ourselves can infer. His doctors will have explained to him that he has a brain injury that has changed his experience of his wife. His common sense should tell him the same. But he cannot use information like visual similarity to update his sense of her. This inability is common in people who, like him, have frontal lobe injuries. He knows the situation is unbelievable, that it is "impossible to understand." But even though he can see, when it is pointed out to him, the absurdity of his situation, he cannot use this knowledge. His impaired frontal lobes simply cannot update the memory headers for his wife. Lacking both sensory and frontal ways of establishing continuity, he resorts to making new memory headers for his wife and family. The headers have now become divided in his mind, leaving him in a state of "bigamy" with his own wife.

Continuity and Morphing

Identities and continuities are central to our experience. Are you the person you were yesterday, 10 days ago, or a year ago? What about further back: Are you the embryo brain that was once in your mother's womb at 20 weeks? Or the cells that preceded them at 5 days? At the moment of conception? What about the atoms that make up your body and brain? Are you they? Were they you for the billions of years after their creation in stars prior to their turning up here on Earth and coming together as the flesh and blood you are now? The transition between caterpillar and butterfly is startling in its metamorphosis. What are the boundaries of our continuity and identity? These are, of course, the sort of questions that drive the inquisitiveness of philosophers and scientists.

Consider entertainment: We take an almost addictive interest in the boundaries between where things are and where they are not. We have a deep-seated fascination with the conjuring tricks of magicians, the visual distortions of cartoons, and hyperreal computer animation effects such as "morphing." The scene in the film *Terminator II* in which a cyborg emerges from the checkered floor tiles has become psychological folklore. Why should we be so fascinated with morphing? We suspect that, in a world filled with discontinuity, we are born with a nose for detecting continuity. Clever brains pick it out as they better aid our survival. Our brains never stop. Our minds therefore are enthralled when familiar continuities are broken, as in image morphing, conjuring, and animation. They draw us, as in science fiction, to the borders of the impossible and a little beyond.

But what if our memory headers are lying to us? Maybe the continuities we take for granted are not what we take them to be. We know that our brains play tricks on us. Consider the experience that things we have never seen before are in fact familiar—*déjà vu*—and its opposite, *jamais vu*—the experience of familiar objects, scenes, and thoughts as totally strange. As Charles Dickens put it in chapter 39 of *David Copperfield*, "We have all some experience of a feeling which comes over us occasionally, of what we are saying and doing having been said or done before, in a remote time—of our having been surrounded, dim ages ago, by the same faces, objects and circumstances—of our knowing perfectly what will be said next, as if we suddenly remembered it."

What grounds do we have to believe that these experiences are rare? Déjà vu could logically be the basis of our experience, but "we" might not be noticing. Suppose we were not really continuous. Suppose, instead, that each day a new person awoke believing he or she was "you," so that really you were a kind of human mayfly, living only 1 day, with your past memories being something inserted into you. Unknown to you, when you are asleep at the end of your day of life, these memories are somehow extracted from you and inserted into your successor. Your memories linking you with your past are, in spite of what you think, nothing but a kind of déjà vu illusion. You think you lived them only because you "hallucinate" the memories of past people as continuous with your experience now. But in truth, each "you" lived only 1 day. Each life adds a separate frame to a "memory film" that is then inserted into the next person. Unfortunately, the operation stops memories of what is happening from being passed on to the person who will be you tomorrow morning, so no one realizes the true short nature of life. These kinds of questions crop up not only in the realm of science fiction. After all, R. K. has lost that sense of déjà vu for the person he had previously been.

Perhaps the man with two wives is not suffering from a delusion but is experiencing a rare insight into the true nature of things. Maybe his wife really *was* replaced by a duplicate. Maybe you and your spouse are not the same people "you" were yesterday. (This is true from a strict physical viewpoint, since on a daily basis you lose a great number of atoms and molecules and thousands of skin and other cells.) Tomorrow, when you pick up this book, maybe "you" in reality will be another, totally different person.

Headers and continuities can apply to places as well as people. You go to a friend's house for the morning and on the way back you get lost. You give your address to a helpful taxi driver, but he does not drive you home. Fortunately, at the house where he takes you, you find your spouse and your daughter. But they are in someone else's home. Bizarrely, it is strikingly like your own—its owners have decorated it with ornaments similar to those you once brought home. Indeed, the coincidences are remarkable, even to having the same items by the bed as you had in your bedroom.[11] But these are coincidences; you know it is not your home. Like the person with two families, brain damage can cause us to experience a duplication of where, as well as who, we are.

How and where might headers arise in the brain? Headers, we suggest, arise out of the brain's prior evolved need to be able to process continuities.

Both places and people change slowly over time in the small details by which we recognize them. New buildings are added to towns, while others are knocked down. It aids us if we can spot the familiar continuity underneath the change. Consider the problems that would follow if a person thought the world had changed every time it snowed. Our brains therefore evolved an area skilled at sensing the hidden continuity of things. Later this ability appears to have been reused for organizing memory.

The Horse-Headed Sea Monster

Put your mind's hands back into your brain. This time put your thumbs under your cerebral hemispheres around where your ears are. The cerebral hemispheres curl in where they join the rest of the brain. This curl is the outer surface of the *hippocampus*. Cut through it and you enter one of the fluid cavities (ventricles) inside each cerebral hemisphere. To early anatomists of the brain, the raised pattern it formed on the bottom of these cavities looked like a horse-headed sea monster; thus the Greek name for this part of our brain: *hippos* (horse) *kampos* (sea monster).

Early nocturnal mammals had to know where they were. The hippocampus evolved to give them a mental map of their nightly travels and so let them know whether they had been to a place before and, as importantly, what they had found there. The hippocampus is made out of special circuitry that (1) detects spatial "continuities" and (2) links them to experiences stored elsewhere in the brain. To carry out the first task, its neurons are "woven" to form what scientists call *autoassociative networks*. To perform the second, the hippocampus and the circuitry of its autoassociative networks have strong input and output circuitry with the rest of the brain.

The price of this special circuitry is that hippocampal neurons are especially vulnerable. A key part of the structure called the CA1, for example, sports long, small blood vessels only two to three times the diameter of blood cells. Not surprisingly, neurons here are the first part of the brain to die when the blood supply to the brain stops. This happens, for instance, when the arteries to the brain are blocked during hanging or strangling. Cut down quickly, a person attempting suicide survives, and so does most of his or her brain, but not the vulnerable hippocampus.[12] Another demonstration of the sensitivity of hippocampal neurons occurs when epileptic convulsions reduce oxygen to the brain. This can lead to the tragic feedback of oxygen lack damaging the hippocampus and thus lowering its threshold of resistance against further epileptic fits, which then cause further damage.[13]

Every mammal, including each of us, has a hippocampus on either side of its brain. Each of our two hippocampi is made up of 14 million neurons and averages about 5 cm^3 in volume, the right one being slightly bigger than the left. Compared to those of other animals, ours are large, providing extra space to help our memory.[14] The three parts of the brain mentioned so far—the amygdala, anterior cingulate cortex, and hippocampus—make up most of the *paleomammalian*, or "old mammalian," brain, more often called the *limbic system*. Some would include the orbital prefrontal cortex. (Birds have something called a hippocampus, but, though its function is similar, under the microscope its structure appears quite different.) Limbic parts of the brain like the hippocampus have a copy on each side. However, it is a convention, and certainly more convenient, to talk and write about them in the singular.[15] Also, the proper term is *hippocampal formation*, in recognition of the several parts of our brain that here work closely together. But again, for convenience, we use the simpler, singular *hippocampus*.

Monster Cartographer

Imagine you see, on a computer screen, the following picture: a rat maze in the shape of a cross. It is colorful. All around, like tangled thread, is a trace of where the rat has moved, in a multicolored line. In some branches of the maze the thread is red, in others blue, and in still others yellow. Indeed, the color of the thread varies with the position of the rat in the maze. Oddly, the colors of the line have nothing directly to do with the maze; rather, they are related to signals picked up in the rat's brain. Scientists have implanted electrodes in the rat's hippocampus. They detect activity in its various "place" cells. When it is in one part of the maze one of these cells is activated and shows as red, when it is in another a different cell is activated and shows as yellow, and so on. Scientists are bugging the rat's own map of the maze.[16]

It is a map that started with our nose. As part of the old mammalian brain, the hippocampus has close links with sniffing and our sense of smell through our olfactory lobes.[17] Before the rise of the rest of the cerebral cortex, the hippocampus mapped the smells (and the associations linked with them) sniffed in the dark world that was explored by the first mammalian noses.

The mapping of the hippocampus is not like the mapmaking done by cartographers. Special neural networks contain "place" cells, but they are not linked to actual places in a simple one-to-one "topographic" mapping. Indeed, place cells next to each other in the hippocampus often respond to different parts of the real world, and each cell responds to more than one place. The reason for this is that the brain, instead of locating places on a grid of longitudes and latitudes, locates them using autoassociative networks.[18] These are specialized neural networks for holding contextual information, the many relationships and configurations of things that identify a location. With such autoassociative networks, the hippocampus can map the webs of associations that interlink things according to where they are.

Even if the brain had the ability to locate things in a simple reference grid, it would not have the time. Besides, by defining places contextually, it gains the dynamic ability to identify places from shifting fragments of information. Such an associative map can adapt. If a few "landmarks" change or disappear, the map can learn new ones to replace those that have disappeared, thus updating the context used to locate a particular place.[19] For example, if the world is suddenly covered with snow, the brain can orient itself using those landmarks that are not hidden. In the spring it can learn new ones to replace those that have disappeared over winter. In this way, if the animal lives long enough, the clues it uses over time can evolve while the places they identify remain the same. The brain can thus flexibly detect continuities in a changeable world, an important skill that, as we will see, can also put to use for other purposes.

Memory Maps

The hippocampus can store information within itself, or it can serve as an index for information stored elsewhere in the brain.[20] Moreover, the autoassociative nature of its networks allows the hippocampus to bind different sources of such information together.[21] This gives the hippocampus an enormous flexibility not only to store but to organize memories.[22]

Memory needs to be organized. Consider the problem of how you create a memory that works. Computer files cannot be stored at random, or the valuable files would be lost among all the unwanted ones. It is the same with the thousands of scientific papers we have drawn upon here: If they were intermixed at random, they might as well not exist. They are useful only when they can be retrieved as needed. Our memory also needs to be structured if we are to recall things.

A computer's memory uses a map structure with directories and subdirectories that contain headers indicating where files are stored. A small amount of memory is thus used to organize a much larger one. We suggest that the hippocampus uses a similar approach. In effect, it creates a map—an autoassociative-based one—for our memories held elsewhere in our brain.[23]

How does the brain use its memory structure? Possible clues come from an ancient technique of memorization called the *Method of Loci*, by which memories are referenced directly onto spatial maps. The Method of Loci originates with a story: An ancient Greek poet, Simonides, was at a banquet given by a nobleman, Scopas. A message was brought to Simonides that two people were waiting outside for him. When he went out to find them, the roof of the banqueting hall fell down, killing Scopas and his guests. Afterward, relatives seeking to bury the bodies could not identify who was who. Simonides, however, could remember the dining places where they sat and so identify them.[24] The Method of Loci takes advantage of our ability to store the location of visual things so that, like Simonides, we can recall things from the places they occupy.

To remember items according to the Method of Loci, first change them into visually bizarre mnemonic associations and place them around a familiar room. (Things to be remembered must be visual and striking, as word descriptions link poorly with places.) If your list is of recent U.S. presidents, for instance, then you may create mental images of a road (or a car) crossing a river, peanuts, a movie, a small tree, a cigar in an improper place in a saxophone, and an even smaller W-shaped tree. Then, in your mind's eye, imagine a room, take these mnemonics, and physically locate them around it. Do so in an odd way. It is easier to remember a human-sized peanut wearing a "vote-for-me" badge sitting on a chair than an ordinary-sized one just lying on a table. When you want to recall the presidents, you reverse the process. You imagine the room and then mentally go around it seeing where you placed the various mnemonics. This aids memory because the problem with remembering things is not storage but recall.[25] The part of our brain used by the Method of Loci is as yet unexplored, but its spatial nature and link with memory strongly hint that it takes advantage of the hippocampus's ability to recall associations linked with places. Indeed, when people explore and learn the positions of objects in a virtual reality maze, the hippocampus is activated.[26] The hippocampus is also activated when people are asked to encode and recall object locations or explore journeys around a map or a virtual or real place.[27] London taxi drivers likewise activate their hippocampus when asked to use their mental map of the city.[28]

Thus, it is likely that while other parts of your brain may have lost the associations needed to recall what you have memorized, this part of your brain has not—at least in its map of the room. Research shows that people with "super" memories often rely on spatial, and so presumably hippocampus-aided, retrieval skills.[29]

The Method of Loci works by using context to retrieve memories. It is limited, however, to memories we have intentionally organized. For most of its remembering the brain needs a more spontaneous means of cuing memory context as needed; it needs to be able to perform what is called *recognition recall*. In recognition recall, the brain scans contexts set up for retrieval, into which memory headers are semantically, conceptually, and otherwise located.

As we've seen, internal access through memory headers is achieved with the aid of the prefrontal cortex.[30] The hippocampus is linked by at least two pathways to the prefrontal cortex.[31] (It is also massively linked to the temporal cortex that surrounds it.) PET scans show activity in the prefrontal cortex when people retrieve or encode memories. This suggests that the prefrontal cortex–hippocampus links enable the internally cued retrieval of memories—recall. Supporting this concept is the finding that those with injured frontal lobes have problems with free but not externally cued recall of memories.[32]

The files that headers retrieve need not be in the hippocampus. For the most part they are scattered around the cortex.[33] Those concerned with personal details of our life are thought to be found in the polar areas (the frontmost parts, nearest to the frontal lobe) of our temporal lobes.[34] The prefrontal cortex is strongly linked to these areas.

The prefrontal cortex uses the context in which a memory was laid down in order to retrieve it. If you try to recall whether you have met someone, you do so by going over the circumstances in which you might have met. The associations surrounding a memory need to be encoded along with the memory itself. Making these associations is the role of the prefrontal cortex, not the hippocampus (though, of course, the hippocampus readily stores them). The prefrontal part of the brain lights up on brain scans when people learn lists of word pairs. Try remembering a list of word pairs, such as *horse-seat, clock-louse, hack-wheat*. Then try learning a similar one such, as *horse-wheat, clock-spider, hack-saber*. The associations of the first list interfere with the learning of the second one. It is hard work mentally encoding the second list so that they are not confused. Brain scanners find that trying to learn them activates the prefrontal cortex.[35]

Consistent with this, people with prefrontal cortex injuries do not use context in recall: They may remember something, but not when or where they learned it.[36] They lose the sense of order in which they learned things.[37] They are impaired in what psychologists call "source memory." This could be linked with a failure to use context in recall, but part of the problem must also be a failure to file memories along with the associations needed to aid their recall. If their problems lay entirely with use of context to access their memories, then it would expected that if they had recalled a memory, they would also be able to remember its circumstances. They do not.

Memory Loss

What happens if the memory structures used to locate our memories are lost? Early versions of DOS, the operating system for the PC, had a command called, of all things, RECOVER. It removed the directory structure that organized files and put them all into one single directory with new file names, such as FILE0001.REC, FILE0002.REC, and so on. While it did not delete the files, it might as well have done so, since it destroyed the means of locating them. Something similar happens when people have an injury to the hippocampus. Near-suicides quickly cut down from their nooses lose their ability to remember new and old events. They even lose the memories of their attempt to hang themselves. Just as when you mislay a paper in the wrong file, it is, unless you chance upon it by accident, as good as torn up.

Damage to the hippocampus affects memory in three ways. First, people with such injuries cannot form new memories. They have brains without directories and the ability to make them. They experience information but have no way to put it in a place from where they might recall it; they can't create a header or retrieval context. For them life ceases to be a series of episodes stretching into the past. Instead, it is confined within the few minutes of the short-term memory that acts as a small buffer to temporarily hold things like telephone numbers. Beyond that a gap exists until they can reach the memories laid down in their cortex prior to their injury. Not that such people cannot learn. Psychologists find they can acquire new skills. For instance, someone who had used only a dial telephone could learn how to use one with buttons. They can learn associations between ideas—called *priming*—so they recognize things better when they are repeated. But that is all; they cannot remember making a call or even having learned to use the new telephone. These kinds of memories—short-term, procedural (learning), and priming—are carried out elsewhere in the brain, in the basal ganglia and the cerebral cortex, and do not need organization to retrieve them.

Second, people with damage to the hippocampus have problems recalling some of what they have already learned. Before their misfortune, they would have learned many things. Some they can remember, others they cannot. If the injury affects the CA1 part of the hippocampus, memories more than a year to two years old can be recalled. But those laid down within the last year or two are lost. This loss may stretch back decades if their injuries affect other parts of the hippocampus and related areas of the brain, such as the *parahippocampus* and the *perirhinal* and *entorhinal cortices*. The CA1 part of the hippocampus keeps the headers and other retrieval contexts for memories more recent than 2 years. Lose them and you lose access to the memory files. Longer-term memories, stretching back a life-time, are recalled with other parts of the hippocampus and brain. Damage them and memories going back decades are blotted out.

Third, if you cannot find old memory headers or make new ones, you cannot refresh them. Not only do you lose your memory of the past, but also of the present. If you are unable to update, nothing can be experienced as old or as having been before. Everything, however often it is repeated, is as new as when first experienced. The mind has become a sieve. However many times you meet a person, he or she remains a fresh face. The hip-pocampus lets us experience "now" because it allows us to know there has been a "then." Without its workings, even the oft-repeated seems brand spanking new.

Why past memories are less affected than recent ones by hippocampus injuries is not fully known. Neuroscientists explain it by the *consolidation* process, mentioned in Chapter 2, that during sleep memories move out of the hippocampus into the cortex.[38]

This explains why memories leave the hippocampus. As noted above, headers need to be updated, but only when what they recall changes—and not all things change. If an old school friend moves next door, we remember him or her both as he or she once was and as he or she is now as a neigh-bor. We do not need to update our school-day memories of him or her, since after all, they are mental records of how he or she once was. Only his or her ever-changing present identity needs to be kept up to date. Thus, old head-ers do not need to be updatable; in fact, we do not want them to change—we want them to be memories. Indeed, old memories may not even need headers; the associations that arise over time within the stored memories may be sufficient for recalling them. And this means that old memories need not depend on the hippocampus.

In contrast, memory headers for people and things in our everyday lives need to be constantly refreshed, and for this they are dependent on the hip-pocampus. If all memories were redirected out into the cortex, they would be turned into "archival" memories. Because it destroys the brain's ability to update memory headers and retrieval contexts, damage to the hippocampus impairs recent memories more than old ones.

The hippocampus's mix of autoassociative networks and widespread connections is likely to be used to enhance many aspects of memory and cognition. Some skills require that different parts of the brain share information that is spread around the cortex. For instance, driving a car requires the ability to interface your skill in seeing things—movement through a landscape and people about to walk into the road—with your motor control over your hands and feet. The brain can easily form the links needed to do this, but it takes time, perhaps several months. The brain needs a "prepermanent" way of linking these abilities in the cortex. But how? One idea that has been proposed is that such prepermanent links are temporarily routed through the hippocampus.[39]

Another skill we need is spotting whether we have encountered something before. Here the hippocampus works with the prefrontal cortex to give us our capacity to be surprised. Continuity and expectations need to work together constantly if we are to note what changes around us. The prefrontal cortex may spot the unexpected and the novel, but first the hippocampus, working with other parts of the brain, must lay down the background against which we can experience the bright figure of surprise—familiarity.[40]

Our ideas about what the hippocampus does should thus not be too dogmatic. It does not do any single task. The brain is not like a chip with its various parts purposely designed for carrying out specific functions. Its neural networks can take on many roles. The brain is flexible and resourceful. It is in the business of survival, after all, not of obeying design plans. The neural networks of the hippocampus therefore serve, and are open to, many uses.

Moreover, we should be wary of present theories of memory. They look as if they ignore as much as they model. Contemporary notions of memory are dominated by the notion that memory is *episodic*, or dependent in its recall and preservation on a *spatiotemporal context*. For many memories, where and when the events happened does serve as a retrieval context for their later recall. But this is so only with some, and they are not our most important everyday memories. The focus on episodic memory has more do with how scientists investigate memory than with the nature of memory itself. The problem is that in order to control what people memorize, researchers must use information that is new to them. People are bound to memorize such information, as they lack any other retrieval context for it. Almost by definition and default, then, this forces such research to study the episodic memory. What is ignored is that memory itself is the most important context in which people remember things. The critical thing about something new is often not that it is novel but that it slightly changes an established memory or identity. But such revised memories are hard to investigate, because they are individual to each of us, having grown over many years. Updating memories nevertheless is as much part of memory as laying down fresh episodic ones.

Here we can see the overwhelming problem of studying the brain. Not only are we in the dark about how the hippocampus functions, we also do not know many of the tasks it used for. Introspection does not tell us, however good we are at it. Excellence in a skill is to a surprising degree compatible with utter blindness as to its components. Consider trumpet playing: For years, teachers of the trumpet encouraged their students to learn to put as little pressure as possible with their mouth on their instrument's mouthpiece. They even gave them special exercises to help with this. Now, you would have thought that these teachers, being skilled trumpet players themselves, knew what happened between their lips and their trumpet. But it took scientists at the University of Strathclyde, in Glasgow, to show that their ideas were totally wrong. They measured the forces that professional trumpet players put on their instruments and found they were substantial.[41] All those exercises to keep pressure uniform were worse than a waste. But if expert trumpet players can be so ignorant about how they play their instrument, what else might people misunderstand about their own skills? We face the boundary of ignorance all the time. If we are ignorant here, we are ignorant everywhere. We could well be strangers to even the simplest things we do, perhaps even more than to the complex ones.

Frameworks

The patient was smart, but he was more than a little lost: "I know what you tell me, but it seems to me I'm in Paris." The previous day he had thought he was in a large hospital outside London. The next day he said, "Do you think the doctor will come to Arizona to see me?" In the following days he said: "It's a luxury hotel, somewhere in the Far East, probably Tokyo"; "It's a hotel in China"; and "Concord, Massachusetts"—and so on across the globe, from Chicago to Baghdad, from Denver, Colorado, to Timbuktu. In fact, the patient was in Massachusetts General Hospital in Boston.[42] He was not demented or confused. He was an architect and could discuss a paper he was writing with a colleague. But following a stroke, he had lost the ability to orient himself on the map in his brain that located him in the world.

We follow our inner maps. Most of us are good at navigating our houses, our home town, or indeed, the world. It may be only going from our bedroom to the car, or from JFK, New York to De Gaulle, Paris via London Heathrow, but we can find our way around. More important, we know where we are on these inner maps and so where we are in the real world.

After some kinds of strokes people can lose half of the world. Ask them to imagine themselves facing east at the western end of Constitution Avenue in Washington, DC, and they can describe the buildings on their right but not on their left. Their world will include the Lincoln Memorial and the Washington Monument but not the White House. Ask them to imagine themselves toward the western end, and the world they see will change: They will note the Federal Reserve and the White House but not the buildings they previously mentioned.[43] Ask them to imagine themselves at the front door of their house or the top of the road where they live, and they recall rooms or houses to the right. Suggest they are at the back door or the other end of the street, and they reverse what they recall and note the other rooms and houses. Ask them for directions, and they give right turns and leave out most left ones.[44] They live in the right half of the universe; the left (it is usually the left) has disappeared from their awareness. Their loss, called *hemineglect*, might affect their imagination, but not what they see or hear. Or it may be deeper and cut into what they perceive and do, in which case they draw half a clock, eat only half the food on a plate, and dress only on one side. Such people live marooned in only one side of the world. The other half of the universe, or rather the brain's processing of it, has ceased for them.

Our brains place themselves in two types of framework. The first one uses our body for an anchor: It is focused on itself and sees things in these terms (this is the framework injured in cases of hemineglect). The second anchors itself on landmarks beyond us (this is the one injured in the architect). The first, body-centered framework is called *egocentric* (*ego* is Greek for "I"). Your egocentric map has a landmark—your body. It may be what is to the left or the right of you (or up or down, or forward or behind). Mentally, the brain maps the world around you as a hub with spokes radiating from your body. The egocentric map usually provides the framework for our own bodies and things near us. Damage it, and the part of the world around you, including your body, ceases. You find yourself living in only half of the world that the rest of us know.

The second framework is *allocentric* (*allo* is Greek for "other"). It tends to start where the egocentric map ceases. It underlies our sense of the layout of a room or of the whole globe. Whereas egocentric maps are useful for orienting us to things nearby rather than far away, maps anchored outside us are useful for moving about in the external world. The hippocampus helps make not only the allocentric framework but possibly also the egocentric one.[45] Evidence suggests that the latter is also tied to the *posterior parietal cortices*, particularly the right one.[46]

We live in two frameworks. We sense our world, act upon it, and recall what it was. And we are also oriented and located in it. The hippocampus lets us experience a world of continuities and identities, but our physical relationship to them is always changing. As we move around, things are sometimes to our left or right, behind or in front of us. The hippocampus maps our place in the world, and the posterior parietal cortex (with the help of the hippocampus) maps our particular orientation in it. If something is behind you, you may be to the north of it as well. One can change without the other's changing. If I turn around, something that was once behind me will now be in front of me without my ceasing to be north of it. But things can be the other way around. The tail end of a plane may be south of me one moment and then to the north the next without ceasing to be behind me. What in our brains sorts out the links between the two?

Remember plunging your fingers into your brain to feel your cingulate? What, you may have wondered, did the cingulate in the back of your brain in the parietal lobe do? Here we have its possible function. The anterior cingulate cortex links the emotions of our amygdala and the actions done by our motor system, and the posterior cingulate cortex links the maps in our hippocampus with the sense of orientation in our posterior parietal cortex. What the anterior cingulate does for motor control and emotions, the posterior part may be doing for orientation and location. "It might," neurologists write, "participate in the transformation from a parietal representation of space based on a body-centered frame of reference to a parahippocampal representation based upon a world-centered frame of reference."[47] Here, if they are right, the mindmaker of continuities joins with that of our personal orientation.

And, there's more. Just recently both PET scans and studies of people with brain injuries have suggested that the posterior cingulate is involved in encoding memories.[48] Perhaps this is not surprising: It is well linked to the hippocampus and, like the anterior cingulate, to the prefrontal cortex. Its links with the parietal cortex could join its mapping of experience with that which takes place in the hippocampus. It has also been suggested that it underlies the feeling of "familiarity" irrespective of what it applies to—a person's voice or looks—whatever it is that is lost in those with Capgras syndrome.[49] Frustratingly, we do not know the details, but, as elsewhere in this book, we mention these tentative findings to point out the borders of our knowledge. They may be hot tips as to where the science of the twenty-first century will make its big breakthroughs.

New Maps of Self

We define social and personal relationships in terms of spatial ones. Social language is full of location descriptions—top people and underlings, in- and out-groups. In negotiations, people take "positions." We all have our "boundaries" on which people "intrude." This may reflect the human creative use of metaphors, but it may also demonstrate something about how the brain orients itself in the social world of others. In an Internet search, we typed *allocentric* and found not only a few papers by neuroscientists about the hippocampus but also quite a few by social psychologists. They have fixed on the same word to describe the locations of people in their social worlds. Some people and some societies are more allocentrically or egocentrically oriented than others.

Could it be that our sense of where we are in our *social* world gets processed in the same parts of the brain used to orient us in the physical world? Probably. No obvious alternative sites exist for it to happen, and the brain is sufficiently flexible to make multiple uses of the same neurons as parts of different networks to do entirely different things. Moreover, the hippocampus is well wired with the limbic system that underlies our emotions. (Indeed, it is part of it.) Both the amygdala and the orbital cortex are well connected to it in both directions. The hippocampus could easily take on the processing of both emotional and social experiences of continuity.

Orientation? Mapping? Memory? Experience and sense of continuity? Here, in each of us, lies a potential as flexible as language—not just for familiar cold cognitions but even the hottest emotions. There is still a great deal that we do not know. But even in our ignorance, we know one thing: the potential of our brains is plastic, superplastic—putty reworked throughout human history as well as in our individual lives. The brains of our species may have operated to change the very entity we were to become, and how we were going to become what we became. Can we dare say how? We now must turn to the trickiest problem of all—*consciousness*.

12
What Are We?

It is all odd and not a little unbelievable, the story of this book. Here we all are, touching physical things, the pages of this book or the seats we are sitting on. Scientists tell us that although we feel solid, we are in fact made of trillions of atoms. They tell us also that we live on a spinning planet, not the static, flat world we see with your eyes. Your body, they go on, is not the flesh you feel with your hands but is made up of millions of cells, each of which holds strands of information, DNA, the blueprints of your life. And further, in your skull there exist 100 billion (10^{11}) cells intricately wired together—your brain. All this is overwhelming; nothing you intuit about yourself or the world is true. But perhaps nothing is so bold and beyond belief as the idea that that brain feels this astonishment! How can your consciousness be made of matter?

It is a problem as hard to pin down as it is to answer. Here are a couple of quotations to hint at it, the first from Massachusetts Institute of Technology computer guru Marvin Minsky:

"There's something queer about describing consciousness: whatever people mean to say, they just can't seem to make it clear. It's not like feeling confused or ignorant. Instead, we feel we know what's going on but can't describe it properly. How could anything seem so close, yet always keep beyond our reach?"[1]

The philosopher Daniel Dennett says that it "is both the most obvious and the most mysterious feature of our minds. On the one hand, what could be more certain or manifest to each of us than that he or she is a subject of experience, an enjoyer of perceptions and sensations, a sufferer of pain, an entertainer of ideas, and a conscious deliberator? On the other hand, what in the world can consciousness be?"[2]

Hold your braincase and dip your fingers into it again in your imagination. Fondle your neocortex and do some wondering. Go touch your anterior cingulate, palpate your hippocampus, and tickle your frontal lobes. And ask yourself: What do these neural organs have to do with this feel, so immediate, intangible, and elusive of being "me" and alive? How could science make physical this incessant feeling that we are not physical but quite the opposite, something that is definitely not part of the material world? There seems to be an unbridgeable gap between the physical and what it is to experience consciousness—not just on first sight, but however deeply we think about it. All there is in the brain are neurons, plus the information their synapses store, plus the totality of their neural network interactions. How could anything mental arise out of them? Science might find the most extraordinary things, but it cannot discover magic, not even "neuromagic." The alchemists tried to turn base lead into gold. Are we not seeking to do something similar: turn matter into mind? And even if this is possible, what kind of theory could imaginably let us understand and explain it?

The earlier chapters of this book were not written with the intention of giving an answer to this question. We have sought to understand our origins, not the fact that we are conscious. Indeed, we would rather not write this chapter and so enter the heated fray about this, the biggest question about the mind. But there is a temptation to go beyond looking at the workings and odyssey of the mind to examine this link. To omit it would, in any case, leave these chapters devoid of something. And like it or not we have, without seeking to, begun to offer a hint of an answer.

What are all these mindmakers, these parts of the brain, discussed in the previous chapters? Let us look at them again, in terms not of what they do but of how they add to consciousness. Alone, none of them seems to us to satisfy the notion of "mind" or "brain." All stand instead, in some way, partway between the mental and the physical. The activities of the mindmakers are more essential to our feeling of self than other, more familiar brain-directed skills such as sight, hearing, and the ability to move; we may be born (or become) blind, deaf, or handicapped but still feel fully ourselves. Mindmakers give us an intimate sense of who we are. Indeed, they are so fundamental that we are not ordinarily conscious of the gruntwork they do in maintaining our feeling of self.

The unified mind—our sense of self—is, we believe, most likely an artifact or illusion, the seemingly singular result of what are in fact multiple underlying processes. Consider the sight-brain link. It takes up to 32 different areas in each cerebral hemisphere somehow working together to produce what we experience as sight. However far apart they are in the brain, we experience vision as a unified phenomenon. The same is true of the seven maps of our body's sensations; we experience those seven homunculi not as seven bodies but as one. This suggests something quite profound. However much the functions of our brain are parceled out, the experience they give us still has a sense of coherence. The mind, we suggest, is experienced likewise. As with vision, it is not quite the unity of experience that we imagine it to be; under it lie many different mindmakers in numerous areas spread throughout our brain. Individually, they do not make our mind, any more than those individual areas of vision can create our experience of sight on their own. But collectively they may. Together they create the feeling that Minsky and Dennett observe as so indescribable and difficult to pin down.

More mindmakers await discovery. Some parts of our brain have been named—such as the *claustrum* (found below the temporal lobe) and *habenula* (on the inner side of the thalamus)—but we have few hints of what they do. And there are other uncharted territories. Deep in our brainstem are groups of neurons with odd names: *nucleus basalis of Meynert* ("Meynert's base nut"), *locus ceruleus* ("blue place"), *raphe nucleus* ("seam nut"), and *ventral tegmental area* ("belly covering area"). These areas send axons up into our cortex, which secretes neuromodulators—brain chemicals affecting how neurons fire. The names of these chemicals are nearly household words from books on psychiatry and psychoactive drugs—*acetylcholine, norepinephrine, serotonin,* and *dopamine.* Even if you do not know their names, you have surely heard of the drugs that mimic them or stimulate their production: nicotine (acetylcholine), beta-blockers (norepinephrine), LSD (serotonin), Prozac (serotonin), ecstasy (serotonin), cocaine (norepinephrine, dopamine), and amphetamine (dopamine). They obviously touch the very essence of what underlies experience. All these aspects of our brain may therefore be key to who we are, yet we cannot quite grasp in what ways. Fortunately, our brain is an area where science is making rapid advances. In future years our understanding of its unknown parts and its neuromodulators will no doubt sharpen, but there will be a wait. Until we have developed the generation after next of brain scanners (and perhaps even the generation after that), what we do not know will vastly outweigh what we do. At present, all we can seek to do is stretch our imaginations in considering what hides within our skulls. We are as people were at the beginning of the sixteenth century in regard to the physical world. The New World had just been discovered. The map of Africa showed little more than a rim of coastline. Australia, Antarctica, and the vastness of the Pacific Ocean might as well have been on a different planet.

The full exploration of the globe was to stretch centuries into the future. Now we are in the same position with regard to our minds: We have begun to see the outlines of the vast continent, the slippery and fascinating and wildly inhabited mindscape beneath our skulls.

Embodiment

The best place to start investigating consciousness is with our bodies. If nothing else, each of us is a body.[3] We feel our emotions in our bodies. Where do we feel sick or disgusted? Usually in our stomachs. Fear is felt as a bodily freeze rather than a mental thought. It hits us where we act. If we do anything, it is our bodies that do it. Minds by themselves never do anything physical—telekinesis has never been shown to exist. However, every minute of our lives our brains move and do things with their—our—hands and feet. Without our bodies we cannot live. They are yoked to our minds as constant companions, continuous with us from birth to death. Our names may change, we may move, lose our closest friends, and replace our lovers. But our bodies never leave us. No one can divorce them. They do not mysteriously and disloyally leave us only to unexpectedly return. Nor are they like cars that we can sell, borrow, exchange, and then leave in the scrap yard. However much we may daydream about it, we cannot hire for a few days, to try out as our own, the body of Arnold Schwarzenegger or this year's top supermodel. Body swapping is out. Even if our consciousness is lost during sleep or when we are anesthetized, our bodies remain much the same. You will never wake up with the body of someone else. It is one fear we never entertain.

But there's a problem. We are not our bodies. Remember leaving the dentist with an odd feeling in your mouth after you had a local anesthetic for a filling? Perhaps you never thought much about it, but your mouth's numbness presents a minor brain puzzle. For a start, what could be more real and part of you than the feel of slightly bloated and tingling cheek and gums? It is a feeling of "me-ness"—though a little odd—in your mouth. For a short time the local anesthetic stops input into your brain that comes from the nerves of your teeth and mouth. But that means your brain is *not* experiencing that area of your mouth, as its nerves have been knocked out by the local anesthetic. But if your brain is not receiving inputs from it, what are you feeling? Nothing? But what you are feeling is something. Oddly, what you are experiencing is a phantom, a neural extension of feeling.[4] It is usually a short-lived inconvenience, but many people suffer persistent phantoms after a dramatic life event.

Following an amputation, "the patient often wakes up from the anes-thesia and asks the nurse when he is going to be operated on. On being told that his arm or leg has already been removed he may not believe it until the covers are removed."[5] Input to the brain does not necessarily cease after nerve blockage. Cut off a limb and you cut off the information that it once sent to the brain, but that does not end a sense of its existence. A leg or an arm that has been surgically removed still feels as if it extends from the remaining stump.[6] Sometimes the feeling is vague, but most often, in spite of some "tingling," it still has the feel of the limb that is no more. Over the years this will change. At first a phantom leg feels as though it is made up of a foot and a knee positioned like a real foot and knee but with gaps—vacuums—between them. Gradually, the parts tele-scope together. Indeed, after many years the foot withdraws up into the stump. [7] (These perceptions are probably related to neural plasticity changes in the maps of these parts on the brain.) But in spite of these changes, the phantom still feels like "me." Sometimes phantom limbs are felt as static extensions of "me," and sometimes people, such as Paul Wittgenstein (the pianist mentioned in Chapter 3), sense that they are movable. Phantoms happen in the brain. Remember motor alpha, or mu, activity? It disappears when people move their limbs, not only real ones but phantom ones as well.[8] They do not need their bodies to be able to feel them, or at least their neurons do not.

Like a real limb, a phantom can feel that it is burning, excruciatingly and exhaustingly cramped, or in other ways severely painful. But unlike pain in a real limb, it is unhealing pain. Worse, it is a hidden suffering. It is easy to get sympathy for a burned limb, which is visible, but not for pain in a limb that no longer exists except in one's mind. Its pain is often related to the time of loss, as is the position in which it is experienced. A soldier, for instance, might feel his hand holding the bomb just prior to the moment that it exploded prematurely.[9] A phantom may also perpetuate the more mundane sensations of the former limb. A person may still feel an old bunion; as one reported to his doctor, "I feel the ring on the finger that isn't there."[10] Others feel watches keeping time on wrists that are no longer there.

It isn't just legs and arms that become phantoms but also noses, tongues, and breasts. One in four women experience the phenomenon after a mastectomy.[11] The nineteenth-century neurologist Weir Mitchell noted briefly the report of a case of a phantom penis that sometimes became "erect."[12] Some people with severed spinal cords report orgasms, during dreams, in sexual organs no longer linked to their brains.[13]

You do not need to lose part of your body to experience phantoms. Have an accident that breaks your spine, and you will mostly likely be quadriplegic for the rest of your life. Not only will you be paralyzed in every limb, but your brain will be cut off from the sensations coming from them and from the rest of your body below the neck. Yet your sense of your body will not go away. In place of your limbs, you may feel phantoms of them as they were just at the moment when your spine was injured. The paper[14] from which we obtained these details contains illustrations of people's accidents and the positions they now feel their phantom limbs to be in. A person thrown by a bull feels that his legs are forever splayed above his head. People sitting with crossed legs just before their car turned over feel that they remain so. Their embodiment has come apart from their still-surviving bodies. Curiously, if a person was unconscious at the time when his or her spine was broken, no phantoms arise and they lose any sense of existing below the neck. Instead of having a phantom body, they feel that they exist bodilessly, as only a head and shoulders.

This is all rather mysterious and shocking, a side of surviving accidents many would prefer not to know about. But it is overwhelming evidence that our brains invent the sense that our bodies are real and with it the surety that "we" are real. We know they do this because there can be a separation— as with phantom limbs—from actual physical embodiment. Our everyday experience is thus wrong: Our sense of being a body does not rise directly from our physical self but from our neurons. It is a conclusion with profound consequences for how we understand the nature of the relationship between our brain and our experience.

If you doubt that the brain creates the illusion of physical embodiment, then consider the following. It is a phantom movement illusion discovered by Vehe Amassian of the State University of New York (SUNY).[15] Amassian stimulated his motor cortex with rapidly alternating magnetic fields. This triggered it into sending two sets of signals, one to make a motor movement and another—to the parietal cortex—to tell it the fingers were about to move. He then cut off inputs to and outputs from his hand using a tourniquet on his arm, so that his hand went numb. The signal triggered in his brain was thus unable to produce any body movement, and the brain could not tell whether any had been made (at least by feel). But the motor cortex had also sent signals to the parietal cortex to tell this part of the brain that the hand and fingers were about to move. Now, if our sense of existing is purely neural, then Amassian should have felt movement in his fingers. Indeed, he (and various others who have gone through this unpleasant procedure) did feel his fingers move when the magnetic fields were applied, but if he had looked, he would have seen that they had not. Thus, whatever happens to our bodies afterward, it is the initial transmissions in the brain that give rise to a feel of "me."

Action Extension

Our brain is so flexible it actually allows us to experience ourselves in the artifacts we use. A surgeon feels extended to the tip of her scalpel. An operator handling radioactive material using remote-controlled "hands" feels embodied in his robotic arms. Perhaps when seated behind a steering wheel you have felt a physical sensation, a kind of wince centered in your head or spine, in anticipation of an automobile scrape; we have. Such body extension occurs even with phantoms. Among those who have lost legs, some feel the phantom—even if it has shortened into a stump—extend into an artificial leg.[16] Some people with such phantoms embodying their artificial legs even report being able to feel coins or the shape of the ground underfoot. They not only feel it but can incorporate feedback from it into their motor control. Then there is the Nielsen illusion.[17] You put your hand in a conjurer's trick box that contains a window through which you can "see" your hand. Of course, it is not your hand that you see but, through the clever use of optics, the hand of someone else hidden by a screen. The surprising thing is that you embody what you see, even when you attempt to move "your" hand and find that the hand that you are looking at remains motionless. Logically, you should realize that what you see is not your own hand. But instead, you experience a feeling that your arm is paralyzed. You have embodied yourself into the visual feedback generated by the sight of a stranger's arm.

A variation of this phenomenon can be evoked using mirrors so that you see your right hand when you think you see the left one (or vice versa). That is not very interesting if you have two arms, but the effect can be enormously beneficial for those with a phantom arm. Recall that many phantom limbs are painful because the arm is in a twisted or impossible posture and so suffers "clenching spasms." Shown their "real" arm in a mirror, people felt their phantom being touched when they saw "it" being touched. Some who had never been able to move their phantom found that they could, with visual feedback from the mirror. Some experienced a paradoxical effect in which the sight of "their" lost arm caused them to lose their phantom sensation, as if their brain needed them to see the missing limb as real in order to reorganize itself to let it "disappear."[18]

Why should this be so? The reason is that our brain's experience of existing in our body does not arise from our body's consisting of pieces of attached anatomy but through our brain's ability to do things with them. This results in a "body schema" built up using the daily feedback from our body. As noted, people may feel a phantom leg existing, but only in the parts of it that move. (The internal organs—bladder, womb, and rectum—in which we can have phantoms might be thought to be exceptions, but they are muscled, even if it is only to let us empty them.) People are more likely to feel phantoms of the parts of the body that stick out. The sensation of a phantom breast or nose often will exist only at its tip, where there is most physical contact.[19] We may not be able to move these parts, but our brains need to know they are there so that they can avoid bumping them.

Embodiment, therefore, does not directly map that which lets us move—bones, sinews, and joints. Instead, it arises from the activity of populations of neurons distributed throughout the brain, using feedback that guides our movements in the external world. This is logical from the brain's point of view, since the brain has no direct knowledge of exactly what our bodies are made of. The brain is very knowledgeable, however, from sensory feedback, about the ability of its bones, sinews, and joints to change position, articulate, and do things.

The part of our brain that guides the motion of our bodies is called the *motor cortex*, but it might more properly be called the "motor-control-under-tactile-supervision cortex." As the neurologist Edward Evarts makes clear, injuries to the primary motor cortex particularly affect those movements made "under guidance by somatosensory inputs."[20] Supporting this link is the fact that brain scans of people discriminating by feel with the right hand the length of objects (but not their shape) show they activate their primary motor but not their somatosensory cortex. Oddly, it is the motor cortex of the right and not the left hemisphere that controls the right hand.[21] Our primary motor cortex is thus also a "somatosensory cortex." The premotor cortex (found in front of the primary motor cortex) is likewise not a motor cortex but one that guides and organizes movement under visual and other sensory feedback.[22] Neurons in the F5 area of the premotor cortex in monkeys discharge both when a monkey performs a hand action and also when one sees the same action done by another.[23] PET imaging detects activation in the caudal part of the left inferior frontal gyrus of the motor cortex when people look at hand actions. It thus processes not only feedback about its own limbs but also that of others.

The supplementary motor cortex, another part of the motor cortex, guides our movements using inner scripts and plans.[24] Further, there is no sharp division in the brain between the cortex which receives sensory input from our bodies and that which sends motor signals; they are all part of a common process, differing only in degree of specialization. The somatosensory cortex, which is usually seen as the cortex that receives touch input, also has, for instance, its own projections to motor neurons in the spinal cord.[25] These projections are functional: Cool the motor cortex and the sensory cortex can take over the control of movement.[26] Our sense of touch is, therefore, intimately bound up with motor control in both the somatosensory and primary motor cortices.

What is conspicuous by its absence in the brain is anything like a "muscle cortex."[27] No cortical neurons have been found that act upon individual muscles in the way piano keys activate the movement of piano strings. Instead, all motor neurons in some way map what can be done through the muscles.[28] Thus, it is through our doing things with our body that we get a sense of being a body.

Weird Bodies

We also sense feedback related to our movement as it happens in the space around our body. The brain's sense of this physically nearby space is made in our parietal cortex, and it is part of our egocentric orientation to the world. Interfere with the working of the parictal cortex—as can happen in migraines or epilepsy—and people experience a distorted sense of embodiment in the outer world. During attacks or seizures people might feel themselves as very small or large. It is called "Alice in Wonderland syndrome"[29] after Lewis Carroll's book. Charles Dodgson (the real Lewis Carroll) suffered from both migraines and epilepsy, so it is likely that Alice's experiences of growing tiny and huge were based on his own experiences of size change during attacks of migraine or epilepsy, or maybe both. Jonathan Swift, the eighteenth-century satirist, is also thought to have had epilepsy, so size change experiences might have influenced him to write about Lilliputians (miniature people) and Brobdingnagians (mammoth ones) in *Gulliver's Travels*.[30] Such size change experiences are linked with disruption, particularly to the right posterior parietal lobe.

This suggests not only that neurons create our sense of embodiment but that disturbances to them can change how we feel in our bodies. To take another example: Changes at the neuron level can affect the physical experience of sex. A person with a phantom foot can feel it as an extra "sexual organ" during intercourse. One man reported "that his erotic orgasmic experience 'actually spread all the way down to the foot instead of remaining confined to the genitals'—so that the orgasm was 'much bigger than it used to be. . . .'" The reason for this is that the map in our brain for our sexual organs is next to those for our feet. (It is believed to be a developmental "fossil" from the time when the brain first laid down body maps in the embryonic stage. During this period the genitals, due to the way the fetus curls up in the womb, are next to the feet.[31]) Remember the example mentioned in Chapter 3 of people who feel that water dripping on their face is also dripping on their phantom fingers? In such cases, due to neural plasticity, the face map has invaded the hand map, which, due to the amputation of the hand, is no longer receiving hand input. In the man described here, it seems that for the same reason his genital map has started to invade the nearby one for his missing foot! The scientists who reported this genital-foot link suggested that neuron activation may also spread in those with intact legs; they commented, "It has not escaped our notice that this may provide an explanation for foot fetishes."[32] But do not think of cutting your leg off in order to have more interesting sex. Sexual excitement is not the only thing that spreads to the feet from a person's genitals. Those with phantom legs also find them stimulated—often painfully so—when they urinate.

Distortion of embodiment can take even weirder and more frightening forms. After suffering multiple strokes, people may claim that they have two left hands, or even three heads and six feet.[33] They may say they have a nestful of fingers under the bed sheets. One man, following a right-hemisphere stroke, when asked about his left hand explained, "My mother has it in a suitcase and there are at least three pairs of fingers in there, and they're all functional." "How did that happen?" "We brought them in through customs." "And where are they now?" "My mother has them. There should be a leg, and there should be three pairs of fingers . . . from the left side."[34] One woman complained of having an extra hand. Once, in response to a query concerning her left hand, she said, "That's someone's hand, someone forgot it—that's funny, you read in the paper about people losing purses but not a hand."[35] She persistently complained about being kept on a neurological ward when her only problem was her hands.

Embodiment can cease to be tied to our bodies. Goethe, after he left his fiancée, wrote: "I saw myself, not with the eyes of the body, but with the eyes of the mind."[36] Such a visual body-image delusion is called *autoscopy*, or *out-of-body experience*. Hallucinations of the self are not uncommon in near-death situations, such as when our heart stops, or in emotional crises, such as Goethe was going through at that time. Certain people are prone to them. They are characteristic of "schizotypy," which describes the personality type of those who tend to be reclusive, suspicious, and prone to "magical thinking" and experiencing visual illusions. (Schizotypy is badly named, since although schizophrenics score high on tests for it, so do many other people.)

Embodiment, in spite of being a product of the brain, is felt as totally real. While no necessary link exists between it and our bodies' real extension, it is still a remarkably powerful "me" experience. After all, we feel that we are our bodies. There is no doubt or hesitation about it: Hurt your hand, and it is "I," not some scientist's neural network model, that feels the pain.

Indeed, this feeling turns out to be more fundamental to us than our knowledge that we are extended. Merely knowing that we are attached to a limb does not make it part of us. The neurologist and writer Oliver Sacks tells of a young man who had found a "severed human leg" in his hospital bed. The only way he could explain it was as a "rather monstrous and improper, but very original joke." "Obviously one of the nurses with a macabre sense of humor had stolen into the Dissecting Room and nabbed a leg, and then slipped it under his bedclothes as a joke." He tried to throw it out of the bed—but he was attached to it. It was no good explaining to him that it was a part of him. "A man should know his own body, what's his and what's not."[37] People in such a confused state will try to attribute the alien limb to the doctor examining it. One American woman in the 1930s, after two strokes, denied that her paralyzed limbs were hers. When asked whose they were, she said, "Yours." A three-limbed doctor made more sense to her than the idea that that "thing" was part of her body. Shown that her arm merged with her shoulder, she observed: "But my eyes and my feelings don't agree, and I must believe my feelings. I know they look like mine, but I can feel they are not, and I can't believe my eyes."[38]

These and other such cases lead us to one conclusion: Our physical sense of being is made by our neurons. It is not just our sense of extension but also the sense of "me" that goes along with it. Here we have come halfway to answering the problem of consciousness. Embodiment may not be consciousness, but the brain, in making it, also makes this inseparable sense of "me." If our brains can do this for our physical bodies, might they not also be able to create a sense of "me" in a nonconcrete reality? While such a question does not answer the problem of consciousness, it does suggest a new approach.

The approach lies in answering a rather simple but overlooked question: Do our brains give rise to a sense of "me" in more than our physical extension? We have shown above that embodiment arises from our brains' doing things with our bodies. Are there other things in which the brain might feel we exist that are not physical? If we look, there are several things done by the brain that could be "embodied" with a feeling of "me-ness." Here, starting with sociability, we shall discuss them, stretching and challenging our quest for an answer to the question, "Who are we?"

Social Embodiment

Evolution made us not out of clay but from a fission-fusion ape. We inhabit not only an external, physical world but also, as argued in previous chapters, a social one. As William James put it, "A man's social me is the recognition which he gets from his mates."

The sociability that gives us this recognition, of course, does not arise magically. The brain has to work to get recognition from others. Moreover, human bonds are not passive or fixed, as they have to be actively kept alive with simple, often overlooked actions.

Do you not chat? Do you not smile and laugh with your friends? You don't do this mechanically. While sitting alone in a cafe or on public transport, try some people watching. Just look at the human species as an alien would. Look at how people greet each other. One moment there are two dead faces, and then suddenly they burst into life. As they say "Hello," their eyes, faces, and hands become a duet of responses that echo between them. Our faces are brilliantly animated, skillful, and sensitive social contact organs. Unfortunately, in psychology, it is taboo to marvel at them. We are not supposed to be awed at our ability to be social. Perhaps the camera is partly to blame. In magazines and advertisements, we are surrounded by expressions that stare out of paper—they could be of waxworks. Photography falsifies our awareness of how alive we are. In reality our faces are never static and dead but interact with others continuously with split-second timing.

We also express our being with others through body movement, the tone of our voices, and the sensitivity of our hands when we touch and hug. All these can powerfully connect us with others. Indeed, the rich expressiveness of much music may be an extension of such human connection with melody, beat, and tonality.[39]

All these forms of expression are actions—social actions, done with great sensitivity, sending and echoing in a chamber of social and hoped-for social recognition. After all, we do not smile at brick walls. Our expressions seek an audience, some kind of social reply. We smile to other faces—ones that smile back. (Musicians likewise need to play to listeners.) The use of expressions gives our brains a means to keep alive our presence in our social world. If no one responds, if a group stonewalls you, then you are out of their social world. You are not one of them. Our expressions fight against this to keep us part of others' lives. We wield our "me-we" sonar. We try to echo the smiles and expressions of others.

Our sociability has a goal: to let us know we are not alone. People respond to us, and we learn how best to socially interact with them so that they do. Sociability is an essential link made between our brains and others'. There is no such thing as negative publicity, say the media. No solitary animals, we like to get noticed, preferably favorably, by others of our kind. And doing that requires plenty of skilled brain work.

Thus, as much as motor actions have sensory feedback, so is sociability guided by feedback. Touch the movement of your face when smiling spontaneously with others. Doesn't part of the feeling of being a "me" lie in it? Suppose your face turned into a wax mask and your hands and body turned into one of those clever automatons animated by hidden mechanisms found in "dark ride" exhibits at amusement parks. And what if your voice were changed too, to become a synthesized deadpan computer monotone. With a waxwork face you could not make even the slightest hint of a frown or smile. With automaton limbs you could, robotlike, get a cup of tea but not wave, pat anything, or offer a handshake. Nor could you, with your monotone voice, intone a subtle hello, or laugh. Such a condition is imaginable, but it would be a psychological hell. We could do without our legs and hands, but could we do without the expressiveness of our faces or of our voices and gestures? Without expression, we would be cut off from that which makes us what and who we are.

If physical actions performed in the physical, three-dimensional world give us a sense of physical extension, could not those performed in the social world likewise give us a sense of extension—social existence? Expressing ourselves to others through our faces and otherwise is crucial to our reality, not in the physical world so much as socially, embodying our identity and presence. It gives the brain a strong sense of "me."

Subjectivity

The brain still goes on existing even when alone or in sensory deprivation, which means that there must be other kinds of "me" embodiments beyond the physical and social. We have memories of ourselves and others, not discontinuous ones but ones that flow from past experience to join with the present. Things, places, and people, including ourselves, may change, but as we have seen, with our memory headers, we are skilled at experiencing the continuities and identities below surface alterations. The hippocampus and associated limbic areas in the temporal lobe seem to orchestrate continuity, organizing our memories and our sense of existing through time.

In its limbic parts, a brain knows something apart from its body and its senses; it has a feel for life, a continuous sense of embodied "me" throughout the chaos. According to the neurologist Paul MacLean, "without a cofunctioning limbic system, the neocortex lacks not only the required neural substrate for a sense of self, of reality and the memory of ongoing experience but also a feeling of conviction as to what is true or false."[40] Here, perhaps, lies the neurological center of our subjectivity, the feeling of "me" that is not that of our body but of our existence and being.

But do our physical or social embodiment and our subjectivity make up consciousness? They may be thought to cover various of its aspects, but consciousness, as the Dennett and Minsky quotes at the start of this chapter suggest, is a fickle thing. We are still left with the question of why our experience seems so unlike that of being matter. Being an embodied "me" comes from the experience of doing (or having done) things in the physical world (and, we suggest, the social one). Subjectivity is passive—it is something sensed. But we actively feel we are conscious. So what is the source of this sense that we embody intentions, actions, thoughts, and feelings, and how does it link to the sense we have of being a "me"?

Inner Maestro

We have a world within that exists because our brain organizes its actions and thoughts with internal cues. Some philosophers, however, deny the existence of such an inner place. According to them, our inner feel is a "beetle hidden" in a box we cannot open and so is not meaningfully there. To them, to see consciousness and mind as things inside us is to see ghosts in a linguistic mirage generated by a misuse of words. Perhaps they are right within the context of their philosophical reasoning. But it would seem that they ignore the prefrontal cortex and its vibrant life of inner cues. Philosophers never see any need for the brain to make its actions and reactions independent of the outer world, for they imagine us to have mushy brains, not assertive ones with inner cues. They are the neuroscience equivalent of medieval scholar-monks counting angels on neurons, blissfully ignoring twenty-first-century science and its discovery of the subtle logic of our biocomputers.

As you think, recall, and imagine, you are, in a sense, your inner cues. They may not be the actions taken in the outer world, but it is through them that we act, if only in the inner world of our memories and imaginations.

Perhaps the act and the actor are the same? If the brain can create an intensely "me" sense of embodiment in limbs that no longer exist, then what of mental actions orchestrated inside us? They may not offer us three-dimensional embodiment, but, as shown above, extension is not needed for the "me" feel of embodiment. What is needed is some control-feedback relationship. And, as with social presence, the relationship need not be physical. It would seem that the inner cues guiding actions, recall, imagination, and thought are part of our sense of being a "me." Here are a multitude of control and feedback processes and cues flipping motions, memories, images, and ideas in and out of existence.

There are in fact clues that the preparation done by our brains before we act is linked with consciousness. One thing prefrontal inner cues do is initiate thoughts and actions—we are anything but vegetables. We are constantly doing things, if not with our bodies then with our minds. But few actions and thoughts arise fully formed. Before we voluntarily take even the smallest action, our brains prepare.

Such preparations have different durations. Some, taking half a second or less, happen in the parts of our brain dealing with movement. Before we move, our motor cortex draws up programs as to how to act in a complex process that involves linking motor memories together into sequences, or *motor programs*. And few actions happen without feedback control: Ongoing sensitivity to feedback requires subtle preparation so that our actions can be integrated and so guided by sight and touch. This is all brain work, a labor done silently by our motor neurons in the half-second or so before we act. And it is not done alone. Overseeing this work, the anterior cingulate cortex attends to the consequences of our actions, focusing up to 2 seconds before we act. And before that, other neurons, in the prefrontal cortex, may start up one or many seconds earlier, depending upon the task.[41] They ask when and where the movement should start, and under what conditions. Is this the time to act? Such prefrontal preparations come not only before actions but before we reach a mental conclusion, or face an expected event or punishment. Our minds are always looking ahead and anticipating. There are whole families of processes being loosely summarized together here. But they share a brainwave "signature." The details of how they do this are just being discovered. What we know is that before we act there is a general shift in the electrical activity of our brains. Temporally extended action requires the slow potentials described in Chapter 5 that are under the prefrontal cortex's control.[42] They are also required for intentions that are never actually carried out, arising not only before we try to move but also when we seek to relax.[43] This is where our sense of willing things may come from. Involuntary actions—tics, for instance—are not preceded by such *readiness potentials.*[44]

A person senses the conscious decision to move his or her little finger about a third of a second *after* the onset of the motion's readiness potential.[45] It is a negative potential linked to the preparation made by our supplementary and other motor cortices before an action. It is hardly a major act of will, but it is an act of will nonetheless. But if the consciousness of making an act arises with the act itself, what of the other brain preparations? Do not the other potentials tied to our prefrontal cortex also give rise to a sense of consciousness as we think ahead and prepare—intend? After all, this part of us is focused on making and supervising the inner cues organizing our actions and thoughts. Scientists can see this on PET scans: Blood surges into part of the dorsolateral prefrontal cortex[46] when people will actions— and only when they will them. This does not happen when our actions are guided from outside, such as when we copy movements. We do not "will" such actions.

Another possible link between consciousness and the focusing and willing of our brains is our gamma (40-Hz) oscillations.[47] We experience our senses as a unity even though the brain does not process them as such. That unity seems to come from the linking done by gamma. But gamma is not only found in our sensory cortices; it is also found when our prefrontal cortex guides our focusing on touch and when we prepare to do things.[48] In these cases, gamma, instead of joining our senses, binds the various processes that let us attend and do things. So gamma may unify not only perception but also our otherwise varied senses of doing—intention and will. Some evidence for this comes from anesthetics.

"The consciousness was terrifying. . . . The . . . terror of trying to signal one's conscious state to someone, but unable to even twitch a bloody eyelash."[49] To wake up during an operation is a nightmare worse than any other. (Fortunately it is very rare; you are more likely not to awake at all after the operation.) But very, very exceptionally it does happen. Anesthesiologists seek to give us the lowest effective dose of an anesthetic, since the drugs can kill and the safe dose range is small. Once in a while they are overcautious and underdose the patient. Added to some anesthetics are drugs to stop involuntary movements by the patient, which might cause problems for the surgeon. At too low a dose, these paralyzers may work but the anesthetic itself may not. It is a nightmare: paralysis and consciousness on the operating table. The anesthetist needs a way to know when a person becomes conscious even though paralyzed. The easy clues, such as heart rate and blood pressure, are not reliable. But one thing in our brains seems to be—gamma activity. The drugs that have been given the patient paralyze the body at the level of the muscles, but they don't stop the initiation of thoughts in the brain. A person knowing that the surgeon is operating has a brain alive to that fact. The binding of its thoughts and experiences can be monitored. When gamma responses weaken and disappear, consciousness vanishes as well.[50]

All this adds up to a picture of gamma's linking to consciousness. As with attention-to-action and the sense of "me" buried in our thoughts and intentions, gamma seems to be a primary mindmaker. As two leading brain scientists, Rodolfo Llinás and Denis Paré, suggest, "Those aspects of brain function which form part of our consciousness must occur at the same time, most probably with 40-Hz activity."[51] Francis Crick, the codiscover of DNA, similarly asserts that such activity's transiently binding fleeting attention to short-term memory makes for "vivid awareness."[52]

Inner Freedom

While it binds our attention, our memory, or even our preparation to do things, gamma itself might be just a correlate—a shadow, not the substance—of consciousness. Other brain activity may be inseparable from consciousness itself. At issue is what it is that gamma binds to create a "me."

Our brains are constantly animating an embodied private life. When we are blindfolded, earplugged, and at rest, our prefrontal cortex still uses more energy than other parts of our brain, indicating that the mind is highly active.[53] What is it doing?

It may be busy embodying a "me" feeling created around an inner world of questions about where we are and what is happening, or going to happen. Our brain is born to be constantly alive with such insistent queries. The world is perpetually changing around us. We must keep up with it: What does that comment mean? That tidbit of information? Or sound? To survive and learn, brains—we—must continually attend to the changes happening around us, which might be to our advantage or not.

Inner cues have a life of their own. Ideas play actively with each other, coming together in statements and questions. Consider the main inner cue used not only by your mind but in this book, indeed, in all books—words. Words empower us to describe and articulate, anticipate and question, better than we could with, say, images. They help us work out expectations and focus our concerns. Here, in the sketchpad of our thoughts, we hold court about what is happening within ourselves.[54] If the prefrontal cortex enables the brain to organize its awareness by internal cues, many of them come from this inner conversation. We sense these cues as a voice—our inner one. Without speaking aloud, you hear yourself say "I." Who is speaking?

It is you. According to the American philosopher of the mind Daniel Dennett, our inner voice is linked with consciousness. He suggests that this is a place (which is not really a place) where the brain tells itself stories about existing. To use one of his phrases, we have a "narrative center."[55] It is a sort of bulletin board or workspace that emerges from neural networks as the brain tries to keep track of its plans and concerns. According to Dennett, our continuity as a mind comes about as these self-told narratives unfold. We tell them to ourselves in inner speech. They organize and structure our actions and ambitions, the stories about ourselves that we tell others. They—we—are the inner prose our brains use to tell themselves and others what kind of person they are and want to be.

Here embodiment, subjectivity, and inner voice come together. As much as with our physical extension, we do things in our inner world so that we embody our inner voice with a sense of "me." Gamma, binding the various threads of our inner voice with what is happening in the rest of our brain, may well be involved. Here, in doing this, the brain does, feels, and knows it exists. It acquires a first-person experience.

Part of the experience of consciousness is not only that this "me-ness" exists but that it acts as a free agent. Perhaps this reflects the brain's concern with control. A brain that is awake to its opportunities and restrictions, after all, must always be attending to questions about the causal environment in which it acts. We live in a world of doers and done-to's, causes and results, antecedents and effects. We must spot how things happen. What follows my actions, and what determines them? Am I a causer, or am I caused? How can I gain control and escape restrictions? We seek the boundaries of our choice and our limitations. We attend to the scope of our intent and volition, and not just our own but those of other people as well. Social psychologists and those studying apes and monkeys find social position is determined by who can do what to whom. Low ranks are controlled by higher ones, never the other way around. We need to see causation for our welfare and survival.

Discovering how to make others respond to us (which also often comes down to learning how to respond appropriately to them) also enables us to socialize. Think of 8-week-old babies. Although they can hardly manipulate the world, they can smile, laugh, and move their heads from side to side. Malcolm Watson, a psychologist, placed a mobile above the cots of 8-week-old babies and observed their movements.[56] Watson found that they laughed and smiled at the mobile, even before they had laughed and smiled at their own mothers. As infants grow up, they constantly seek ways of mastering their environment and engaging with things. We learn to play games like peek-a-boo. Finding islands of predictability gives us a sense of control even as it keeps us on the lookout for further surprises.

Some things clearly shape our actions. Take, for example, the laws of physics, the knife to our throat, the dictates of tyrants, poverty, and social obligations. But many things are within our control, if we wish to make them happen. We can move our hands, focus on the whiteness of this paper, plan a meal, cook it, and invite guests with whom to eat it. A previous generation might have sought such control in magic. We value it in the modern conveniences by which we have mastered our environment, such as the remote control, the private car, and the mobile phone.

Our thoughts are constantly focused on those things that might block our freedom and on how we might overcome them. We seek liberty of action, space in which to do whatever we want. To the degree we find it, we feel free; to the degree we do not, we feel trapped. Freedom affects our emotions; a stressful noise that we can turn off is not as stressful as one over which we lack control.[57] A child feels fear of a toy when it cannot control it, but pleasure in it when it can.[58] Children, not surprisingly, have a strong urge to gain a sense of mastery of things.[59] They feel frustration when things that were controllable stop being so.[60] As adults we get frustrated over the aggravations and hassles of life. We bear them if we chose them; if not, we resent them or we try to gain control over them. In this way, our brains are steadily sensing out and, if possible, enlarging our "elbow room."[61] The prefrontal cortex is making its inner cues, after all, for a purpose—to give itself freedom from being limited by other people and what goes on around us. Here the brain searches out how to make things go along with its plans and desires. Thus, we wish the world to be contingent on us, not us on it.[62] We seek to do our own thing, not be the means to the ends of others. We desire to be the supreme causer in our affairs, not a puppet of events, pulled and pushed by necessity.

We can experience control through our prefrontal cortex's internal cues. The outer world may frustrate us, but here inside, hidden from it, we are embodied in a "me" that feels at total liberty. We—our brains—therefore feel ourselves as agents in the world, even if it is only privately.

You can, for instance, think any thoughts you wish. You are entirely free in your mind. The only limits on your inner voice are your sense of logic and your imagination. You may lack the wings of the birds, but if you close your eyes you can be up in the air with them. Maybe your imagination is not always free—if you stub your toe, pain pulls your attention constantly to it, however much you seek to focus your mind elsewhere. That is a reason we dislike pain: It rules our attention! But when pain-free, we can focus with great liberty on such things as planning a date or writing a book. And we can do something else: Our minds can engage the senses to focus on the inputs into the brain's experience. For instance, we can stop and attend to the whiteness of this paper or the blueness of the sky outside. Philosophers call this *qualia*—the *what*ness of experience. Your prefrontal cortex does this by manipulating and tuning its links to shift the attentive processes by which your visual cortex experiences what is before your eyes.

As we live through our inner cues and brain modulations, we can feel free and independent of the physical world outside the brain. But this interest in freedom is not only about physical limits. This brain experience we call "me" is as active in questioning its own constraints on its knowledge as in testing those it encounters in the physical world. It seeks to find freedom in the models and stories it tells itself. We tell stories that emphasize how we overcame restraints and how we determined what we did. We love tales of David against Goliath, Papillon escaping Devil's Island, heroes who fight against the odds and succeed. In our lives, we play down how things shaped us. We may have been slaves to fortune, money, and others' dictates, but we would rather tell ourselves stories in which we were not.

Is this true only of the stories of our everyday lives? Is it not also true of those with which we orient ourselves in the wider world of human knowledge? Within its embodied inner reality, the brain wants to tell itself stories that it lives in a "metaphysical" world, one beyond nature. Here lies the threat we feel from those 100 billion cells in our skull. We fear that our inner volitions, in some distant way, are merely those of their matter, making us contingent to the physical world and its laws. No, we shake our heads, no, we—our brains—are separate, and somehow different, from matter, and so free. Material explanations of mind are experienced as traps; they threaten our prefrontal cortex's embodied sense of having inner freedom. Our brain would rather not know that beyond its immediate senses it is merely another physical thing in the universe, that the restraints that limit and rule the outside world also, in a hidden way, limit and rule it. Our brain would rather tell itself stories that something exempt from outside influence makes it a free "me" or "I."

This need to be free of the physical makes our brains sensitive and threatened by life's end. We see loved ones decay in their brains, go demented and stop being the people we knew. We see them die, and know that the same fate awaits us. Here lies the horror that each brain faces in decay and death. Our freedom may have an end. It is a story our brain would prefer not to hear.

Beyond the Prefrontal Cortex

The prefrontal cortex cannot be the whole story of consciousness. People can injure their prefrontal cortex and still exist. They may lack empathy or an ability to plan or focus. They may not be inner driven and instead be tied to the world around them. But that does not necessarily mean they are not conscious. It might mean that they have a different experience; they may be less conscious but still have a kind of consciousness.

Also, as noted in Chapter 5, meditation puts the prefrontal cortex on standby without stopping consciousness. The calm awareness of meditation slows and halts its incessant activity. Yet here, with the prefrontal cortex turned down or off, consciousness still exists. Obviously, it is a different kind of consciousness. Indeed, it may be better in some ways, richer in its attunement to the external experience to which normal consciousness gives short shrift.

Allegedly, practiced meditators can go beyond such calmness and experience transcendence. You might think that a book like this should not talk about such things, but some research requires that we should. The experience, according to meditators, goes beyond words, so it can only be hinted at. Gurus wave their hands, suggesting it is something like the knower, the known, and the process of knowing becoming one. They claim that ordinary experience is distorted and that only in meditation do people become truly aware of things. The problem, according to them, is that our lives are full of petty cares. While they are the necessary stuff of living, they also blind us to what exists beyond them.

Curiously, something happens to the brainwaves of meditators during "transcendence." The prefrontal cortex does not stay turned off. When meditators who are wired to record their brain activity have signaled their entry into "transcendence," gamma activity returns over their prefrontal cortices. In some meditators the activity appears not just in the prefrontal cortex but all over their brains.[63] Nobody knows what to make of this, but it suggests that a still unknown link connects the prefrontal cortex, gamma, and what Buddhists called *nirvana*.

The Brain's Enigma

Is anything mentioned here or earlier in this book really you? In some ways all these phenomena seem to be. But it could be that they all touch just a little upon what it is to be, so that while none of them individually makes our minds, each makes its own, subtle contribution to consciousness, all dovetailing into a unified experience of being alive. As the fragmented visual cortex appears unified in our vision, so it may be that the various activities of the mindmakers come together in the "I" of our mind.

Our minds must be distributed around our brains. It would seem rather odd if scientists were to announce they had found a square centimeter of our brains—the "me cortex," say—that was solely responsible for consciousness. The individual processes involved are very diverse. We have been wholly ignorant of many of them until recently, and many more are yet to be discovered. But simply knowing they exist demands we reverse philosophy's understanding of how the brain relates to consciousness. Many philosophers hold, for instance, that the key fact of our experience is its apparent unity. Using this as a starting point, they investigate the nature of our being. But this could be a trap, misleading us into thinking that we are seeking one mysterious link between mind and brain. There may be no one such link. If anything, the problem is turning out to be one of too many mindmakers.

In the past, philosophers were just not in a position to have any deep insight into who we are. That may sound arrogant, but think of our bodies and the speculations of ancient doctors about blood, cholera, phlegm, and black bile—the four humors—before modern physiology and anatomy. Ancient doctors were hopelessly wrong. Until recently, philosophers were paddling upstream in the same boat with regard to brains. Medieval philosophers thought mind was in the brain's ventricles. Descartes saw free will in the pineal gland. Taking an opposite approach, behaviorists denied consciousness existed; some twentieth-century philosophers even attributed it to an artifact of language usage. Without scanners to picture brains as they think and feel, how could anyone have started a serious investigation of what underlies our sense of who we are? The crucial information as to what went on in the brain was simply not there. But now lights are beginning to shine. As little information as we have, it dwarfs the cumulative knowledge of previous centuries. Embarking on a quest to the gray continent of the brain without this knowledge is as foolhardy as trying to make sense of MRI scans of the body using Galen's theory of the four humors.

To understand consciousness, we need to freshen our imaginations and free ourselves from the old stories about who and what we are. After all, it would not be the first time. To take one example, when we think that stars are made of matter like our Sun, we do something people 3000 years ago could not have grasped. For them, they were gods and spirits. It took the Greek Anaxagoras (500–428 BC) to break with this and suggest that the

Sun might be a burning stone and that the Moon might have a landscape of hills and ravines.[64] Old views of what is material and what is not have changed, and we must be prepared for them to change again.

It is only now, after the turn of the third millennium, that humans can fully grasp what a wonderful thing the biocomputer in our skulls is. We are, in many ways, the first people in a position—thanks to neuroscience—to probe the key question of what it is to exist. But we need to be prepared to change some of the ways in which we expect that question to be answered.

13

Of Human Bonding

When Julius Caesar crossed the northern Italian river Rubico into Cisalpine Gaul in 49 BC, the Roman Senate regarded it as an act that could never be taken back, an irrevocable act of war. It is from this act and the river's name that we derive the word *Rubicon*, meaning a line or border that, once passed one way, can never be passed back the other way. Was there an evolutionary Rubicon—a point of no return—that apes crossed millions of years ago to put them on the journey to becoming us? We believe they did take evolutionary steps from which there was no turning back.

To see what kind of steps they took and what they crossed requires us to look at what makes us different from other apes. This is easier said than done. Our obvious differences from apes and monkeys, such as the relative hairlessness of our bodies, may not be the most important ones. This holds also for our social natures. As noted in Chapter 7, friendships, lifelong bonds, romantic makeups, and deceptions mark the social lives of apes and monkeys. In a way, this shows how similar to us they are. It is not only humans that have close feelings for friends or form alliances with each other; apes and monkeys have been doing so for millions of years. They need attachments as much as we do.

But as we begin to see our closeness to apes and monkeys, we also get a better focus on what makes us unique, clues to the location of that evolutionary Rubicon. We are a social ape like none other.

The Social World of the Human Ape

Apes and monkeys live in many kinds of social worlds.[1] Some live alone (orangutans), others spend their lives in small isolated family units (gibbons, gorillas, and a few species of monkeys); but most primates (including us) also live in a larger social group—our troop, band, tribe, or village. People universally live in family groups that are themselves part of a larger community. The only apparent exceptions to this rule are a few religious communities, but they get their members from outside.

Primates differ in the nature of the relationships they have with each other. Most monkeys have rather simple social lives, but the social lives of some baboons and apes (and us) are more complex, made out of alliances and friendships. It should be noted that alliances are not the same as friendships. *Alliances* are coalitions of two or more individuals who work for a short time together toward a common goal. They need not be friendly; indeed, under the surface, they may be keen rivals. In contrast, friends, human or not, rather than working toward planned goals, enjoy each other's company; they are, to paraphrase the philosopher Immanuel Kant, ends in themselves, not means. Nothing here totally divides us from other social primates; we, too, have our alliances and friendships.

So how do we differ from the other social primates? In 1991 a group of four anthropologists and primatologists identified the nature of the great ape-human divide—that ancient Rubicon. Lars Rodseth, Richard Wrangham, Alisa Harrigan, and Barbara Smuts argued that our uniqueness lies in the *duration* of our bonds.[2] We are like chimps in having friends, alliances, and a fission-fusion way of life, but the duration of their relationships is much shorter than that of ours; and their relationships do not survive extended physical separation. Chimpanzees may form alliances, like us, but they last only several months, or a year or two at most.[3] Our bonds are tougher, sometimes lasting our entire lifetimes—and we all have them.

But that, by itself, does not make us unique. Some other primates, such as hamadryas baboons, also have long, tough bonds.[4] Such bonds, extending in some cases over 6 years, enable baboons to work together even better than chimpanzees. Baboon blood relatives live in the same place (related baboons sleep next to each other) and are each other's constant companions. We would probably find their lives stiflingly boring; unlike us, baboons have a small circle of long-term bonds at the cost of not having friendships outside them. Baboon troops do not show fission-fusion—the breaking and reforming of groups within the larger one. They combine long-term bonds with social sterility.

Our uniqueness is a mix of the best of baboons and chimpanzees. In the day, we are like chimpanzees. We meet many individuals of our social group, with whom we socialize, play, and work. These relationships change regularly, and they are not with the people with whom we have our deepest bonds.

But at night, we are like hamadryas baboons. We sleep and live with the people with whom we have our long-term ties—our family. All people, from hunter-gatherer bands to industrial states, have a circle of close friends and kin, with, outside them, a social circle of less close but useful friends. A little further out we find the next circle, people with whom we have not had the chance to get so well acquainted but whom we still take to be members of our social group.

Humans are thus unique in combining long-term bonds that are resilient against separation with a fission-fusion style of social existence. No other primate does this. It is a powerful formula: The social richness and dynamism of our species comes directly from this mix of close kin bonds, social networks, and group identity.

How did our unique sociability arise? Did our species take the chimpanzee's ability to form ties outside the circle of blood relatives and upgrade them to make them longer lasting? Or did we extend the long-term bonds of hamadryas baboons to nonrelatives?

Given that chimpanzees are the species closest to us, our bonds probably arose from something like theirs.[5] But how? It is a story about women, daughters, marriage, and in-laws. Long bonds are powerful, especially if they let us bond families across communities.

There is one way in which we are similar to chimpanzees but not other primates. In all primates, some adolescents leave the group in which they grow up and move into another group to spend their adult lives. The general rule is that the males leave and the females stay—*matrilocal residence*. But in chimpanzees, it is the daughters who leave home.[6] This creates resident groups of half-brothers, cousins, and other related male chimps—*patrilocal residence*.

This rare pattern is also found in our species. In most human societies, daughters move away from their parents. Where sons do, anthropologists ascribe it to unusual circumstances in the wake of a group's migration to an already inhabited area. Further, those rare communities in which men leave home are not like those where daughters leave home. Daughters, when they move, usually move to a new village, but when a son moves, he "moves, so to speak, across the street."[7]

In spite of our vast differences, these statements are true of all societies. In all human societies, some blood kin (especially daughters and mothers), however far apart, try to keep in touch and aid each other. We are the only "long-distance family" species, the only one to help our children and other relatives when we do not have daily contact with them. Chimpanzees do not usually visit their relatives. After a young female chimp leaves her mother's band, she does not go back. Her problem, in part, is that her bond with her mother cannot fully be decoupled from her everyday social meetings. Chimps need daily, or near daily, contact with those to whom they are attached, because *they lack a means of being emotionally close in spite of being physically apart.* This is crucial: The uniqueness of the human ape is to have found a way for families to be attached to each other without regular physical closeness.

Kinship links may begin with the mother-daughter bond, but they are strengthened through shared caring for children. After all, the children of any unions in a group would contain genes of both their mother's and their father's families. Brothers and brothers-in-law might not share any genes directly, but both have one-quarter of their genes invested in their common nephews and nieces. Likewise, both sets of grandparents share a quarter of their genes with their grandchildren. That makes both families indirect kin. So while the members of the two families are not blood relatives, reproductive unions create a blood bond in the genes they share in their children. That indirect genetic link matters, as it gives such relatives an interest in aiding the children's survival.

Communities that are linked through kinship into larger units of cooperation enjoy many advantages. Severe droughts dry up the water holes used by the !Kung of the Kalahari Desert about once every 4 years. Fortunately, different water holes dry up in different years.[8] As a group, the !Kung can survive only if they can cooperate and share resources. That cooperation comes from the bonds forged among them through marriage. !Kung with water (often at hardship to themselves) welcome !Kung families without it because they are in-laws. Ties bonded through marriage therefore give these hunter-gatherers a safety net that they would not otherwise have against local droughts and famines.[9] Alliances thus make it possible to venture into new, riskier environments.

Even for us Westerners, lasting attachment to our kin is our deepest means of feeling close to others. We might see some relatives only once a year, yet they are far closer to us than other people we see every day. Such attachments tie families together, even when they are not especially warm or pleasurable. Indeed, meeting relatives at Christmas can be a stressful experience. Nonetheless, next year we meet them again. And even when we cannot get together, when our brothers, sisters, and children are thousands of miles away, we keep "in touch."

Sexual Bonds

Unlike other people, we modern Westerners do not need complex webs of kinship for survival, since we have other resources—pensions, health and property insurance, welfare—to fall back on. We can thus afford to treat in-laws as a source of jokes. If we get along with our in-laws, that is a plus, but it is something we do not necessarily expect—presuming, of course, that we get married in the first place. People who cohabit do more than forgo a wedding; their partnership is also unlikely to lead to in-law relationships between their families. But then, modern relationships have evolved toward the person and away from the family. In our society, people marry out of romantic love, a phenomenon mostly of the last few centuries. (Some medieval historians argue that modern "romantic" love first arose in the thirteenth century with the troubadours and poets of Provence, now part of France.) Western parents today do not attempt to arrange marriages for their children to link up with the right set of in-laws. No modern daughter would stand for it.

The bond between two married people is strengthened by sex. Both men and women report love for each other is more intense after sexual relations have begun.[10] But why? As noted in Chapter 8, after sexual orgasm in both sexes, the brain releases vasopressin and oxytocin. Oxytocin, as noted, is also secreted after bond-forming events such as the birth of a child and during suckling. One study found a 10-fold increase in oxytocin levels in some men after sex.[11] Vasopressin and oxytocin, involved in sexual bonding in one monogamous rodent, the prairie vole, are now also suspected to be at work in one of the monogamous primates, the marmoset.[12] Might true love, and the culturally privileged fidelity of monogamy, have a chemical basis?

Vasopressin and oxytocin, through their receptors, prime mammal brains to evolve pair bonding. They induce biological "marriage" after the prairie vole's first nuptials. When a young pair first meets, they mate for 30 to 40 hours. Thereafter, they form a lifelong bond, with the father helping to look after his offspring.[13] The monogomous prairie vole is in most ways like the polygamous montane vole, except in the distribution of its chemical receptors. Receptors for oxytocin exist in different parts of their amygdalas and *prelimbic cortices* (part of their orbital prefrontal cortices). Oxytocin receptor distribution in the marmoset mirrors that in the monogamous voles.[14] Likewise with vasopressin in the *ventral pallidal region* of the brain—if vasopressin receptors are artificially increased in this area, male voles readily form mate attachments by bonding with a female without necessarily having sex.[15] Thus, evolution, in the layout of these receptors, would appear to have a switch by which to turn promiscuous animals into monogamous ones.

Sex itself also establishes bonds. Male rhesus monkeys often mount each other. This was once thought to signal dominance, but closer study suggests that they do it as a form of recreational sex whose enjoyment aids social cohesion.[16] Similar use of sex has been seen in all-male groups of gorillas. Whether

oxytocin is behind this has not been studied, but it seems plausible. In bonobos, known for their free-for-all forest orgies, sex has been coopted for cooling social tensions. Both within a group heterosexually and homosexually
when members of strange groups encounter each other, sex is used to make
friends, or at least peace.[17] Males stand back to back with one male rubbing his
scrotum against another's buttocks; hanging from branches, they engage in
"penis-fencing." Not to be outdone, females engage in mutual genital rubbing,[18] which empowers them against males by helping them to join into
alliances against them.[19] It even seems possible that such same-sex activity,
by reducing social stress, could have paved the way for heterosexual bonding.

The sexual bond was an evolutionary breakthrough, for it allowed two
parents (and so their families) to bond[20] (as well as allowing the formation
of bonds between in-laws). The fact that humans sometimes mate outside
marriage does not affect this suggestion; monogamy is never 100 percent in
any animal. In "monogamous" mammals, such as the above-described prairie
voles, and in over 100 kinds of birds with pair bonds, females have been seen
to have sex with males other than their partners.[21] But while bonding is not
totally exclusive in birds or in us, it is sufficiently so that most males are
fathers to the children they help bring up. Where open promiscuity exists,
no biological father can know for certain which children are his, and thus his
relatives cannot know which young ones are their grandchildren, nephews,
or nieces, which may keep them from acting like kin.

A male that does not know who his children are will not invest his energies in the young and the mothers of his group. He is blind to which offspring have and do not have his genes. We see this with chimps. Rather than
aiding the survival of genes he cannot see, a male chimp will gamble that
he will fertilize a female by chance and be promiscuous. The more times he
hunts for, finds, and mates with females in this way, the better are his odds
of getting his genes passed on. This strategy, of course, has the drawback that
his energies are wasted in male competition, but he is caught in a trap. Not
only might he be wasting his efforts if he helps to raise children he can't be
sure are his, but he might be helping to pass on the genes of a rival. Gene
blindness keeps fathers from helping—at least in a big way—with parenting.
In evolutionary terms, it is a missed opportunity.

The evolution of big brains needed more than one-parent families. Big
brains are too large to grow to adult size in a mother's womb.[22] If they did,
the baby's head would not fit through the birth canal. (As it is, a human
baby faces a tight squeeze[23] even though the pelvic outlet increases up to 30
percent in area during birth[24].) So our brains must continue to grow in the
first year outside the womb. Because they are born underdeveloped in this

way, human babies are helpless for the first 2 years. Newborn, they are too immature to grasp. Only at 10 months can they stand without help, it is more than a year before they can walk, and it takes up to a further 10 months before they can run.[25] A young ape or monkey, by contrast, can grab hold of its mother's fur from birth if need be and so free its mother to get food. Human babies need to be carried, which means that the mother has to look after them constantly, rather than supporting herself and her family. That was hard work—especially before strollers. A mother in a modern hunter-gatherer band might carry her child nearly 5000 miles in its first 4 years[26]; it is unlikely that our earliest hunter-gatherer ancestors carried their children less. A child's helplessness also keeps a mother from defending herself. On top of all this, the young baby's growing brain is an energy guzzler. An adult brain at rest might take 20 percent of the body's energy, but a newborn infant's brain trying to catch up on what takes place in the womb in other mammals takes 60 percent.[27] A mother and her child might have been able to survive without a mate in good times, but each year in the savannah there was a dry season when things were tough. With what is, in effect, a 21-month pregnancy, the young energy-demanding brain would have had to survive two stressful seasons. Thus, because of big brains, human mothers needed someone willing to support and protect them and their children through such hard times. Female friends might help a mother in good times, but not in the hard ones when the welfare of their own children would be their only concern.[28] What she needed was a sexually bonded partner with as much at stake as she in her child's well-being.

Humans have more to learn than other apes. A child needs to pick up tens of thousands of words, social roles, and many life skills, necessitating a long apprenticeship period of childhood and adolescence. And as any human parent knows, that long period of immaturity takes up much of their time and energy. It is a major investment. Unless more resources were available to human mothers, child development could not go beyond that in apes. Fathers had to help out, but for that to happen, a father needed a way to "see" his genes. A reproductive bond solves the problem. A husband, knowing he is the exclusive reproductive partner of his wife, can know that her children contain his genes as well as hers. And, of course, once fathers know their children, family support can come from his relatives. Then a mother will have two sets of people interested in helping her.

The marriage bond has to include trust. Parents need to be able to separate and still be loyal; they are, after all, living a fission-fusion life. Until our species evolved, the only other way a male ape could be sure he fathered children was to stop other males from having access to his harem of females.

Gorillas and hamadryas baboons live this way. But having harem groups prevents the social mixing that leads to coalitions and friendships. For that to arise, there had to be fidelity between husband and wife.

Our Bonded Species

The marriage bond allowed other bonds to arise within the human species. No chimp is a racist or a nationalist. Nor does any chimp think that his or her way of life is better than that of any another chimp. But humans do. The ability to decouple everyday bonds and carry them in our heads allows us to connect to our groups in new ways. Thus, our identities are linked to abstract ideas via nation, race, or religion.

The human revolution ushered in still more bonds. A bond that needs to be maintained by everyday contact also needs the other person to be alive, but our social bonds can last beyond death. Indeed, in most societies, people love their children and parents beyond death. We say most, as many hunter-gatherers are scared that their relatives may haunt them after death as spirits. (They take special precautions against this, such as avoiding mentioning their names lest they hear them and return.) Here we have a human universal. There may or may not exist a life after death, but our cares and feelings for those who have died live on. In other primates, mothers may try to nurse a dead infant for a short time, but soon they desert the body. Maybe they think about the departed for a while—we do not know.[29] Only humans, however, bury their dead and grieve openly and long. Other animals merely pine; they do not seem to distinguish between the physical absence of another and death. Even knowing that someone will never come back, however, we humans may grieve for years, decades, or even the rest of our lives. Think of soldiers and survivors of concentration camps treasuring faded photographs of their loved ones. Small, private things like that matter a lot to us. In the end they matter more than other physical possessions because they link us to what exists beyond them.

The human revolution led to a new experience of life—and death. Humans, having uncoupled bonds from daily contact, could experience the presence of their special ones in a nonphysical world. Bonds could now give apes an experience of a world beyond that in which they lived. Thus, the human revolution created the possibilities of religion, faith, the experience of a soul, and the promise of a life after death. Whether there is a God or not, the human revolution made an ape with a mind that could pray to one.

14

The Symbolic Brain

Grown men bare their left breasts, knees, and elbows. Ceremoniously, a blindfold is put over their eyes and a hangman's noose around their necks. Stranger things then follow for the next 1½ hours: a dagger's point is put on their left breast; oddly worded questions are asked; people promise that they will not repeat what they are hearing on the threat of having their "tongue torn out by the root and their bowels fed to the birds."[1] To outsiders, a bizarre and meaningless ritual is taking place, but those involved take it seriously—and so should we.

The modern Masonic movement dates from the seventeenth century (though it has earlier roots in medieval craft guilds). Masons trace their own ancestry further back to near the Biblical date of Adam, 4000 years BC. But in a way, their origins are even earlier, at the dawn of our species. Masonic initiation rites contain a vestige of the thing that enabled our ancestors to make long the short bonds of their ape ancestors. Whatever we may think of the Masons, they have found a way to bridge the divide between those with blood ties and those without them. Not for nothing are they called "the Brotherhood."

Strange Mortar: Symbolons and Symbols

The Greeks had a word for it: *symbolon*. Its modern equivalent is *symbol*,[2] but to the Greeks it meant something slightly different. Symbolons are social symbols, the tokens or insignia by which people who are linked can spot each other. Symbolons show we are part of a group. The "symbolon" indeed comes from the Greek word *symballein*, which means "to throw together." Symbolons are a "glue" that can join nonkin with the emotional strength of blood. With them, people could form bonds that would last until death, even if they lived in different communities. They thus freed people to decouple their relationships from the necessity of everyday meeting. That created the adhesive that let people bind together in large, complex social units. Symbolons turned nonrelatives into brethren and brothers-in-law; they provided the bridge across the evolutionary Rubicon that was to turn apes into people.

A symbol works by replacing one thing with something else—it throws things together. A ring stands for a marriage, a handshake means trust or understanding, and a title signifies social position. Symbols are extraordinary for being able to do so many things. They need not be social; indeed, nonsocial ones are much more familiar to us in the modern world. You need not go far to see some. Look at the letters that spell these words. No mere sequences of visual patterns, they also stand for sounds. You may never have read *symbolon* before, yet you know how to say the word. You can "hear" the word symbolon by sounding out its letters.

Let us look at some familiar symbols in our lives. Our music would be very different without the notations that allows musicians to see and so "hear" the pitch and duration of notes printed on a score. What would the blind do without Braille? Here touch patterns, not visual ones, stand for sounds. And sounds are not the only things that can be represented by other things. What would nineteenth-century telegraphers have done without Morse code, in which sound itself can stand for visual images—the dots and dashes replacing letters? Or still more up to date, consider emoticons, the smile :) and the wink ;) and >————-<———<—-<;:—<-{@ the thorned rose of email messages.

It is still only partly understood, but language—names and other words—can take the place of "meanings," and so let us communicate. People once took language and speech to be the same, but now we know better. Researchers find that sign languages used by deaf people are linguistically equal to spoken ones. Moreover, sign language is not restricted to visual gestures; the deaf who also go blind communicate signs by touch. And sound-based languages can use more than mouths to make themselves heard. In Africa, languages exist that can be drummed or whistled and thus "spoken" across greater distances than would otherwise be possible.[3]

In China, logographs are not only spoken but gestured, and serve as the basis of the sign language used by the Chinese deaf. The associations symbols make may be radically different for different people. Black in Western societies stands for death, while in China white is the fatal color. And although the red of blood is associated with danger by nearly everyone, the committee that decided that the coloring for the "live" wire in a three-pin plug should be brown apparently thought otherwise. And what about signs used on roads, flags or icons on word-processor screens for saving and opening files. Using symbols is one of the richest things we do. It is unlimited: We keep developing new ways to extend what we use them for and how we do it. Some clever person decided that these punctuation marks: @;-9 represent Elvis Presley, complete with wink, songster croon, and pompadour. Symbols know no limits except our ingenuity.

Symbols can serve as a secret bonding language within a group. To the early persecuted Christians in the Roman Empire, Jesus was represented by a fish. It holds a hidden meaning. The first letters in Greek for "Jesus, Christ, Son of God, Savior," ICQUS, spelled the Greek word for fish, *ichthys*. Roman pagans at the time would have been lost unless it was explained to them. Freemasons have their secret symbols. The ropes, daggers, white gloves, aprons, and blindfolds used in the initiation ceremony have meanings that only Masons are aware of, as do certain words such as "Jahbulon," "Tubalcain," and "Boaz." They are the mortar and cement that bond Mason to Mason. Like the early Christians, Masons bind themselves together by what outsiders cannot (or should not be able to) grasp.

Such symbols of identification need not be visual. They can be as much ways of doing something, like pressing the spaces between knuckles (a Masonic handshake), or whether as a Catholic or an Orthodox Christian you make the holy cross with a leftward (Catholic) or rightward (Orthodox) stroke, or the use of an animal call—for instance, the donkey cry used by a Sicilian Mafia group in the film *Godfather III*. Outsiders will not always spot these subtle signs, but insiders will quickly notice them. Organizations must have a way of letting fellow members identify each other. In *Godfather III*, the donkey cry warns someone that he is the target of an ambush. During the 1939–1945 war (and perhaps during their recent civil war), when Croatians invaded a Serbian area, they asked people to cross themselves. Those who did it the "wrong" way were shot.

As social symbols, symbolons stand for the emotions that bind us in attachments. People use symbols to affirm and maintain their relationships. The most familiar example comes from marriage: The wedding ring is a socially defined symbol of one person's share in a marriage bond. A gift likewise acts as a symbol between people. How you treat it reflects your bond, or lack of it.

Why should symbols be so powerful? Remember those children who had to wait for a treat? Those who could wait held images—symbols—of the chewy marshmallow in their minds. Here we can see the mental mechanics by which social symbols stretch bonds. Symbols let us hold relationships in our head, not just for a few days, but for years or even a lifetime. Symbols become invested with the emotions that bond us and so can become a substitute for meetings we might no longer have with each other. In our thoughts and in the topics of our conversations with others, they—and our bond—are always with us. This process starts early and enables children to separate from their mothers. Instead of seeking comfort with her physically, children inventively take a piece of cloth (such as the security blanket Linus carries around in the comic strip Charlie Brown) or a mannerism (thumb-sucking) to stand privately within themselves for her and her comfort. Blanket-attached children in a strange situation explore and play as if they were in the presence of their mothers—the blanket functions as a mother substitute.[4] Inside their brains, it is an early symbol for her presence. When they part as adults, children and parents value odd little things—a home-made Mother's Day card, a tiny tattered photograph, or a handwritten letter. Such little things can mean a lot.

We need not be conscious of using symbols as stand-ins. Remember that when we work out whether a hand in a picture is a right or a left one, we do so using our own experience of right and left hands, and we do it without being aware that this is what we are doing. Likewise, few of us realize the importance of that odd letter, or photo—until it is accidentally thrown away. Symbols thus enable our bonds to exist even with those whom we see rarely. The symbol—physical or mental—stands for the absent relationship: a piece of blanket for maternal protection; a ring for marriage. And in a more complex way, we recall bonds when we use names and tell stories.

Symbols also help create the possibility of the sacred. Indeed, the symbolism most familiar to Western people takes place in the Christian church between worshippers and their God. Eucharist, or communion, is perhaps the best example. (We use Christian examples only because they are likely to be familiar to most readers.) Eucharist is the ritual of breaking bread and pouring wine. The act centers around the symbolic relationship by which bread and wine can become, for a Christian, "the body and blood of Christ." Christ left this symbolism at the Last Supper, in which he linked his body and blood to a ritual eating of broken bread and wine. For instance, "Take and eat; this is my body," Jesus said after breaking bread. And about wine: "This is my blood" (Matthew, 26: 26–28). With these words Christ drew on symbolism linked with Passover and familiar to Jews like himself and his earliest followers. But he gave them a new significance. Jesus's words have been understood in different ways (some churches take them more literally than others). But it is reasonable to suggest that one of the key aspects of these words, however understood, is the making of community among Christians, with each other and with their God. In communion, Christians create an emotional and spiritual space in which they can discover the event of Christ in themselves and with their fellow believers through shared symbolism.

We often take the symbols of thought and writing as evidence of our advancement over other animals, but this is unlikely to be why they arose. The symbols (apart from language) we associate with our lives are recent. Writing and numbers are less than 6000 years old, first arising in China, Egypt, and Mesopotamia; most others are far more recent. The symbols that drove human evolution did not spread, in our opinion, because they were such great aids to thought; rather, we believe their popularity can be traced to their role in facilitating the earlier human need to bond. The use of symbols led to success in ape social worlds. It is part of the brain's hardware inheritance: Chimpanzees and other apes can use symbols, but they do not do so spontaneously.[5] In contrast, children denied the experience of language invent their own.

Gestures

The main remaining social symbols in our lives are gestures. People greet each other in ways that differ across the world, though they express much the same thing. As a gesture of welcome, Indians put their hands together on their chests in what, to Westerners, is a praying position while making a slight bow; Arabs sweep their right hand upward touching their hearts, foreheads, and beyond; the Japanese bow; the Maori of New Zealand rub noses; Tibetan tribespeople stick their tongues out at each other. These gestures may seem odd, but are they really any odder than giving someone a firm handshake and looking them straight in the eye like Texans, or squeezing them in a bear hug like Russians?[6] They are all symbols made in movement. It is, of course, arbitrary that one action in one society stands for hello and in another it has a completely different social meaning or no meaning at all. Gestures can even mean opposite things. To the French putting thumb and first finger together in the shape of a circle means zero or worthlessness; to the Japanese it is a sign for money. Do it to a North American and they take it as a sign of "OK," but South Americans take it to mean "Screw you!"[7] Would-be American hitchhikers should be advised that the latter insult is also conveyed in Sardinia by sticking out the thumb.

Gestures matter. Though arbitrary in what they stand for, they play a key role in our lives. Scientists treat them as a form of nonverbal communication, but to talk of them as a form of communication ignores how they "speak" our social feelings for each other. People communicate two kinds of information: relationship and nonrelationship. If we mention that the fridge is broken, it is usually pure information. The problem happened by chance, and it has no greater importance beyond that. But we also do and express things with gestures that convey the emotions that make up a relationship. When you leave someone, you hug or make some other movement. You may say goodbye or use some equivalent verbal communication. If you do not do this, and give no explanation (such as "I'm sorry, I don't want to give you my cold"), the other person may become upset. Their attachment with you will leave them feeling that something is wrong. To refer to gestures as non-verbal communication ignores their vital role in the health of our bonds.

Names

Names are an interesting social symbol with additional practical uses. Our names associate us with our kin: Surnames can be sire-names and so identify our fathers and ancestors. Women usually adopt their husbands' surnames, and children often get their grandfather's or grandmother's first name. *John* is the name of John Skoyles' father's dead brother. *Robert*, his second name, is the name of his great-grandfather (in one of whose houses he drafted these words). *Skoyles* comes from the name of the small village, Scole, on the Suffolk/Norfolk boundary, after which a distant ancestor of his was named in the fourteenth century. Names link us.

Monkeys identify each other by sight and body odors. A powerfully equipped nose can discern millions of easily identified variations of smell. Our pets navigate through a world where each place and each person has his, her, or its own olfactory ID. Police dogs can take one whiff of a suspect's coat and then trace the person over long distances—the record according to the *Guinness Book of Records* is 100 miles. Even humans have been shown to be able to identity their children and themselves by the smell of the T-shirts they have worn.[8] Hellen Keller, who was blind and deaf, could identify people by their smell.[9]

But there is a problem with using smells as identifiers. Smells are limited to the present. We cannot call up in our minds at will the smell of the sea or a garden rose in the way we can imagine a sound or an image. As Proust noted, smells can stir up memories very powerfully, but only if they are reencountered. Only if we are given a sniff of the sea can we find ourselves awash with memories of being a child on the beach.

Smells vividly evoke the feelings that underlie our bonds, but they trap our experience of them in the "here and now." Dogs left by their owners pine because they cannot be comforted by a thought of their return. They are trapped in a mere awareness of absence. Language, however, frees emotions to ignore the absence of others. It lets us update our feelings about people. Consider how you feel when you hear about the death of someone whom you have not seen for many years. Those few words—"Fred is dead"— can give you the most intense emotion. You do not have to see the lifeless corpse. What you have been told is sufficient. Gone is the mental feel of that person's existence that had comforted you in his absence. You know you will never see or touch him again, and you grieve. Or consider the experience of writing a letter to loved ones. They are not physically with you, but you feel as if they were.

Names free our thoughts from the present. Using names, we can talk and think about people in the past, in the future, and in their physical absence. Names are real magic spells that can conjure up a specific person, place, or thing. As such they have practical advantages over other ways of telling "who is who."

The Uniqueness of Our Symbols

Masonic handshakes, bread and wine at communion, words in speech or sign language, numbers, musical notation, the Arabic incantation "alakazam," and the letters on this page are all symbols. They are the fabric of our lives and history, both in what we feel and in what we think. They enable us to do all the mental gymnastics that make us the unique ape. If no chimps get married or play chess, it is because their minds on their own have a weak ability to do things with symbols. In contrast, we are spontaneous geniuses, masters of the code. We manipulate thought and action in many different ways using many different symbols. Indeed, the history of our species is the history of symbols that we have acquired and then abandoned. The earliest people did not know of nonsocial symbols, whereas they are the fabric of our modern thinking minds. And curiously, we gained our nonsocial symbols at the same time that we lost most of our social ones.

But if symbols are a cultural glue, what then is its chemistry? Why do we have symbols? Other apes, after all, do not have them unless taught by humans. And how it is that symbols such as road signs or the letters on this page can so readily exist as part of our modern lives? Symbols are a royal road for the rise of those skills that now make us modern adults. If we could not learn nonsocial symbols, we could not safely drive or read these words. We would be shut out of the modern world.

And this is only the beginning of our questions. How does our brain process symbols? What has driven the rise of intellectual ones and the fall of social ones? How we understand symbols shapes all the answers to how we understand ourselves. We could not even ask such questions without using them.

Symbols versus Iconic Signs

In rather simplified terms, symbols are a kind of sign, a thing that stands for another thing. There are many types of signs, but here we will focus on the iconic and the symbolic.[10]

An icon looks like or in some other way resembles what it refers to or replaces. To throw away computer files, we click and drag them toward an image that looks like a wastebasket. A chain of such associations links the icon with its purpose. A symbol, in contrast, lacks any chain of associations linking it to that for which it stands. The links of symbols are arbitrary. An icon for a wastebasket looks something like a wastebasket, but the symbolic sign for a question "?" does not look like anything but itself. It is arbitrary. Historically, it evolved from the abbreviation "Q̥" for "quaestio" that then became simplified over time into our familiar squiggle and dot.[11] Spanish speakers go further and use an inverted "¿" at the start of question sentences as well as putting a "?" at their end. It is simply a matter of convention. Letters are also arbitrary, though they did not start out that way.

The difference between icons and symbols has until recently been of interest mainly to philosophers. Our understanding of the difference between them is changing, however, with the computer simulation of neural networks. These now let us understand signs and symbols in terms of processes. Indeed, it is only with advances in neurology and computers that we have begun to see their key functional difference.

Philosophers had a problem with icons. They worried that icons may look like what they refer to only because we are familiar with the association. For instance, it was argued that an icon sign for a bicycle on a traffic sign does not actually look like a bicycle; we have just learned to see it as one. The resemblance is subjective. Martians visiting our planet might see road signs for bicycles and see actual bicycles and be puzzled as to why we link the two.

But the computer simulation of the brain's neural networks reveals otherwise, suggesting that Martians would have no such problem. Not only do icons look like what they stand for to us, they also do to computer simulations of how our (and presumably Martian) brains recognize things. The resemblance is objective.

The problem started with Socrates. As recorded by Plato in his dialogues, Socrates insistently tried to find the definitions of things. What does justice have to have to be justice? Or virtue to be virtue? Courage to be courage? What are the necessary and sufficient conditions for meaning? These questions drove Western philosophy for 2500 years and shaped the questions people asked about resemblance and recognition. People slipped into the habit of thinking that recognition, like definitions, must use necessary and sufficient conditions. They thought a bicycle could be detected only if certain necessary and sufficient details had first been recognized. But computer simulations show that this idea is simply wrong. Neural networks simulate our recognition efficiently, using a soup of nonnecessary and nonsufficient clues. Explorations of populations of neurons in the inferior temporal cortex (discussed below) suggest that they collectively recognize images in this way.[12] Socrates put philosophers and, later, psychologists on the wrong trail.[13]

Our ability to identify signs with only a few of the clues we normally use lets us spot icons. A neural network trained to identify real bicycles will, without any more training, identify signs for them. A person can easily link an abstract sign for a bicycle with his or her own visual experience of them. Live in a world of bicycles, and you get for free the ability to recognize bicycle icons.

This is not so with symbols. As with icons, there is a link between a symbol and what it stands for. But unlike that with an icon, the link has to be learned, because of its arbitrariness. You may know what a question is and be able to see and remember the squiggle and dot "?"; but unless you are taught, or guess, the abstract association between them, you will not link the two. Symbols are learned associations; icons are not. It is a fundamental difference, and it leads to a problem.

Symbols and the Brain

How does the human brain learn symbols? Remember that their associations are arbitrary. With symbols, one visual pattern is linked with another, so that "&," "and," and "AND," for instance, are all visual symbols for one spoken word. To understand how we recognize symbols requires us to understand how our brains learn arbitrary links.

Scientists at Oxford University looked at how a select group of individuals learned associations between abstract images. Some of them found the task difficult, but with effort they learned them. Others failed entirely—this was because the scientists had removed part of their brains (the individuals, by the way, were monkeys). Though the researchers did not realize it, they had discovered the mechanism by which our brains learn and use symbols.

The *inferior temporal cortex* is the part of the brain used by monkeys (and us) to recognize images and learn arbitrary links between them. Not surprisingly, the monkeys lacking an inferior temporal cortex could not learn to make arbitrary associations between letters. But there's more. Another group of monkeys had working inferior temporal cortices but damaged *uncinate fascicles*. This "barbed bundle" links the inferior temporal lobe and the dorsolateral prefrontal cortex.[14] Without this link with the prefrontal cortex, monkey brains cannot learn new arbitrary associations. If the monkeys with intact connections with the prefrontal lobe had found learning symbols hard, it should not surprise us, as their prefrontal cortices are between one-fortieth and one-fiftieth the size of ours. They can thus show only a hint of what has become fully developed in us. What this research shows is that to learn arbitrary associations, we need our prefrontal cortex—but only to learn them.[15] Once learned, the use of symbols is not lost if the prefrontal cortex is injured.[16]

In letting us learn symbols, our prefrontal cortex makes us who we are. In general, it works to upgrade the rest of the brain, using its working memory and neural plasticity to neuroprogram it with symbols. The processing of letters, for example, uses a network that spans several different parts of the brain, those involved in sight and hearing.[17] These several parts will never, on their own, learn a link between phonemes and visual patterns of lines. That requires an outside programmer that can make the cross-sensory associations needed to pair letters with sounds and hold them in memory buffers so that they link with each other.[18] The same is true of all arbitrary associations. They do not spontaneously come together, but the prefrontal cortex, with its working memory, can ensure that they do come together; it is an artificial inducer of *symballein*. In this, it is a bit like a catalyst.

A chemical catalyst can bring together two molecules that would not normally meet. It brings them into temporal-spatial conjunction by presenting them with a shared substrate. Now they can react. Likewise, many associations in the brain would never spontaneously occur, because the neural networks governing them do not normally function together. But the prefrontal cortex can change that. Phonetic sounds, for example, would never normally pair with the visual images of letters, but the prefrontal cortex, their catalyst, their matchmaker, brings them together. Once it does so, they can stay linked on their own. We know little about the nature of the networks that support these associations, but they seem to involve gamma synchronization. Gamma not only increases between areas supporting an association but it is more coherent[19] between areas that get fused together in networks generating an association. This is particularly so when the association is first being learned.

The brain's discovery that it could teach itself symbolism was an immense revolution. With the rise of self-taught symbolism, the brain could upgrade and remake itself. This event was a watershed as momentous for the mind as catalysts were for the rise of life.

The Great Explosion of Symbols

First, consider the power of arbitrary links to change what the brain was doing. Before them, the brain had used iconic and other nonarbitrary links. For instance, it could tell whether a picture of a hand was of the right or the left hand using its own internal experience of them (see Chapter 3). But the brain could not go far beyond this and so possessed only limited ways to process information. This restricted what it knew and could do. The prefrontal cortex changed this by upgrading these associations to include arbitrary ones, such as numbers, and rules for manipulating them, like addition and subtraction. This vastly increased processing power without the need for any new evolution of the brain.

But this was only the start of the brain-upgrade story. By themselves, arbitrary associations exist only as a kind of private software that each brain learns in isolation. Once these associations became encapsulated in symbols, the prefrontal cortex could copy the links discovered and used by other brains. Symbolic associations could be communicated. That was a revolution—brains learning from other brains. The software of the brain was now in the public domain. This turbocharged what brains could do, even without the rest of the brain needing to change.

Still, it was only the beginning. Now brains could combine symbols into sentences. Symbols until then had been simple things. Now they could build larger units. For our minds it was a revolution as big as that which ended in the Cambrian jump 543 million years ago, when simple multicellular creatures discovered the tricks to make themselves complex ones such as barnacles, butterflies, robins, and even you. Now a doubly new and entirely unexplored world was opened up to brains. First, they could communicate with great subtlety with each other and so start to live complex lives together. But second, they had the possibility of an inner world; now the prefrontal cortex had a powerful tool with which to organize the rest of the brain: inner speech. It could use it to organize the rest of the brain even more thoroughly. The brain catalyst of symbolism would soon start catalyzing itself in many odd and unexpected directions.

Brains could now go on to invent worlds that existed only in symbols and language—mathematics, literature, and science. Indeed, one philosopher, Sir Karl Popper, suggested that these constitute a different world, "World 3." According to him, it exists autonomously, apart from the real one in which we physically live (World 1) and the subjective one of our inner thoughts (World 2).[20] World 3 is the world of symbols we communicate and store, from the works of Euclid and of Shakespeare to the World Wide Web.

Now the fragmentary patches in our evolutionary jigsaw puzzle are coming together, and we can see what the picture is. We can trace the steps in our brain's odyssey. We need to take a deep breath here, for this is nothing less than the missing link that evolutionary studies have been looking for.[21] We have been searching for a new fossil, but the missing link is a brain fact. Neurology is needed as much as, or more than, paleoanthropology to understand our evolutionary past. More pieces, many more pieces, will be added in this puzzle, some of them unexpected. But in essence, here, in the brain's ability to make symbols, lies the secret that turned our brains from those of apes into those of humans.

Mindware

Symbols make mindware possible—not of a single kind, but of a nearly infinite number of varieties, limited only by human imagination. Our brains had contained many ape skills that were ripe to be changed in what they did. With the right symbolism, they could be upgraded to do things no other ape could do.

As we saw in the initial chapters of this book, the new physics of the human brain—neural plasticity—suggests it is not fixed in what it can do. Our prefrontal cortex, we argue, has the ability continually to empower and rework that potential to do new things. The middle chapters told of the potential of the ape. With the ape brain, evolution had created a biocomputer with a wide range of mental skills that was ready, without further physical evolution, to do totally novel things. These skills had originally evolved to aid the social survival of a fission-fusion ape, but now the brain could potentially rework them to entirely new ends and purposes. It could extend and strengthen its social bonds, but, as important, it could do other, entirely fresh things. However, for this to happen, something had to unlock its potential. Symbols were that key of liberation. Symbols made the mind of the human-ape unlike that of any other. But that was not all. Symbols gave brains the ability to pass that change on. Mind change could thus be virulent and undergo an evolution of its own.

We can summarize what happened in the following two formulas (where, again, NP stands for neural plasticity and PC for prefrontal cortex).

NP + PC + fission-fusion ape social skills = human social symbols

NP + PC + ape sensory and motor skills = human nonsocial symbols

Symbols can change the workings of nearly every part of our brains. The monkeys in the Oxford experiment used the associative skills in the inferior temporal cortex, but many other parts of our brain (possibly all of them) are programmable by symbols. Brain scanners give us clues as to where the brain forms our most important associations. Though the processes involved are widely distributed and vary slightly among people (a point that cannot be overstated), the main networks that symbolically link a word's meaning and its sound exist chiefly in the middle part of the left superior and middle *temporal gyrus.* And then there are written words and meaning, believed to be tied together by networks in the posterior part of the left middle temporal gyrus[22] for English words. For Japanese logograms, or *kanji*, it is the left posterior temporal gyrus.[23] (Word meanings might themselves in part be located in a part of the left posterior superior temporal gyrus usually identified with the *Wernicke's area*.[24]) The area used for musical notation is around the superior and posterior part of the supramarginal gyrus of both the right and left inferior parietal lobes.[25] Such symbolization of the brain could greatly change its functioning. Once there was speech, there was a broad ability to recognize and create patterns and an ability to draw them. After the invention of writing, people linked these patterns and started to read and to write. The adoption of symbols demonstrated the endless versatility of the human brain.

Limbic Symbolons

Symbolization changed the parts of our brain dealing with social ties. These were, perhaps, at first, the most important changes made to the working of the brain. The prefrontal cortex is well placed to use symbols to extend our capacity for relationships. It has links to all the parts of the limbic system— the hypothalamus, insula, amygdala, hippocampus, and anterior cingulate. It can upgrade them so that they can substitute symbols for relationships that in other apes can exist only through daily or near-daily contact. No longer would a person's limbic system make him or her pine in the absence of another. It would now transfer the emotions and feelings of the other's presence into symbols—symbols of attachment. To give a modern example: Your spouse may be away on business, but you are still married. The wedding ring proves it, to yourself and others.

Once emotions had been extended, human evolution had struck gold. Most of the brain can be enriched socially by symbolism. To extend kinship, the brain could use symbols to tap into its attachments, linking kin and extending the relationship to in-laws. Similarly, to make kin through marriage, the emotional experiences between family members could be transferred, using symbolism, to the marriage partner. Thus, the prefrontal cortex, through symbolism and ultimately neural plasticity, changed the brain's capacity for kinship and attachment—and so crossed that Rubicon from being an ape to being a human. One idea we need to explore is whether social symbols mirror—though in a new way—the information processing that happens in the parts of the limbic brain into which—for lack of a better word—they hack. One conjecture is that symbols extend what is done by the orbital prefrontal cortex. In apes, they underlie behavioral "don'ts." In us, they may have made possible the human experience of morals and etiquette. The experience of many people is full of things to fear and be wary of—magic, pollutants, unlucky numbers, evil eyes, sacred places, ritual ceremonies, and superstitions. Might the fears and anxieties of the amygdala become symbolized? And look at how we symbolize our bodies. We wear rings and powder and paint ourselves. Performers, from Muppeteers to strangely dressed rock-and-roll musicians prancing around the stage, distort and enhance their bodies and faces with costumes, props, and makeup. Is the insula or the somatosensory cortex being symbolized? Perhaps the allocentric and egocentric frameworks in which our social nature exists are symbolized. And what about the links between them? Is the posterior cingulate programmed with symbols? Mindware changed everything.

No doubt the idea that social symbols extend the limbic system is too simple. It is possible that symbolic culture is spread around the brain, with no part of a ceremony or ritual being located in a specific part. Maybe no "amygdalan" symbols or "cingulate cortical" symbols exist. They could instead be encoded in widespread networks across the brain. Maybe people differ in the way they symbolize—and thus in how they experience—culture. We may even go so far as to suggest that perhaps some individuals, in their brains, symbolize themselves more than others. Here we suggest that while we know few of the answers, we are beginning to ask something like the right questions.

Our mindware view challenges present anthropology by suggesting that beneath our cultural trappings lies a brain story. Social culture, it argues, is the direct result of a skill we did not realize we had, the ability to symbolize our brains. Since everything we are and do is made possible by our brains, it makes sense to us that the findings of neuroscience would eventually click together somewhere with what has long been known by anthropology. The ability of the brain to symbolize itself may well explain why people live such highly symbolized lives.

These ideas, however, do not fit easily with anthropologists' present notions of what makes symbolic culture. Maybe they need a dialogue with what we are finding out about the brain. There is room for integration. Cultural anthropologists talk of *signifiers* and *signified*. What people do with symbols means something, but it is never clear exactly what—though it has to happen somewhere. Given the brain's role in making our experience, it would be reasonable to suggest neurons. Thus, we suggest, when anthropologists talk of signifiers, they are in general talking of symbols; and when they talk of the signified, they are in general talking of the processing done by neurons, particularly given their role in emotion, those in the limbic system.

This conjecture that anthropologists really mean symbols (as we define them) when they talk of signifiers and limbic processing of neurons when they speak of signified still leaves many questions unanswered. Cultural symbols nearly always link with rituals, and all rituals have certain things in common. Those involved in a ritual take it to be something special. They are moved to do odd things they would not be seen doing in ordinary life. They feel they are encountering a powerful hidden reality. As one anthropologist has noted, "in the performance of the rite a native has made that small contribution, which it is both his privilege and his duty to do, to the maintenance of that order of the universe of which man and nature are interdependent parts."[26] As another anthropologist, Stanley Tambiah, puts it, "it is possible to argue that all ritual . . . uses a technique which attempts to re-structure and integrate the minds and emotions of the actors."[27]

Rituals are emotionally intense. Initiation rituals, including Masonic ones, have hints of death and passage back into ordinary life through rebirth. Are these details of rituals incidental entertainments, or are they a key part of the process by which brains open themselves up to be encultured by symbols? Rituals must happen somewhere in people's brains. Moreover, they must do something, since, after all, nearly every society finds occasions on which to carry them out. We do not know how, but it could be that ritual charges up the limbic system to embody social symbols. Intense emotional experiences are known to be encoded differently by the brain than nonemotional ones.[28] Perhaps rituals reorganize our brains by playing on the same processes in the amygdala that underlie posttraumatic stress disorder, but in a constuctive way, so that what would have been ordinary objects, images, and behaviors become wired with intense emotions—not flashbacks but feelings of relationship, specialness, and the shift of reality into what people call "sacredness." Such an idea is speculative, but the technological revolution that has given us brain scanners makes it at least testable in principle.

This link between symbolic culture and the brain is not an afterthought to the earlier chapters in this book. When reading them, you might have thought we wrote those chapters to explain our minds. We did, but there was a deeper purpose as well. They were prologues to prepare you for this chapter's conjecture that culture rests upon what happens in the brain. After all, this is what really makes our minds. It may seem lopsided—several chapters for these few paragraphs. But in fitting together jigsaw puzzles, more time is spent finding the pieces than clicking them into place.

The discovery of social symbols, we suggest, was the big step we made across our Rubicon. With it, evolution opened up future possibilities. Increased in size, our prefrontal cortex now had the power to skillfully hack symbols into the working of the rest of the brain. It did this with the limbic system to symbolize attachments. In doing this, the human ape found a brain programming language to bond across time and place—symbolic culture. This was to change forever what it meant to be a brain. Now the human mind could live in thousands of varieties of life. Brains could discover how to share a new world among themselves. With this came a vast leap in what brains did. With shared symbols, the human brain retooled its mind to make all manner of new bonds, new selves, new societies, and indeed, if we include the sacred and Popper's World 3—new worlds. With their mindware humans set themselves apart from other animals and the rest of nature.

However it happened, the use of symbols liberated the brain. No longer was it merely a skull-encased organ programmed with instincts to survive and reproduce genes. Symbols freed brains to experience new things in life. Just look around and you can see and feel them. Brains now had the means to ask the big questions: Who are we? Why do we exist? Why are we conscious? At first they found answers in myth, telling themselves stories about gods, creators, and ancestors. But the questions started a process that gradually led to better answers. It led to science and books like this. Neurons with symbols started us on a new journey, one that is far from finished—or perhaps even truly begun.

15

Lucy and Kanzi:
Travelers to Humanity

Nearly a billion suns have risen and set over apes with brains larger than those of chimpanzees—humanlike apes. A billion days is just over 2.7 million years, a short time in the history of our planet and of life. But it is roughly the time since the dawn of brains like yours. It was just after that— 2.5 million years ago—that we started to acquire those large, symbol-using frontal lobes that make us human.

Why did this ability to symbolize arise? It did not have to. No law of nature says big frontal-lobed apes like us have to exist. Something must have happened. Perhaps we should not ask such a question. After all, we are never going to understand the origins of our prefrontal cortices in the same way we can understand how the brain uses them. For that, we need only to investigate our brains, and the tools for doing that are becoming better every year. But to understand where our bigger brains came from requires us to see back across many hundreds of thousands of lives to the dawn of the first human apes, and beyond. And it is a safe bet that time machines will never be invented.

The only clues we have to go on are a few stone tools, bones, and bits of skull, and what we know of ourselves and other apes. Thoughts and feelings do not survive in the fossil record, nor do the situations or problems that propelled their evolution. Neither does the soft tissue of the brain. But the calcium phosphate brain cases called skulls and the body supports called skeletons do survive. Indeed, just enough remains to give us hints as to why our brains arose.

Lucy

The story begins a little over 3 million years ago in Africa with a young woman and the Beatles' song "Lucy in the Sky with Diamonds." The young woman died in circumstances that preserved most of her bones, a fortuitous bequest to us, her descendants. In the 1970s, some young scientists in Ethiopia found her skeleton. Usually only single bones are found, but in this case most of the skeleton had survived. It was an extraordinary find and, at the time, unique. The night after she was unearthed, those who had dug her up, dazzled by their luck, listened starry-eyed to the Beatles song playing in the background. In their jubilation, they called her Lucy. The name has stuck; her scientific Latin name and specimen number, *Australopithecus afarensis* A.L. 288.1, are a bit of a mouthful. And while half a million years would go by from the time she lived before the real action in the human odyssey would begin, the story had to have a prologue. It did, and her name was Lucy.

We may lack time machines, but Lucy's bones and those of other *Australopithecus afarensis* give us valuable clues to how our line separated from that of chimps.[1] They tell us that Lucy and her relatives were chimplike in their upper bodies. Indeed, her hands still resembled a chimp's, suggesting that, like them, she still spent much of her time in trees. But the remains of Lucy's lower body tell us that Lucy was beginning to walk on two legs. Her hips, legs, and feet, while not quite like ours—she and her relatives still had some distance to go—were more like ours than like a chimp's. With her hands freed from walking, she could carry objects and do things with them. Though she was probably not yet crafting stone tools, her arms were free to throw stones and wood missiles at any approaching lion. Lucy's legs might have arisen as much for freeing her hands and standing for self-defense as for walking. But those two legs were also used to move around. In Tanzania, some footsteps left in fresh volcanic dust by a woman and her child have survived. They are not Lucy's footsteps, nor even necessarily from her species, but they suggest that legs were evolving for standing upright and being walked on 3.5 million years ago.

Despite her legs, Lucy was not fully human. If you saw her on the subway, she would look like quite the bag lady. Taking account of their different body sizes, the brains of Lucy and her relatives were only slightly larger than those of a chimpanzee. They showed some hints of human shaping in their lobes, but these differences are slight.[2] Lucy a human? No. A superchimp on the way to becoming us? Yes!

Lucy is a rare find. The best guess is that apes like Lucy lived 5 million, maybe 7.5 million, years ago. A few fragments of an even more chimplike

"Ramid" (root) Lucy have been dated to 4.4 million years back.[3] Lucy had ancestors and cousins. No doubt if we could get into a time machine and go back to before the dawn of humans, we would find many Lucy-like ape species. But time has washed away evidence of them. Neither have remains of ancient chimpanzees been found; the only chimp bones known are modern.[4] But their ancestors, as much as ours, must have been living in Africa for many millions of years. Nearly all evidence of our past has been erased. It was by an extraordinarily lucky fluke that Lucy's skeleton survived. No wonder the scientists who found her were jubilant.

Lucy's discovery, however, leaves us with a mystery: Why did our evolution start feet first? What led an ape to start walking and later evolve superbrains? A brain—a social symbol-using brain—perspective on our evolution is needed. We should warn the reader that ours is only one viewpoint among the many possible. Many pieces of the jigsaw puzzle are missing.

Our brain approach may anger those who study and argue over the bones of our early ancestors. But we must get a perspective on what we do not know. Consider what would happen if we were missing other pieces in the jigsaw puzzle. Look, for example, at the vast diversity of primate life living in Africa today. It teaches us much about the kind of animal from which our ancestors evolved. Yet we could easily not have known this, since it is only by chance that Africa has preserved its wildlife. If agriculture there had expanded to its present extent 1000 years earlier, human activity by now would have killed off all that continent's primates. If that had happened, we would now be totally in the dark as to all the subtlety and richness of the knowledge we take for granted about monkeys and apes. We would have no idea that such a near relative to us as the chimpanzee had ever lived. And if we were lucky enough to find an odd bone of Figan or Goblin, no one holding it could have ever guessed the subtle complexities of their lives as recorded by Jane Goodall. Goodall was only just in time. Had she not made the Gombe chimps internationally famous, Gombe would have become farmland, not a national park. This is the pickle we are in with Lucy and other human ancestors. Nearly all of them, like chimps, have lived and gone without leaving remains. Lucy is a lucky exception, not the rule. We know nearly nothing about our ancestors' behavior, and we cannot pop Jane Goodall into a time machine to go study them.

Yet we should not give up. While much is lost, many unexpected sources of evidence are turning up. Until 1952, few people had heard of DNA, and now it provides clues to, among other things, when chimps and humans diverged from each other. What we are learning about the brain likewise could well give us clues about our past. Not only fossils but present lives provide a window into the past. And in some cases looking at the present can be better; while fossils give us details of structure, looking at modern organisms gives us a clearer idea of the actual processes, physiological and behavioral, that went on in their living ancestors. What follows are a few neurologically inspired stories to stretch our imaginations.

Lucy Goes Home

Recall those female chimpanzees mentioned in Chapter 13 who left the band into which they had been born. They no doubt would have liked to keep alive their bond with their mothers and kin. Chimps are emotionally dependent. As Jane Goodall noted, Figan, "even as a late adolescent, often ran to touch or briefly groom Flo [his mother] when he was upset." Goodall notes of those daughters that stayed (not all leave the band in Gombe National Park that she studied) that they "sometimes search for their own mothers for hours at a stretch, whimpering from time to time."[5] So chimps have the emotional need to see their mothers.

But a female chimp would have problems going back to her mother. First, she would have made friends and alliances in her new band. When she entered it, she would have had to settle herself in socially—females already there might resent her presence at first and be aggressive toward her. If she went home, she would lose some of that social position she had worked to establish. She would also have emotional problems leaving her new friends. As much as she would miss her mother, she might miss them as well. Not using social symbolism, chimps are vulnerable to problems after separations. They are highly aroused after a prolonged absence. On reunion, if emotionally close, they embrace, pat, hold, and groom each other. Chimps who compete with each other become aggressive after separation.[6]

Any chimp who tried to return to her mother would need a suitable "passport" to enter her old band. Chimps normally attack other chimps moving between bands.[7] There is one exception, which is a female with genital swelling. The males in the group she wants to enter can see that she is, or shortly will be, in estrus—ready to become pregnant. Her passport for safe travel is not respect for herself but the opportunity her body offers for sex.[8] It is a passport limited to young females. Older females with children lack this passport, and males will attack them.[9] We stress that this keeps females in their bands once they become mature mothers and thus stops them from doing something we do, visiting kin. The revolution that happened between Lucy and us was the discovery of a new passport that could allow families that had split between different communities to meet again without violence.

Around 1972, the Gombe band Jane Goodall was studying split into two, a northern group and a southern one. One female, Madam Bee, went with the southern males, and one of her daughters, Honey Bee, stayed with her. Her other daughter, Little Bee, moved in with the northern ones. In spite of hostility of the males, mother and daughters attempted to meet. Madam Bee occasionally visited the northern group during 1972. But she was not welcomed, gradually finding boundaries put up against her. In August 1974, she visited but had little social contact. A week later she tried again but was put off by the display of a male consorting with Little Bee. After this, the mother stopped going north, but Little Bee still came south to visit her and Honey Bee. The males now tried to stop this. In September, mother and daughters were attacked, Madam Bee suffering a deep gash in one of her legs. This was the first of at least three assaults. Then, on September 14, 1975, friction turned to crisis as the males became more aggressive. They caught Madam Bee and stamped, hit, and pounded her. Among the most frenzied was Figan, who dragged her around for yards, shivering and too winded to scream her terror. When she tried to stand, he and others smashed her to the ground again and again. They did this in front of her daughters. It was murder. Madam Bee received bloody wounds to ankle, knees, wrist, hand, and back. She lived another five days, nursed by her daughter Honey Bee.[10] She had paid the ultimate price for trying to keep in touch with separated kin.

Ultimately, those who killed Madam Bee were defeating themselves. Madam Bee could have been the grandmother of their children. After all, Little Bee was seen consorting with Figan, who is believed to have later fathered her son, Tubi (born in 1977).[11] If Figan had had a brain that could have realized that he was attacking his future child's grandmother, he might not have done it. Indeed, he might have come to her aid. As a grandmother to his son, she could have helped his son's and so his genes' survival. Given the challenges chimps face in surviving and later gaining social position, they need all the help they can get. But Figan's son never had that extra help. In murdering Madam Bee, Figan made a bad move in the struggle to perpetuate his genes.

But why weren't selfish genes coming to the aid of Madam Bee and giving her a passport? If Figan had had a gene that would have caused him to help Madam Bee, it would have helped him pass more of his genes (including the helping gene) on to the next generation. It would have been an evolutionary winner. Something must have gotten in the way. We suggest this something was his emotions. Figan could see Madam Bee only in the short term. She made it harder for him to consort with her daughters, because she was a competitor for their attention. Mentally trapped by this viewpoint, he could not see her long-term value to him as someone who could aid the survival of any child he might father by her daughters.

The problem is that chimps are emotionally shortsighted as to who are their kin. It's not just with their mothers-in-law; it has been reported that even some brothers and sisters, after their mother dies, fail to continue to act like kin.[12] The reason is that they lack the mental spectacles that let us know our kinships; because they don't have names, they can't express simple relationships such as mother-of-so-and-so. Without a means of seeing who is kin and who is not, a male chimp's emotions cannot recognize the existence of long-term benefits from kin and possible kin relationships (such as visiting mothers). Their brains fail to link their emotions and the future advantages offered by kin. Limiting themselves to a narrow range of short-term advantages, they exploit only a few of the many possible advantages kinship offers. An evolutionary pathologist (if such a profession existed) might diagnose their condition as "limbic temporal shortsightedness." The first symbols, we suggest, changed this. Lucy-apes discovered social symbols to mark off actual and potential relatives, and this enabled apes to relate to kin in terms of longer-term advantages. It put them on a new road of evolution, and of socializing.

Social symbolism, if they had it, would not only make apes wise to kinship but also give them the mental abilities to exploit it. Not only did Figan need to know that Madam Bee was of advantage to him, he also needed some mental skills to change his behavior toward her. He would have been troubled by his hot, aggressive reactions to outsiders such as her. Just knowing she was a "good" outsider would not be enough; he would still need a means of inhibiting his natural aggression against her.

Symbols acting as inner cues would have allowed him to do just that. As we saw with children waiting for sweets, some symbols can cool and aid the control of hot emotional behavior. A symbol of kinship could have taken the place of their strangeness, thus cooling Figan's emotions toward Madam Bee. Perhaps he might have even used symbols to make arrangements for Little Bee to share her time between him and her mother. Without them, he could not see such possibilities. Symbols would have freed Figan's mind to treat Madam Bee like his own mother.

A dotty idea? Recall that those children able to wait for sweets using symbolism were different from those who could not. They were more successful at school and more socially competent. But there were other differences. A child less able to use symbolism also tended to be under more stress, more easily rattled, and more disorganized. These children were rated by their parents as more shy and reserved and more stubborn, whereas the ones who could use mental symbols to delay gratification were rated by their parents as playful, able to concentrate and express their ideas; they were curious, resourceful, and open to new ideas. Not being able to wait is linked with what psychologists call lack of ego resiliency. Insufficient internal symbolism leads to insufficient inhibition of impulses. By contrast, symbolism gives a child's prefrontal cortex a way to control his or her arousal, behavior, and emotions.

Still, there is a problem with our suggestion for the chimps. The use of symbolism requires both brains able to use it and symbols for them to use. Even the less able among us are far more equipped to use symbols than are other apes, because we have big prefrontal cortices. Chimps have some prefrontal cortex, but is it enough? The evolutionary snowball of prefrontal enlargement did not roll uphill; it had to start somewhere. Could the great snowball have been started rolling merely by the rise of some kind of symbol-aiding ego resiliency on meeting kin? Did Lucy-like chimps have the brains to learn kin symbolism to welcome their in-laws? Studies demonstrate that chimps have the potential to use and be shaped by symbols, and this suggests that our mutual ancestors may indeed have possessed the bare essentials to get that snowball rolling.

Kanzi

One of the most hotly disputed areas in science today concerns the divide between us and other animals. The issue touches upon key issues in ethics and religion. A line has traditionally been drawn at our unique ability to speak and so also use symbols. It was believed that other animals might have feelings, but they could not talk. But is this really true? Other animals do speak to each other in simple and limited ways, such as the "this-is-my-territory" bark of a dog and a purr of pleasure from a cat. But can they do more? Do they have within them the hidden potential to communicate and use symbols as we do?

Until recently, this has been an area where anecdotes have reigned. A person's prejudices shaped how he or she saw the issue. But all that changed in the mid-1980s. Since then researchers have rubbed out that line between us and other animals, at least for chimpanzees. These apes can unquestionably learn simple language and symbols. If brought up in a world where people talk to them, they can talk back at least as well as a 2-year-old child does. They thus have the potential to be much more like us than we realized. This in turn, we suggest, changes how we should view our own evolution.

The potential of chimp brains has been missed in the past. The problem was not just that people wanted to preserve that divide between us and them, it is also that adult chimps are difficult to work with because they are so strong. A young man has the strength to pull about 175 lb, an adult chimpanzee four times that. Thus, for a long time, scientists could study only young chimpanzees. And they could not start to teach young chimps with their mothers around, since the mother might want to make her contribution.

Washoe and most of the other talking chimps of the 1970s, moreover, were motherless. They were captured in the wild when they were very young, and their mothers were shot. Considering the potential importance of mother-child bonds in primate communication, these chimps were not ideal subjects for study. It would be reasonable to suggest that, like young humans, young chimpanzees learn best when they have a stable bond. Work in the 1980s, however, has turned to "well-adjusted" chimps. And these chimps have shone. In short, some of the failure of earlier attempts to compare chimps with us was due to the circumstances in which they were tested, not their true abilities.[13]

There is yet another reason. Researchers have learned from past attempts to teach chimps language. In the 1970s, Washoe's ability to talk with signs made her a star. Unfortunately, it was one thing for Washoe to use language and another to show skeptics that she did. That required ruling out alternative explanations for her behavior. Some scientists dismissed claims that she could speak. In retrospect, their arguments appear weak, but they killed most research into talking chimps for a time. However, some researchers carried on, learned from the criticisms, and took advantage of new technology.

Washoe learned to talk with her trainers, Allen and Beatrice Gardner, using American Sign Language. Chimps lack the vocal abilities needed for making speech sounds—speech requires a skilled coordination between breathing and making movements with the larynx that chimps lack.[14] (In the wild they rarely make sounds unless they are emotional.[15]) But they can "talk" with their hands. With sign language, however, it is difficult to judge whether a chimp is truly making a sign to communicate or is just imitating the sign movements made to him or her. So, more recently, chimps have also been taught visual signs, or *lexigrams*, displayed on a "keyboard." To communicate, they press a geometric sign, which lights up and at the same time, through a speech synthesizer, says the word. While trainers "talk" words through the keyboard, they also pronounce them in English. The electronic keyboards used were not always reliable, so there was also a plastic pointing board.

Several chimps have learned to "speak" with this system. The most famous of the new generation of "talking" apes have been Kanzi (born October 28, 1980) and his younger sister Mulika (born December 22, 1983). Kanzi, a bonobo chimpanzee, appears to be remarkably intelligent.[16] He started on his road to learn language through the training given to his mother, Matata. Researchers tried to train her using a keyboard, and he was with her while they were doing so. He took an interest in what interested her, an example of the social referencing mentioned earlier. At around age 18 months, he started to play with the symbols on the keyboard during his mother's training. Indeed, to his mother's and the experimenters' annoyance, he would try to grab signs as they lit up. Ironically, Matata never mastered the keyboard, but Kanzi did—and spontaneously. He even practiced on his own, something children also do when learning language (psychologists call this "crib talk," since it often occurs at night). Thus, Kanzi went on to become an active learner, trying to understand what others were saying and get them to listen to him. By 10, he had a "spoken" vocabulary of some 200 words. He did something else: While using the keyboard, the trainers spoke the signs; Kanzi started to hear and understand their words. While he could not speak back to them, he learned to understand what they had said to him.

The extent of Kanzi's abilities was tested by comparing him with a 2-year-old girl, Alia. When both of them were out of sight, they were given simple instructions in spoken words, such as: "Go to the microwave and get a shoe," "Go get the snake that's outdoors," or "Can you make bunny eat the sweet potato?" It was a task Kanzi could do better than Alia. He could distinguish the different order of words. Asked to put the pine needles on the ball, he would do that and not the other way around.[17] It was a task in which Kanzi made fewer errors than Alia.

Kanzi did this when he was 5, and Alia was only 2. But it was not really a fair contest. Alia was learning not only to understand spoken speech but also to speak, something that would provide feedback on her comprehension. Since Kanzi could not make speech sounds, he was working under a handicap when trying to understand spoken English. It is remarkable that he could understand single words, let alone the short sentences above. Interestingly, while Kanzi will never, for anatomical reasons, be able to speak, he does have a far wider range of vocal sounds than other chimps.[18] Perhaps with a more articulate vocal apparatus Kanzi would be able to talk. He might wish he could, since as an adult, he has preferred the company of humans to that of other chimps.

Kanzi's case suggests that language is not unique to us, but we should be careful what lesson we learn from it. His case has rubbed out one line dividing us from other animals but put another in its place. Kanzi shows that while chimps may have the potential to learn language, they require a "gifted environment" to do so. Kanzi was surrounded by intelligent apes with PhDs who spoke to him and gave him a stream of rich interactions. They gave Kanzi's brain a world in which it could play at developing its ability to communicate. As we have seen repeatedly, the brain is molded by its learning experiences. Adolescent monkeys in rich environments, for example, increase their prefrontal cortex activity by a third compared to those lacking such experiences.[19] Therefore, as much as in his brain, Kanzi's skills lie in the environment that helped shape it.

Gifted people, after all, are aided considerably by an environment that allows their potential to flourish and grow. Wolfgang Amadeus Mozart might have been a child prodigy, but his father Leopold was a gifted composer himself. He devoted himself to giving Wolfgang (and also his sister, Anna) the opportunities to discover his musical talents. Kanzi, like Mozart, grew up in a world that gave his potential every chance of being revealed. One of the key factors in Kanzi's success was his mother's attempts to use the keyboard. He came into a world where the one individual who mattered most to him was trying to learn to communicate. It jump-started him.

Mind Changers

Kanzi's ability to use language gave him symbols. It changed his mind. While we view language as a means of communicating, it is, as importantly, a mind changer.

Think again of those children who were given a choice of sweets. They could delay their gratification if they could hold onto a "cool" idea of a sweet in their minds rather than a "hot" one. The cool idea was a symbol, an inner cue that children could use to manipulate their reactions. The symbols chimps learn have a similar power. Research was done on some chimps that had been taught symbols for numbers. Like the children, they were given a choice of two groups of sweets, but they did not have to delay to get the higher number of them. Instead, they had to play a game with another chimp. They had to point at the choice they did not want (either three sweets or five), and their selection would be given to the other chimp. So to get the higher number of sweets, they had to point to the lower number. It was a frustrating rule, as they could not keep themselves from pointing to the higher number of sweets. As soon as they did, they appeared to the researchers to have a facial expression as though they were thinking, "Oh no, I did it again. She is going to get more than me!"[20] They never learned how to keep from resisting the higher number until they were offered a choice, not between the sweets themselves, but between numbers—symbols—standing for them. With this choice, they had little problem in selecting the lower number and so getting the larger quantity of sweets. Numbers were "cool" and allowed them, like the children, to control their immediate "hot" reactions. In these experiments, chimpanzees too gained inner cues with which they could manage their impulses.

Learning symbols seems to have improved Kanzi's ability to imitate actions. Seeing one of his human friends using a roller to squash some putty, he picked up the roller and did likewise. Chimps without language did not.

What was happening in Kanzi's mind? Symbols, in our way of looking at things, gave his prefrontal cortex a better inner orchestra. With symbols, he could plan, focus, and imagine how to do things. The prefrontal cortex's skill lies in manipulating representations, but such a skill is only as good as the cues it has available to it. However big or talented the prefrontal cortex, without a rich variety of representations, it is like a conductor of an orchestra of two or three instruments that can play only one or two notes. The prefrontal cortex cannot take advantage of the brain's potential if it is limited to only a few mental concepts. If it lacks language, for example, it will be able to use only the few internal representations it picks up by trial and error. Worse, they will tend to be contextual rather than abstract. Abstract symbols offered the prefrontal cortex a kind of mental Lego. With them, it could mentally manipulate inner cues to do things other than delay. It could fit together the most useful representations and build on them. Indeed, the prefrontal cortex, with its inner cues, graduated from managing representations to managing ideas—thinking.

Symbols offered Kanzi's mind a new, smarter ability to focus on the world. An ordinary chimp faced with someone using a tool might see just a person using a tool, but Kanzi could break down what he had seen. Using symbols for a person, a tool, and doing something, he could reorganize his perception of what he saw. He could focus on the tool being used while ignoring irrelevant details such as the nature of his relationship with the person. Now he would see an interesting tool movement, one he could look at in terms of his own experience of doing things. He could link the movement with something not as yet part of the situation, his use of his hands. Then, when offered the tool to use, his mind would be ready to try to do what he had seen.

The Rise of Gifted Environments

If the chimp brain of Kanzi can learn symbols, so could those of Lucy and other human-apes. Here we have a big clue as to our origins: Kinship symbolism—the passport to cross between bands—did not need new brains to arise. An enriched environment sufficed. That changes radically how we view the story of our brains. It did not—at least to begin with—need evolution. It could have started with apes teaching themselves how to make better use of what they already had. They would then have begun doing the things that later would lead to changes in their brains.

Giving gifted environments this role in our evolution is not entirely speculative. Kanzi is not the only chimp whose brain has been enhanced. Wild chimps instruct their young and so stretch their brain's abilities. Chimps in the Taï forest hammer nuts open to get at the rich kernels inside. They demonstrate a skilled use of stones or branches to pound whole nuts in "anvils," roots with small hollows into which they can be fixed. It takes each chimp about 4 to 10 years to master this skill. It does not arise spontaneously. Mothers must start off the learning process in their children by teaching them the first steps. Thus they offer them a gifted environment, albeit a very limited one.

Consider the skill of Salomé as she helped her young son Sartre, as observed by Christophe Boesch and Hedwige Boesch-Achermann . Salomé and Sartre were breaking a hard nut containing three separate inner kernels. To get at each kernel without smashing it, the nut had to be repositioned exactly right for each hammer blow. Sartre grasped his mother's stone hammer. "After successfully opening a nut Sartre replaced it haphazardly on the anvil in order to try to gain access to the second kernel. But before he could strike it, Salomé took the piece of nut in hand, cleaned the anvil, and replaced the piece carefully in the correct position. Then, with Salomé observing him, he successfully opened it and ate the second kernel." Some other mothers go further. One, called Ricci, was seen in front of her daughter Nina, to have, "in a very deliberate manner, slowly rotated the hammer into its position for efficiently pounding the nut. As if to emphasize the meaning of this movement, it took her a full minute to perform this simple rotation." She then broke a few nuts, giving their kernels to her daughter. Afterward, Nina took hold of the hammer and "always held the hammer in the same position as had her mother and never changed her grip nor the position of the hammer" (though she adjusted her own position and that of the nut). Ricci left her, and she broke four nuts on her own.[21]

Sartre and Nina have a gifted learning environment—at least for the skill of hammering nuts open. For chimps, breaking open nuts is a skill that must be taught. It is worthwhile for a mother to aid her children. She cannot let her children try to learn it just by watching her, or they will get frustrated and stop her. So Kanzi is not unique, as might at first be imagined, as we've seen that gifted environments can arise in the wild.

Could such gifted environments have created us? Here we meet an intellectual roadblock. We know a little about the rise of larger brains. We have fossil skulls and therefore can make estimates of the size of the brains that once lived in them. And in Darwin's theory of natural selection we have a means of modeling their evolution. But we know nearly nothing about what might shape the rise of gifted environments. Processes, like the ones by which Lucy-like apes taught their children, leave no fossils. And even if we could travel back in a time machine, we lack any means of modeling how such teaching might itself have evolved or interacted with culture to change the evolution of our brain. We are in virgin territory, but we know there is an important story here. Kanzi shows that having brains is not enough: Gifted environments are needed to exploit their potential. So how could learning brains have gotten started? What laws helped them sustain their advances, and develop?

These are fresh and, at present, unanswerable questions. There might be an evolutionary science for our genes, but there is not one for our learning environments. We can, however, try to sketch out some possibilities.

One question that arises is whether Kanzi and Mulika could provide a gifted environment for their own young. Several things might happen.

First, Kanzi and Mulika might create a stimulating learning environment for their chimp children, and when their children grow up they might, in turn, give their children a similar gifted environment. Washoe was seen to teach her adopted son, Loulis, to sign. As Jane Goodall comments, "It is as though because Washoe herself was taught, so she is able to teach."[22] Likewise, Kanzi has been seen to aid a second sister of his, Tamuli. She is being taught by her trainer to understand speech. Kanzi, however, does not agree that her trainer knows the right approach and acts out her requests to make them clear. Thus, after her trainer tells Tamuli to put her arm around Kanzi, "he takes her hand in his, and places it under his chin, squeezing it between his chin and chest. In this position he stares into Tamuli's eyes with what looks like a questioning gaze. When Kanzi repeats these actions, the young female rests her fingers on his chest as if hesitating over what to do."[23] Here Kanzi shows just the kind of kindness and sensitivity to another's limited understanding that is needed to teach them language. The gifted environment given to Kanzi and Mulika might therefore prove self-reproducing.

On the other hand, Kanzi and Mulika might not be able to re-create for their own children the gifted prefrontal stimulating environment given

to them. It may be that Kanzi and Mulika would turn out to be less stimulating to their young chimps than humans with PhDs in animal behavior
had been to them. As a result, their children would learn less and give their
children an even less rich learning environment. The environment would
go downhill, weakening with each new generation. It might eventually bottom out at some self-sustaining low level, or it might eventually die out
completely.

But there is a third possibility. In some circumstances the chimps might
be able to pass on the gifted environment, while in others they might not.
A gifted environment represents a major investment in time and patience.
We can easily imagine why a chimp parent might not be able to afford to
create one. One reason is lack of help. Chimps are brought up by their
mothers. While they receive some assistance from males, it is neither enough
nor of the right kind. A lone mother would not be able to teach her young
very much if she lived in a world where she had to spend most of her time
maintaining social alliances or foraging for food. Neither would she have the
patience to help them learn if she were worried and pressured. Researchers
find that the presence of stress in a mother when she forages for food can
affect her infant's later behavior. As adults, such offspring, when themselves
stressed, are both more submissive and more aggressive than those raised
by unstressed mothers.[24] Indeed, there is evidence that their neurotransmitter systems are permanently affected, making them more anxious and
depressed when under stress.[25] If stress has this effect, then it could easily be
causing other problems, some of which might affect learning. Thus, the single chimp mother might have been too stressed to aid the development of
her child's potential as she might have wished.

The difference between a gifted environment that was sustained
through generations and one that was not might be as simple as the availability of help, time, and freedom from stress. What shapes these? The rise of
monogamy might be one thing, but we cannot assume its existence early on
in our evolution. We can, however, push the question back into an area of
research that fortunately gives some clues, the interaction of habitat and
behavior. We need to take a closer look at how chimps live.

Ecological Variety, Chimp Variability

Habitat shapes how chimps live socially together, especially if it is easy for
them to feed together. This, for instance, underlies the differences between
bonobos and common chimps. Bonobos tend to live off a stable supply of
fruit, while common chimpanzees have to live off more fluctuating and inse-

cure ones. This lets bonobos forage in larger groups,[26] while the common chimpanzee, living in a more marginal and food-poor environment, cannot.

These differences in foraging styles are reflected in the way females bond and so underlie the social dynamic of chimp troops.[27] Foraging together gives females the opportunities to form alliances with each other. With them, female bonobos can assert themselves against males.[28] In consequence, males form weaker alliances among themselves. As a result, bonobo bands lack the male-centered dominance hierarchies found in common chimpanzees. Bonobo females, unlike common chimps, tend to have more equal relationships with males. That has an important effect: They are less socially stressed. Thus, something that could easily be overlooked in the environment—the opportunities it gives females to forage together—can change something as apparently distant from it as the power balance between the sexes. That could affect the possibility of their sustaining a gifted environment.

But a reduction in stress is not the only advantage female bonobos gain. Another is that they receive much more help from males. Because males do not bond so strongly with each other, they have increased ties with their own mothers and other females. Further, bonobos often consort, that is, two go off together as a pair to mate. While this happens with other chimpanzees as well, such relationships do not last among them, because the male dominance hierarchy breaks them up. The formation of more stable, less violent relationships between the sexes results in females' being more willing and ready for sex. Common chimps mate only during estrus, but bonobos do so whether in estrus or not. This keeps a male by a female's side, where he can be of assistance in raising their young. Thus, the opportunities an environment offers for collective feeding cast a long and distant shadow, affecting the opportunities for apes to support the development of their young.

Here is a key with which we could unlock our origins, but it is one we may lose, if we have not already. The human species is undergoing a population explosion, and wild chimpanzees are rapidly being squeezed out of their habitats. Those studying animals complain that the wild animals have lost genetic variation, but an equal complaint is that we have lost the natural variation given them by the worlds in which they live. Chimpanzees as we find them now exist in a limited number of habitats—the ones we have not yet taken from them. They are mostly marginal places, since the most productive land has long since been taken for agriculture. Without the ability to study chimps in widely differing habitats, we will never learn the subtle ways in which habitats can shape their lives, and thus we may never gain the clues we need to understand the lives of our first ancestors. Consider how different our ideas of Lucy would be if we knew only the Gombe chimps and had never studed those at Taï.

The chimpanzees studied by Jane Goodall at Gombe individually hunt small animals such as young colobus monkeys, bush pigs, bushbuck fawns, young baboons, and even, in one case, a 6-year-old *Homo sapiens* boy (scarred for life on the face).[29] But they do not hunt in organized groups.

In the 1980s in the Taï rainforest on the Ivory Coast, Hedwige and Christopher Boesch, a Swiss wife-and-husband team, found a band of chimps that organize their hunts. What is most remarkable is that these social hunters share their kills in a manner that ignores the social position of the hunters in their dominance hierarchy.[30] They have created special rules by which they divide the spoils only among those who have cooperated together on the hunt. Not only this, but after the kill, they eat their prey with tools. At Gombe, chimps have been seen to use sticks to "fish" for ants and termites; those at Taï use them to extract marrow from the bones of animals. When they can't get at their prey's brains, they hit the skulls against trees to break them.

The Taï chimps form female coalitions with nonblood band members that parallel those of bonobos.[31] Their rules for sharing killed prey are followed irrespective of sex. As Christopher Boesch observes, "The sharing rules prevailing in Taï have allowed adult females to reach a surprisingly high status during the meat-eating episodes, with the higher-ranking females gaining more meat than most of the dominant males."[32] He further speculates that this might have changed the role of females in other areas of social life. If so, here we have a factor pushing their lives in a direction that could increase the chances of their creating gifted environments. It is a small thing that looks as if it could have a big impact on their lives.

We know little about how this sharing arises in each group of Taï chimps. Christopher Boesch suggests it comes from their habitat, which is less open than the forest at Gombe. (High trees presumably would make it difficult for them to hunt alone.) But the rules could also be something they have learned to pass on from one generation to another, a result of behavioral culture. We do not know. But we do know that chimps display behavioral culture in other areas of their lives.

In each location in Africa where chimps live, they do things in their own way. In Gombe they use tools to fish for ants and termites, but 105 miles away at Mahale they fish only for termites, and at Kibale they do not fish for either. Likewise, chimps at Mahale use a different form of greeting, involving a hand-clasp, from that used by those at Gombe.[33] Even their "pant-hoots" differ acoustically in different areas, suggesting they have different "dialects."[34] The environment is not creating this diversity; the chimps are copying these norms from each other and spreading them directly across the generations.

Such learned behaviors can even be passed between species. Rhesus monkeys, which do not readily reconcile after conflicts, learned to do so when brought up as infants with stump-tailed macaques. When put back with members of their own species, they made up with their opponents three times as often as other rhesus monkeys after fights.[35] If reconciliation can be propagated, what about rules for sharing? If many bands of brainy apes compete among each other, wouldn't variant moralities arise and compete? Such questions are researchable and need to be studied experimentally and with more observation of wild chimps.

The most pertinent question is, of course, could chimplike apes have supported females' efforts to give their young gifted environments? We know that the cooperation needed for hunting at Taï changes how males relate to females and, more interestingly, that these females are also the only wild chimps known to teach their young. If this happens today at Taï, can we rule out the possibility that it also happened millions of years ago to Lucy-like chimps on the African savannah? It would seem arbitrary to assume that it couldn't have.

But we can go further. Very few chimp bands—probably only half a dozen—have been studied to the depth that the Taï have. It would be remarkable if they were unique. If we could study every band, not only those surviving but all bands that have existed in habitats from which they are now extinct, we no doubt would find that other bands had evolved organized hunting, sharing rules, and tutoring, in a great variety of ways. The behavior we find at Taï is but one sample taken at random from a distribution of lifestyles and adaptations of which thousands of variations must exist or have once existed. Some of these would no doubt be less suited for propagating gifted environments, but it is equally likely that some would be better. If one out of the six bands of chimps being studied today demonstrates tutoring and group hunting, what remarkable behavior might we have found if we could have studied the several thousand that once existed? Ditto for the distant past in which Lucy lived, except here we have the added element of untold variations of Lucy-chimp types. We must remember that neither bonobo nor common chimpanzee fossil bones survive from the past: What we have studied are chance survivals. In the past there must have been a vast variety of Lucy-like apes across Africa, subject to a great many more behaviorally creative influences than those shaping Taï chimps.

One would imagine, for instance, that scavenging in some niches might have increased cooperation and with it, as at Taï, the sharing of meat. Could this have led Lucy's cousins, indeed even herself, to live in equality with males as communal killers? If so, could this have let them support the growth of the minds of their children? After all, it would have put mothers in a strong position to focus their efforts on teaching them. And they would have skills to teach—how to scavenge and hunt. Many paths would have been tried, so one could have well existed that let Lucy start giving her children a self-sustaining gifted environment.

But there is a problem. Take the example of Taï nut cracking: Many groups of chimps existing elsewhere in Africa could benefit from it but do not. They live surrounded by nuts that could give them a rich source of food, but they ignore them. The problem is that nut cracking is a self-sustaining skill. Where it has not arisen, it simply does not occur. Something has to start it off.

The same problem exists with Kanzi's ability to make kinship symbols and speak. People with PhDs who are eager to teach chimps how to talk do not exist in the wild, and they never did. The skills they taught Kanzi had to come from somewhere, not just the gifted environment they offered his brain, but also the symbols they employed—or at least the seeds for them. Gifted environments are necessary but not sufficient. If self-sustaining stimulation of brains is a fire, a spark or a match first had to exist to ignite it. What flint might have been struck to ignite the original brain tinder?

Lucy's Words

Monkeys and chimps have a protolanguage. Many of their calls express simple ideas. Vervet monkeys use different alarm calls for leopards, eagles, and snakes. If one of them makes a leopard call, others run into the trees; if they hear an eagle call, they look up; and with a snake call, they look down.[36] Wild chimpanzees have a similar vocabulary. Jane Goodall lists 13 emotions in the Gombe chimps that are communicated through 34 types of call. They ranged from fear of strangers to "sociability feelings."[37] It is likely those 34 calls stand for 34 emotions, some of which, like those of the vervet monkeys, relate to specific events. The problem is that Goodall and her fellow researchers can only guess at what they might mean. However intimate they might be with chimps, they are not chimps themselves.

Hints exist, however, that some chimps can be smart communicators. In the Taï rainforest, the Boesches found that the parties making up a chimp band organized themselves as a larger group by using signals. Their food is spread too thinly for the chimps to forage together as one large group, but they must stay fairly close to each other in case of attack by leopards. Other primates could do this by constantly making noise, but the Taï chimps must keep silent most of the time so as not to attract the leopards. Thus, to keep together, they signal where they are to each other every 10 minutes by pant-hooting and then drumming on a tree. The pant-hoots identify the signaling chimp, the drumming his location. One chimp, named Brutus, makes special signals from time to time, since as the band's alpha male he must lead the separate parties. He keeps their movements tied together by telling them when to change direction and when to take a rest. He signifies direction by pant-hooting and then drumming one tree and then another in the new direction they are to go in. To signal a rest period, he pant-hoots and then drums one tree more than once within 2 minutes. The Boesches suspected that the more times the same tree was drummed the longer the break the chimps were to take. But they never had the opportunity to follow up their observations. They started studying the chimps in 1979. At the start of 1984, most of the males in the group they were studying were killed by poachers. Brutus, with his signals, could protect them from leopards, but not from human predators.[38]

Many animals have ways of communicating with each other, but the fact that monkeys and apes use many kinds of specific calls suggests that they have some capacity for more sophisticated communication. So the question is, why did only one ape—the human ape—turn communication into language? There are two views. First, it might be that there is something uniquely difficult about higher language skills and that only humans worked out a way to break through that barrier. A second view is that while language skills might be easy to learn and use, no other group of primates has ever had any use for them, and so they never acquired them. According to this view, humans developed language simply because we were the only ape that needed to. These two views look at the same situation but see different things. The first attributes the rise of language to overcoming difficulties in making, learning, and using language. The second attributes lack of language to the rarity of situations that prompt it to arise. Most people who are trying to answer how language arose take the first point of view. But are they right?

Kanzi's case suggests we need to look seriously at the second viewpoint. Chimps communicate in the wild, but not much. It is not because their brains lack the potential for simple language. If they did, then Kanzi's brain should not have been able to learn it. The explanation must therefore lie in circumstances. Wild chimps do not talk to each other because they can get on nicely without using their language potentials. They do not have 34 calls because their minds are limited to that number. They have 34 calls because they can get by as well using 34 as they could with 134 or more.

Let's stop here and pull some ideas together. We have seen that chimps can use symbols and, in a certain context, language. It does not require a new brain, only a gifted environment. Whether such an environment can be self-sustaining depends in subtle ways on male-female relationships and how much help, time, and freedom from stress a mother enjoys. We now know that habitat, and probably culture in the form of hunting cooperation and sharing rules, shapes these relationships. And if they shape them today in the dense Taï rainforest, then they almost certainly did the same when Lucy-apes hunted and scavenged on the savannah. But modern chimps in the wild stop here. They use symbols and teach them to their young, but they have not developed them into language because they lack the problems that would make it advantageous for them to do so. We can surmise, however, that somewhere in the distant past, Lucy-like apes with no more abilities than modern chimps could learn language, at least in the right circumstances—if they had the need.

We should therefore search for the origins of language in a new area. It is now the search for the problems that might have caused chimplike brains to start using symbols. We have, if we think about it, already seen some. Cooperative hunting and scavenging with rudimentary language would enable better planning and organization. Even simple things like the ability to signal the location of prey would seem advantageous, as would warnings of predators. Symbols might even be useful when doling out meat. Suppose one chimp cannot be around when the kill is divided up because it has to do something else—say, act as lookout up in a tree. It would benefit the sentry chimp if one of its friends could point to a leg and say it belongs to Luke who will be coming later. If that cannot be done, it is unlikely Luke or anyone else would be willing to be a lookout, however important that task might be for the safety of the whole group.

The First Names

Cooperation highlights the need for kinship names. You do not need full-blown language before you have naming. It begins with personal identities, names that can link kinship ties. All humans have naming systems that encode such information. Many wives are known by their husbands' names, so that we know that a Mrs. John Doe is married to John Doe. Surnames are often "sire names"; Johnson, for example, means son of John. In some societies, everyone has such patronymic names, containing the name of their father. Ancient Classical Athenian names went further and also included the local "deme," the village unit, in which a person was born and registered.[39] To stretch an example, we can extend the name of Kanzi to "Kanzi son of Bosondjo of Yerkes Regional Primate Research Center."

How might such kin name symbols have arisen? Wild chimps do not need personal names, as personal looks and smell are sufficient for them to identify each other. Why and how might apes have learned names when they were otherwise well provided with means of identifying each other?

The Taï chimps' pant-hoots suggest a possible way. The males in each party signal their location by a pant-hoot followed by drumming against a tree.[40] The pant-hoot is a simple ID—it identifies who is drumming. The Boesches were able to identify individuals from the pant-hoot's intonation and vocal quality. The interesting thing is that the chimps using them did not seem to be born with these identifying calls. They learned them as IDs, and if one chimp ceased using one ID, another would take it over. This happened when poachers killed most of the males. Within 3 weeks a young male, Macho, changed his pant-hoot to that used formerly by Le Chinois, one of the murdered males![41] The Boesches suggested he did this because certain patterns of vocalization were better than others for being heard distinctly. This simple observation may be an important clue for understanding how language arose. All social animals have indicators by which others can determine who they are, but these are usually physically fixed to the individual they identify. One chimp cannot imitate the smell or look of another. But that is what pant-hoot IDs can do. They could let apes copy each other's signals. Macho had to do so to take over another chimp's ID. But a chimp that can take over another's ID is just one step from doing something else—using it as a name, a word to refer not only to "me" but also to "you." With IDs, the Taï chimps can do more: They add a message to their IDs. It might merely be "Brutus says rest" or "Brutus says go East," but it is the beginnings of a sentence. The Taï chimps are on the road to making language.

Brutus and Macho are one step from language. What they need to do is start pointing to make the first sentences. Chimps, even wild ones, engage in gesturing[42] and understand the meaning of finger-pointing. Point at something, and they look not at your finger (as other monkeys do) but to where it points. Indeed, chimps spontaneously use finger-pointing to indicate things.[43] Thus, if Taï chimps can make an ID and then add drumming on a tree to signal where they are, then they surely can take this further and make a pant-hoot ID and add a finger direction. They could invent the start of grammar. (Kanzi seems to have invented some of his own rules for putting signs together, suggesting that the beginnings of grammar are there in the chimp brain).[44] All the ingredients for language thus exist in the wild chimp; they just do not have a reason to come together. What stops them is success: Chimps can communicate what they want in most circumstances without investing in learning the sophisticated language skills of which they are at least potentially capable.

But suppose they did have a reason and started to experiment with language just a little. It would catapult them into a new situation. Imagine you buy a car to go to work. It changes your life, since having made that investment, you can do lots of other things for free, like visit distant relatives and take the kids to school or the park. The point is that now you will do those things, even if you would never have invested money in buying a car just to do them. It is the same with simple language skills. If chimps ever had a survival need to learn language to do one thing, it would prompt them do many other things with it as well. It would usher in a new way of life.

16

The Runaway Species

One and a half million years ago on rich floodplain grasslands in Africa an 11-year-old boy died. The rainy season floods washed his corpse face down into a swamp. His body decayed. Large herbivores trod over his remains. His teeth fell out into a hippo footprint. Under the water, catfish and turtles chewed his bones, before they were embalmed in the mud. But in August 1984, Kamlya Kimeu, a colleague of Richard Leakey, found part of his skull, thereby launching the biggest archaeological excavation ever of an early human skeleton. Five years later, over 50,000 cubic feet of soil had been removed from an area of 4500 square feet; much of the work was done by hand.[1] "Nariokotome boy" had been found.

If he had grown to adulthood, he would have stood around 6´1´´—probably taller than you. He was lightly built; he probably would weigh about 150 lb as an adult.[2] And he was bright—his skull held no chimp-sized brain. While only two-thirds the size of yours at about 2 lb, his brain was far bigger than would be predicted for a chimp of his body size. Without precluding what we say later in this chapter, his brain was a smart one. Although not a member of our species, which arose later, he was human, a *Homo erectus* human.

How did Lucy-like apes become humans like Nariokotome boy and us? Most people see the origins of our larger brains in our abilities to speak and do things with our hands. These, no doubt, are important. But to pick up the "social brain" thread of our earlier chapters, there was likely a bigger factor in the development of our brains—families.

As we suggested, the children of Lucy-like apes would have wanted to visit their families. When Lucy's daughter visited her mother and brothers in her old band, she would have taken her children along. In this small act she started a new ball rolling. Those she once would have left behind now could get to know as she bonded with a new set of kin. New relationships arose. Grandparents would start showing interest in their grandchildren. Uncles and aunts (if they had not left) would meet and know their young nieces and nephews. Each small journey back home was the beginning of a great voyage, the symbol-spurred human odyssey. These seemingly minor trips were the beginning of families' extending across the savannah into two or more bands.

In time, the new kin would have found ways to visit each other, not just when young but throughout their lives. As kin, they would matter to each other. Here were the beginnings of bonds lasting into old age. Extended families had existed before in some monkeys and apes, but not like this. With humans, the landscape of Africa became crisscrossed with large extended families. Children found themselves born into a tangled web of kinship networking over the savannah. Crossband kinship was cemented with the simple symbolism of names.

The beginnings were simple, perhaps no more than occasional meetings and simple rituals affirming kinship, not the formal, ceremonial links made by modern humans between communities or even the sophisticated contact that arose later in the Upper Paleolithic, 40,000 years ago, as evidenced by their surviving traded and stylized materials. Though uncomplicated, these early meetings laid the groundwork for culture by forging crossband links and keeping them alive.

We need not think far to imagine circumstances in which having kin in the next band might be useful. A local drought or flood might force members of one band to move temporarily into another's territory. Or individuals who lost out in the politics of their band and were deposed might need a short-term refuge.

And what would happen if two parties came across the same freshly dead animal while out scavenging? If chimpanzees on four legs have territorial fights, then it would follow that groups of two-legged ones would also have them. When chimps fight other chimps, they use the same skills with which they hunt.[3] While scavenging they would likely have carried stone tools for butchery, so they would be armed. A small conflict could easily end in bloodshed. But they had a way out. If there were kinship links between the two groups, then there would be peacemakers, kin acting as intermediaries to avert conflict and find ways to cooperate. Perhaps, when the members of a band scavenged on the margins of their territory, they made sure an individual with relations in a neighboring band went with them. If conflict arose, that individual would be able to calm things down. That would not only aid the survival of the two bands but give prestige to those able to use their kin this way. Having relatives in the next band would be a plus for individual as well as group survival.

Social life would not remain unchanged in these circumstances. If relatives used symbolism, so, no doubt, would nonkin when making alliances. With symbols, non–blood band members could learn how to have the closeness of kin. A smart human-ape would now be one that could better use and manipulate those bonds. Ape brains, as we have suggested, arose as much to survive the politics of social groups as to survive nature. Successful apes had been those best able to form alliances and so win and keep dominance in their groups. Here now were new opportunities for success. It would still be important to work relationships to one's own advantage, but there would now be a more complex social world in which to use such skills. The focus of social success thus broadened from navigating relationships within a band to practicing diplomacy across bands.

Who knows? Some beginnings are so hard to imagine that we will perhaps never understand how they got off the ground. But that does not stop us from noting that, however started, they would reinforce and magnify themselves. A slight ability to communicate kinship would select apes better able to use it. In time the ability would refine itself and become more of a skill. Evolution would have made a force that would now drive the selection of brains best able to symbolize, sustain, and communicate kinship.

This did not take a miracle, only an enrichment of the mental skills apes already had. The ingredients were there. What was needed were the circumstances conducive to putting them together. The changing African landscape was creating the circumstances, and Lucy-like apes would have adapted and taken advantage of them.

Like modern chimpanzees, our ancestors would have lived in a wide variety of habitats that shaped the ways they foraged and hunted, and in so doing their social relationships. If they had not already done so 2 to 3 million years ago, Lucy-like apes now evolved into several kinds of long-legged standing and walking chimps. Many chimplike social lifestyles would have existed. Some of them would have been dead ends, but evolution had hundreds of thousands of years and many habitats across Africa in which to explore them. Nature, in effect, was fiddling with the numbers on a combination lock. Eventually it hit upon the right formula, and one group of apes started to take advantage of their hidden potential for symbolism. It could be that there were many attempts before one became self-sustaining. It would not have needed to be a once-and-for-all innovation. Though we talk of the rise of symbolism as if it were a single jump, it could have been made in many smaller steps, the breaking not of one combination lock but several. The odyssey to becoming us thus may have had many short breaks, but over tens and hundred of thousands of years it carried on.

Symbols would do more than extend the families of these early evolutionary pioneers. They would extend their mental ability to survive nature. A brain better able to plan cooperation would also be better able to plan how best to scavenge recent kills and make tools. Put to different uses, symbolic skills would no doubt have given rise to other symbolic skills. Now able to organize themselves and exploit nature, Lucy-like apes would become increasingly human giving rise to *Homo habilis*—"handy man." They would venture into a greater variety of habitats, which would lead in turn to further changes.

So whatever the story was in its earliest stages, once begun, the process of change would have been self-accelerating. Symbolic culture and brains would have started to coevolve with each other, driving each other's development.

Evolution Accelerates

Then in a short time, around 1.8 million years ago, brain size exploded. Handy man had evolved into *Homo erectus*, erect humans like Nariokotome boy. Ape brains grew to be two-thirds the size of modern humans'. Early humans increased in height until they were taller than most modern men. Whereas the australopithecine Lucy, a very early "human," had been much smaller than her brothers, now the two sexes became roughly the same size. With this new species of human we find a new sophistication of tools. The early humans changed their behavior, and some of them moved out of Africa into Asia. Indeed, remains of erect humans in Java, Indonesia, predate those of Nariokotome boy by 300,000 years at 1.8 million years ago.[4] That date equals that of any remains found in Africa. Humans, therefore, at this stage, might have evolved in Asia. There is no reason why not. Remember that time has washed away nearly all remains of our early ancestors. The story we fit together depends on those fragments luck has thrown our way. When looking back, we should respect the complexity and subtlety in the millions of lives that vanished without a trace. Today, just one new unexpected find from one of those lives could change our entire view of human evolution.[5]

Wherever it occurred, the jump from *Homo habilis* to upright humans was swift and sweeping. Had an evolutionary foot been put down, as it were, on the accelerator of brain selection? Had the rules selecting apes been altered? Did people, some 1.8 million years ago, somehow wrest evolution from the hands of nature and take it into their own?

There are both direct and indirect ways by which sharper minds could have been selected. Some apes, after all, would be better at using symbolic skills than others—the ones with brains more skilled at remembering or at using mental tricks to recall who were and who were not their kin. These abilities could lead to associative mating (those with like genes mating together), creating a positive feedback effect on their own selection.[6]

Consider language. Apes as we find them today have some abilities to communicate. Early hominids would have at least these skills if not better ones. Such hominid communicators would have sought out other good communicators for company. Maybe a smart Lucy found herself spending most of her time with an equally smart Luke, and they chose to bond. Did the males and females with the sharpest symbolic brains, by the social effects of language, find themselves picking each other out as partners?

Or maybe selection was more indirect. Brains that can communicate even to a limited degree would be socially advantaged, the one-eyed kings in the land of the blind. Hearing and speaking, moreover, are not comparable skills. There is an asymmetry between the passive and active roles in communication. It is far harder to talk than to listen. Few write like Shakespeare, though most of us can gain pleasure from the subtleties of his words. It is an asymmetry with a big effect. Clear communication is a skill that aids social manipulation, working out what someone said less so. Humans, apes, and human-apes that could make themselves understood would be skilled in building up and forming alliances and social contacts. Those able only to understand would be their passive partners. Females picking successful talkers would be selecting sharper brains.

This leads us to the fact that Charles Darwin created not one but two theories of evolution: natural and sexual selection.[7] In natural selection, it is nature that winnows the survival of animals; in sexual selection, it is done by the animals themselves.

Sexual Selection

Darwin was led to propose sexual selection because not every trait in nature aids the survival of animals. For instance, the fantastic-sized tail displays of male peacocks and birds of paradise would seem not to aid survival but to make it harder. While females in these bird species have a dull appearance, males stand out so much they should attract predators. Darwin puzzled over this, arguing that outlandish male traits might be caused by animals' influencing their own selection. "The courtship of animals is by no means so simple and short an affair as might be thought. The females are most excited by, or prefer pairing with, the most ornamented males, or those which are the best songsters, or play the best antics."[8] He argued that when we look at peacocks, what we see is the selection of cocks by peahens: "males which [are] decked in the most elegant and novel manner would have gained an advantage, not in the ordinary struggle for life, but in rivalry with other males, and would consequently have left a larger number of offspring to inherit their newly-acquired beauty."[9]

Female birds need to be sexually aroused by males of their own species; one of the ways this happens is by their instinctively spotting the small plumage differences that identify their own kind. But such markings not only differentiate males of their own species from those of others, they also signify "maleness" to the female. Thus, the hen will be sensitive to differences among males of her own species and be attracted to those that stand out as more "malelike." When this happens, such males will be favored as mates. Over time, this selection can cause plumage to become exaggerated.

Sexual selection might have had a functional advantage unknown to Darwin. Plumage is a sensitive marker for health. As much as an ill person looks ill, so does an ill bird. Gaudy plumage is like litmus paper showing up the inner ability of a bird to fend off parasites. Colorful display feathering is expensive but optional—if parasites degrade a bird's health, resources are diverted from it, making the handicap of parasite infestation visible. Females that mate with males resistant to parasites are at an advantage over those that do not, and a peahen that selects cocks with the gaudiest plumage will be choosing parasite-resilient genes for her chicks.[10] Indeed, it has been shown that the offspring of peacocks with the most elaborate trains and biggest eye-spots weigh more and live longer.[11] Thus, sexual selection will select for cocks with even gaudier feathering. And this applies not only to peacocks. Male collared flycatchers, for instance, have a white forehead patch sensitive to a bird's condition. The offspring of flycatchers with larger white patches are healthier than those of males with smaller ones.[12]

Charles Darwin logically suggested sexual selection might have happened with people.[13] The human race varies widely in its looks. Like birds, people go for certain looks in their partners. "We see that with savages the women are not in quite so abject a state in relation to marriage as has often been supposed. They can tempt the men whom they prefer, and can sometimes reject those whom they dislike, either before or after marriage. Preference on the part of the women, steadily acting in any one direction, would ultimately affect the character of the tribe; for the women would generally choose . . . the handsomer men, according to their standard of taste."[14] And, of course, the same would happen with men choosing women by their looks. Once a small preference established itself, it would take over as a cultural notion of beauty. As such it would confer an advantage on those lucky enough to have it and so, over time, shift people's appearance as certain looks become favored in the mating game. Different cultural notions of beauty in this way would have led to the rich variety of looks— skin color, hair distribution, and facial shapes—that enriches our species. (Sexual selection may have magnified skin color differences selected earlier by nature and linked to UV radiation.) People's looks, like plumage in birds, also reflect underlying health. Ill people not only look ill but also physically attract us less than those in the bloom of fitness. Those studying what makes the faces of people attractive link beauty to signs of good health and future fertility.[15]

But was it only our looks that were affected by sexual selection?

Sociosexual Darwinism

Beauty may be only skin deep, but could not our brains and their abilities also have been the subjects of sexual selection? When human apes started bonding as parents, they created a new kind of advantage. In the game of getting your genes into the next generation, it helps if you pick a mate able to succeed in life. Previously, that advantage had been hidden by promiscuity, but now with parent bonding, our human-ape ancestors, in passing on their genes, were not playing a lottery so much as making an investment. Having choices, they could make better or worse ones. After all, when an animal selects a mate, especially if they bond to bring up children, they make a decision that will affect the numbers and success of their offspring. Some of your potential mates will have genes that make your children poor survivors, others genes that will make them winners. Some partners will be good parents, others not. A gene that lets you spot the difference will itself survive, as it will aid its own survival by linking itself with such winning genes. But what sorts of genes would best enable the spotting of such mates?

We have already met one possible kind: genes for bigger prefrontal cortices that aid a person's symbolic skills. The bigger this region of the brain, the greater an animal's ability to control its brain through symbols and language. That will give it more skills to let it spot similar abilities in potential mates. Further, successful selection also involves waiting for a good partner rather than going for the first available mate. Just as abstract thought allows children to forgo immediate gratification in the marshmallow-munching area, so those with larger prefrontal cortices would be better at not falling for the first potential mate who comes along.

Females, according to Darwin, may select "those who were at the same time best able to defend and support them."[16] Did males and females go for the mate who showed the best skills at making alliances? Social skills may, by their nature, lead those good at making them to meeting each other and so to associative mating. So sexual selection may have amplified prior selections made by humans themselves.

Sexual selection is interesting because it powers itself and so, unlike natural selection, can "run away" in a self-reinforcing fashion.[17] Genes for picking males are inherited. If the children of selected mates make up a higher proportion of future generations, the genes responsible for selecting them also proliferate. Thus there will be more such selection. This pushes the selection of such genes further. With bird plumage, the need to escape from predators usually puts a brake on the selection of too-unwieldy feathering. But not so in places such as New Guinea, where few predators exist. Not surprisingly, that is also where birds of paradise live. But something different would have happened with human brains. Once the selection of smart brains for processing symbols had begun, nothing would have stopped it. The prefrontal cortex would have been dramatically pumped up in size, because unlike bird feathers, sharper brains would aid escape from predators, not hinder it.

In suggesting these ways by which such selection could increase at runaway rates in human-apes, we are not suggesting that any single one was responsible for creating *Homo erectus*. Our evolution is likely to have been shaped at different times by quite different kinds of sociosexual selection. The point is that if they existed—and the ingredients are there—these selection processes would have been powerful shapers of our evolution.

The nature of symbols and language, moreover, would have made their selection even sharper and more intense. Symbols offer an opportunity for those best gifted in using them to show off. Humans display their "quality" as potential partners through what they do and say as much as birds do through showing off their feathers. We try to put on a good "show" not just for potential mates but also for future employers and others who might be useful to us. Dates, job interviews, and search committees are all institutions that give people an opportunity to make judgments before they select each other.

How might an early human judge a possible partner's prefrontal skills? Maybe an ability to tell interesting stories was appealing. Since storytelling requires a good memory and the ability to organize ideas and communicate them effectively, it would be a litmus test for these abilities, largely based in the prefrontal cortex, as well. Also, these skills correlate with other ones that determine an individual's later success in the social games of life and surviving off nature. So a gene for picking a good storyteller will pick a better-quality partner and so leave more descendants than average. The story-liking gene will itself start propagating and evolving in a feedback loop. Partners with superior storytelling abilities will be advantaged if they show off their skills, telling more and more riveting stories. If they get selected at a higher than average rate, this will lead to brains with even better abilities. This in turn will feed back into the sophistication of the stories told, thus pushing yet further the sophistication of language skills. The interesting thing is that once such a situation arose, it could be self-sustaining even if the language abilities being selected for were to cease to be of practical use. This is runaway selection. The genes for selecting language skills will not be deactivated but will ruthlessly carry on picking better language abilities.

Those best at selecting language skills in others would be those who also have some skills already. Better language would therefore cause an even more demanding selection of even better language, and the runaway process would have yet another loop amplifying it. Language has often been seen as an adaptation to surviving nature, but once ape-humans started selecting partners, adaptation would cease to be as powerful as people's ability to select themselves.

Such selection would have radically shaped the mental skills of our species. The process of selection itself would leave its mark in our interests and appreciation of what we can do with language. Once it had been selected, there would be no reason for joy in hearing stories to be deselected. Telling stories, making explanations, and being interested in the past would become part of our mental inheritance. We would be left with brains that wanted to hear about what went before us, such as our evolutionary odyssey.

Half-Brained People

It might seem like a heresy to suggest that sociosexual selection rather than natural selection forged our superbrains and language. It asks us to make what on the surface is the weirdest of analogies, that is, one between our big brains and the exotic plumage of some tropical birds. This is not the way those studying humans tend to look at our origins, but it makes a crucial prediction: that our evolution would have given us brains far in excess of what was needed for our ancestors to survive in their natural world. In other words, natural selection alone would have had no reason to push our brains so far. Those studying the evolution of our brains have overlooked a key fact: Brain surgeons have discovered that we don't really need these big brains to function.

You would think that cutting out one-half of people's brains would kill them, or at least leave them vegetables needing care for the rest of their lives. But it does not. Consider this striking story. A boy starts having seizures at 10 years of age when his right cerebral hemisphere atrophies. By the time he is 12, the left side of his body is paralyzed. When he is 19, surgeons decide to operate and remove the right side of his brain, as it is causing fits in his intact left one. You might think this would lower his IQ or leave him severely retarded, but no. His IQ shoots up 14 points, to 142! The mystery is not so great when you realize that the operation has gotten rid of the source of his fits, which had previously hampered his intelligence. When doctors saw him 15 years later, they described him as "having obtained a university diploma . . . [and now holding] a responsible administrative position with a local authority."[18] Nico, an Argentinian half-brained child is not only remarkable for "the richness of his vocabulary and syntax" but also in that to gain a second language, he "attends English classes at school, in which he attains a high level of success."[19] Another person in the United States with half a brain worked as an industrial executive (traffic controller) part time and majored in sociology and business administration at a "prominent midwestern university." The average man has 1100 cc of cerebral cortex, so if you had half your brain cut out you would be left with roughly 550 cc. One and half million years ago, Nariokotome boy had a skull whose cortex, when fully grown, would have been larger than that, at 673 cc.[20]

Apparently we are born with far more brains than we need.[21] If brain volume alone makes us who we are, then the last 2 million years of evolution was unnecessary. Apparently, early humans would have had enough cortex to work in the modern world as industrial executives and get university degrees. (It should be pointed out as well that the cases we have described of people with half their brains removed had already suffered a disadvantage, since they had started life handicapped with severe epilepsy.) Admittedly, these estimates use only back-of-the-envelope calculations and ignore the percentage of the brain that is prefrontal cortex. One cannot deny that our prefrontal cortex has increased greatly in size and that this might be more important than the total amount of cortex. Kanzi's prefrontal cortex was 43 cc to our 320 cc—less than one-eighth the size of ours. But how much prefrontal cortex is needed to live the life of a successful early hunter-gatherer—43 cc, 100 cc, or 320 cc? That modern people can live successful lives with only 160 cc (half a brain's worth) suggests that we could do without a good part of our prefrontal lobes.

Do you disbelieve this? Consider the case of Daniel Lyon, a watchman for 20 years at the Pennsylvania Railway Terminal in New York at the end of the nineteenth century. He could read and write, and according to legal representatives of the company that employed him, "there was nothing defective or peculiar about him, either mentally or physically." Nariokotome boy's brain would have weighed about 2 lb as an adult. But the reading and writing Daniel Lyon's weighed only about 1.5 lb![22] Upon examination, anatomists could find no difference between it and other human brains apart from its size, with one exception: The part of his brain attached to the brainstem, the cerebellum, was near normal size. Thus, the total size of Lyon's cerebral hemispheres was smaller than would be suggested by a total brain weight of 1.5 lb. We do not know how bright he was—being a watchman is not particularly intellectually demanding—but he clearly was not retarded. A pound and a half of brain may not be enough to manage a career as an attorney, a professor of theology, or a composer, but it was sufficient to let Lyon survive for 20 years in New York City.

And Lyon is not alone. It is a medical myth that microcephaly (having a head smaller than two standard deviations (SD) below average circumference) is invariably linked to retardation. In a group of 1006 school-aged children, Clifford Sells of the University of Washington School of Medicine, found that 19—nearly 2 percent—were microcephalics.[23] Of 12 whose IQs were measured, 7 had average or higher IQs for their age; indeed, one had an IQ of 129. Jay Giedd and his colleagues at the National Institute of Mental Health's Child Psychiatry Branch carefully screened 624 ordinary individuals with psychometric tests and a psychiatric interview. Of the 104 who had successfully completed MRI scans, one otherwise normal individual had a cerebral cortex volume of only 735 cc and a brain volume of 888 cc.[24] Of 188 children with microcephaly (of various origins, including Down's syndrome) studied by Edward Sassaman and Ann Zartler at the Child Development Center at Rhode Island Hospital, 60—39.1 percent—were not retarded though below average intelligence and 13—7 percent—were of average IQ.[25] Livia Rossi and her colleagues at the University Medical School of Milan made a genetic study covering 21 autosomal-dominant microcephalic adults and children in six families. Psychometric tests were given to 13 of them and (except for one individual) found them to have nonretarded IQs. One, a 29-year-old mother, referred to as C2, had a brain circumference 4.7 SD below normal (giving it a volume somewhere near 760 cc) and, in spite of this, an IQ of 112.[26] Several people refused to have their intelligence tested at all, because they resented the implication that, because of their small heads, they were not fully normal. The idea that small head size is linked to retardation is not only a medical myth but also an overlooked source of prejudice.

So how well equipped was *Homo erectus?* To throw some figures at you (calculations shown in the notes), easily well enough. Of Nariokotome boy's 673 cc of cortex, 164 cc would have been prefrontal cortex, roughly the same as in the half-brained people.[27] Nariokotome boy did not need the mental competence required by contemporary hunter-gatherers. Hunting has come a long way since his time. There has been a technological revolution between the earliest human hunter-gatherers and the !Kung as big as that between the !Kung and us. As noted in Chapter 2, the first bow and arrow are probably not more than 40,000 years old, if that. Stone hearths with air-intake ditches to cook meat date from around 60,000 years ago and the first controlled use of fire from 500,000 BC. The first lamps were made around 40,000 years ago (thus letting people go deep underground to paint the first surviving cave art). Such a simple thing as hunters preparing tools in advance seems (the evidence is controversial) to arise in the Upper Paleolithic (the later Old Stone Age) around 40,000 to 30,000 years ago. For the previous 2 million years, tools seem to have been usually made for immediate needs. Compared to that of our distant ancestors, Upper Paleolithic technology is high tech. And the organizational skills used in hunts greatly improved from 40,000 years ago to 20,000 years ago.[28] These skills, in terms of our species, are recent, occurring by some estimates in less than the last 1 percent of our 2.5-million-year existence as people. Before then, hunting skills would have required less brain power, as they were less mentally demanding. If you do not make detailed forward plans, then you do not need as much mental planning abilities as those who do. This suggests that the brains of *Homo erectus* did not arise for reasons of survival. For what they did, they could have gotten away with much smaller, Daniel Lyon–sized brains.

Big Heads

Not only is there no apparent need for our big brains, but they are evolutionary big trouble. Big brains must be housed in large braincases—our heads. And if there was anything our evolving ancestors did not want, it was that their heads would grow big.

For a start, big heads make birth long and difficult. The birth canal of modern mothers is only just big enough for a baby's head. Even then, the baby must be turned on its side half way through to squeeze out. A human baby's head is uniquely round, which helps it "roll" on the way out, and its head and body are able to rotate. The mother's pelvis may have a small opening relative to the size of her child's head, but it is shaped optimally to aid its head-first birth. (Breech delivery—head last—in most cases requires a cesarean section.) The baby's cranium is incomplete at birth. Fontanelles, or gaps between its bones, allow the head to mold to the shape of the pelvic opening. In fact, about one in six babies' heads end up getting noticeably changed in shape during delivery. The human ape is the only primate with such adaptations to aid birth delivery,[29] but it is still a difficult process; and human mothers can rarely give birth without assistance.

The problem, however, is not so much that human babies' heads are so large but that human mothers' pelvises must be so small. Our brains grow big mostly after birth, not before. Look at any picture of a newborn baby and his or her mother. Note the sizes of their heads. A newborn baby's head may be big for the birth canal, but it is still tiny—less than three-tenths the size of an adult's. Other apes are born with proportionately much bigger heads—slightly under half adult size.

But if a big-headed genie came out of the narrow aperture of a bottle to grant our ancestors one wish, it would be that their children's heads at birth would be even smaller. The problem is with bipedal running. Whereas most other primates and apes are built for tree climbing, we specialized in long-distance walking and running. That adaptation requires a narrow pelvis and thus a small birth canal, since the wider the pelvis, the less efficiently one can use one's legs.[30] Thus, the human infant's head is the biggest it can be and still be delivered by a mother able to run efficiently. Imagine what it would be like to run or walk with a pelvis wide enough to give birth to a baby with a head the size of a 1-year-old child's! As it is, the female pelvis handicaps women in athletics. World records in track events range from about 6.4 percent (100-meter dash) to 33 percent (pole vault) less for women than for men. This is not due to a general physical inferiority. Females compete on a par with males in athletic events that depend on upper body strength. Indeed, Gabriele Reinsch, the female discus record holder, has thrown 2.7 m farther than her male discus record rival.

Running is more important to our species than we may realize.[31] We take our legs for granted, but the stamina with which we can use them is a bit of a mystery. We are great long-distance runners; if need be, we can keep up a steady fast pace from dawn to dusk. Records in older editions of the *Guinness Book of Records* (before many of its interesting facts were removed to make it feel like less of a "textbook") show that people can run up to 188 mi (303 km) within 24 h and up to 621 mi (1000 km) in a single 136-h and 17-min stretch. And we can do this with heavy loads: Sedan chair bearers walked 30–35 mi (48–56 km) each day. Many non-Westerners, both men and women, are reported to carry "light" loads of 60 lbs (27.2 kg) for 25–40 mi (40–64 km) over rough landscapes. There is a report that one group of Chinese schoolgirls regularly walked 50 mi (80 km) daily.[32] In the manhunt for the outlaw (and Indian runner) Willie Boy in 1909, when horsemen came within 20 minutes of catching him he increased his stride into 5-foot paces. He carried on like this for 15 miles, and the horses were forced to stop and rest. He managed to outrun his pursuers for 500 miles.[33] In Mexico, Tarahumaran runners in kickball races covered 150–300 mi (240–480 km) over 1–2 days—more than six marathons. Imagine a chimp running even a few yards, and you can see how radically our legs have been turned into efficient long-distance running machines. No other primate—or any creature, for that matter—is so specialized for endurance.

If this is surprising to us, perhaps it is because most of us live such sedentary lives. Four in five Americans do less than the daily equivalent of half an hour of brisk walking, and one in four fail to do even the equivalent of half an hour's stroll.[34] We are dependent on our cars, not our feet. Not surprisingly, while our brains might be larger than would be expected for an animal of our size, our hearts are smaller—except for some regular marathon runners. Perhaps our relatively recent existence as nonexercisers blinds us to the power, efficiency, and uniqueness of our legs. We look back upon evolution as a story of increased brains, but then we are all acutely aware in our lives of the importance of sharp minds. This should not keep us from seeing that in the animal world, our running ability is unique, just as our mental capacities are.

Humans specialized in endurance running for a reason. To understand this, consider cheetahs and how they run. They are speed-sprinters that can outrun their prey in short dashes of 60 mph (100 km/h). But their physiology keeps them from doing it for long—just under 1 min and 0.6 mi (1 km). The maximal use of muscles increases their metabolic rate 50-fold. A human runner doing just over 5 minutes to the mile is burning energy at a rate of 1500 W (1.5 kJ/s). Movement takes up only 30 percent of that energy; the rest is turned into heat, which a cheetah's body has no effective way of removing as fast as it is made. That excess energy could kill it in the same way that an engine without a cooler will overheat if it is run for more than a few moments. If cheetahs kept running at 100 km/h they would cook themselves. Cheetahs avoid that fate by running only in short sprints, but that is not the only solution. Developing highly efficient means of getting rid of waste heat is another, but that requires a biological radiator. Though we do not take it as one of our most unusual features, we have turned the skin of our bodies into an efficient heat transfer system. Look at yourself—the human ape lost its fur and gained a cooling fluid. Every time you bathe, thank evolution for giving you a sweaty, naked body. Our sweat glands, when we are hot with exercise, secrete far more fluid than those of any other animal. A human marathon runner can sweat over 5 qt (5.5 L) in a run. Evaporating from bare skin, sweat can rid our bodies of 2450 kJ (583 kcal) of heat per liter and so keep a runner cool.[35] People doing heavy exercise may lose two and a half times as much fluid in sweat as in urine.

If you think about it, running for long periods is rather an odd skill. Other animals run to escape predators or to catch other animals. They must move quickly. What is emphatically not needed for them is a means of running for long durations over long distances. But that is the ability we have evolved. The answer may be that such a running ability could be of advantage to an intelligent, bipedal predator. Standing high on two legs, humans could see farther than their prey and, if need be, climb up trees. Thus, however fast the animal they pursued might run, it could not keep dashing out of sight when persistently trailed by a clever, upright hunter built with the endurance to fatigue it, literally, to death.

This is not speculation, at least for modern people. Though it may not be widely known, people worldwide hunt animals by running them down over 1 or 2 days. Anthropologists record that in Africa, the San (Bushmen) run down wildebeest and zebras. In Australia, aborigines used to run down kangaroos. In ancient Syria, around 400 BC, the Greek writer Xenophon recalls starving soldiers running down bustards for food.[36] In the American Southwest, it has been claimed, the Navajo ran down one of the fastest of all animals, the pronghorn antelope.[37] This animal can go 53.1 mph (88.5 km/h).

While Lucy's legs were not quite like ours, Nariokotome boy's were. *Homo erectus* fossils cannot tell us of the miles they ran and walked and why they did so, but we can tell from the structure of their leg bones that Nariokotome boy and his kind exercised their legs well. Human evolution could not foresee the advantages of having genes for competing in the New York and London marathons, so what were they for?

We cannot deny that early humans hunted. Taï chimps show us that some smart apes would already have been doing that. Further, modern people show that endurance running enables people to hunt even the fastest of animals. On what grounds can we exclude our earliest ancestors' doing likewise? We would have to assume that they failed to exploit an opportunity for which their bodies were specially adapted. This argument, of course, is not evidence. We will never know for sure whether early people used their legs to outrun prey. However, our bodies are evidence that something critical to survival caused us to evolve endurance running. The most likely story is that Nariokotome boy's people were hunter-runners who depended on their legs to chase down their prey.

Whatever use running had back then, it came at the cost of a narrower pelvis. Thus, our lower body and our enlarged cerebral hemispheres had been put on an evolutionary collision course, and our need to run should have won. Our brains thus arose not because of our legs but in spite of them. Evolution solved the big-brain/running conflict by delaying most of our brain growth until after birth to keep pelvises efficient for running. Thus, unlike other primates, our brains keep maturing for a year or more after birth. Our brains have a 21-month gestation, not a 9-month one.[38]

As noted, the first year of a human infant is one of total helplessness. After all, a human baby should really still be in the womb. It is so immature that its mother must constantly hold it and give it care. She is trapped into full-time caring for her newborn child to a degree not faced by any other primate mother, and she cannot be alone. She needs help and support from relatives and friends; such commitment was the price of big brains. Big brains would have arisen only in a situation in which a mother's extended family could "see" their own genes in her children and so be willing to support them. This is one reason for linking the rise of *Homo* so strongly with the above-discussed rise of family and symbolic kin networks. Only these would have given the needed help. And, of course, since these networks required recognition of male as well as female blood links, that would mean we must have been pair bonded and practicing sociosexual selection.

So those Lucy legs had to come before our brains. Some of the descendants of leggy apes would have been forced down the path of extreme infant dependency, pair bonding, and therefore sociosexual selection. That offered a tougher selection of brains than any made by nature[39]: the choices made by brains themselves. So it was not nature, as apart from humans, that selected the brains in your head. They were expanded by the sociosexual winnowing done by other brains. And our emerging language competence accelerated even more the runaway process.

This process did not stop with *Homo erectus*. About half a million years ago our species arose—*Homo sapiens*, the "wise humans." But they were not yet us. Our species' evolution continued, forming several subspecies. We are not quite like the first *Homo sapiens*. Such early humans had small bone differences from us, such as heavy brows and a small or absent chin, but they were large-brained. The best known of these extinct *Homo sapiens* sub-species, the Neanderthals, *Homo sapiens neanderthalensis*, may have had even larger brains than ours.[40] Only in the last 100,000 (perhaps 200,000) years did *Homo sapiens sapiens*, anatomically modern humans, arise. We are more lightly built, fully chinned, with smaller teeth. This subspecies of *Homo sapiens* includes all people alive today regardless of skin color or "race." As Darwin suggested, skin color and other human differences are due to appearance selection—it is a surface difference. Under such things we are all the same. The selection of our looks went in different directions, but our brains could only go in one.

But there is still a problem: Did we really arise with our brains? What if we could thaw an earlier human, say a Neanderthal child accidentally preserved in a glacier? Adopted into a modern family, would his mind be like ours? Or would there be something lacking? In other words, how far back did brains with the potential to be you or me arise? We do not know. For convenience, we suggest 100,000 years, the lower end of the 100,000–200,000 year range for the rise of anatomically modern people. But maybe the first *Homo sapiens*, some half a million years ago, had that potential. Or maybe it arose even further back—1 million or 2 million years ago, with erect and handy people. Kanzi seems to do remarkably well with a chimp-sized brain. And while we tend to link retardation with small brains, we have seen that people can live completely normal lives while missing pieces of their brains.[41] Brain size may enhance intelligence, but it seems we can get away without 3 pounders. Kanzi shows there is much potential in even 13 oz.

Maybe searching for who we are in our brains is leading us to some wrong questions. In the mirror, we look and see our big heads full of super-bright brains and attribute our intelligence to them. We think evolution evolved our brains for that intelligence. But there could be another story. Our brains might have evolved for purposes other than those for which we now use them. What we do with our brains now might be something added on to what had already evolved. If human evolution went into overdrive and pushed for unnecessarily big brains, the human species would have been left, like a peacock, with something of little practical use. But there is a differ-ence: A peacock tail does not lead to anything. But superbrains are flexible. Many generations after they arose, they might have done something novel. They may have changed what they did, and with that, our species. There could have been a second human evolution, not of its neural hardware but of what was done with it—its mindware. Selected for one thing in a billion days, humans may have found themselves on an odyssey in their last billion hours (roughly 100,000 years) to become a radically new kind of animal. As much as Kanzi needed a gifted environment to learn new skills, we suggest we needed one to become the people we are now.

It seems an unlikely story, until you remember that your ability to read this was never evolved. It is something the human species learned to do only a few thousand years ago. And this is true of most of the things you do. We now turn to the rise of the gifted human environment and the ascent of mindware.

17

The Billion-Hour Journey

The word *symbol* suggests simple images like flags or emblems, but the symbols that empower our brain stand for processes and operations, like +, ×, %, and ÷. What the brain learns in terms of symbols is complex. Its theoretical knowledge and technological know-how consist of whole symbolic systems. We need another word for this learning and what it gives our brains: *mindware*.

To get a feel for the nature of mindware, consider for a moment its computer analog: software. Software exists at many levels. A basic machine code directly links symbols and the operations of the computer's hardware. But this code is difficult to work with, so programmers use languages like BASIC, Pascal, or C++ to turn the hard-to-follow symbols of machine code into something with which people can easily write programs. But such computer languages are only the start. They are used to write our familiar applications, like databases, drawing packages, spreadsheets, word processors, flight simulators, and games. A further refinement is made as we customize and in other ways personalize our copies of computer software with the colors, keys, skins, and icons of our own choice. Such an ascent by symbols is powerful. Merely looking at the instructions written in the machine code, it would be hard to imagine them creating mock typewriters, filing cabinets, paintbrushes, and Cessna aircraft. But thanks to programmers, a small number of symbolic operations can become complex, multilevel worlds of skills and structures.

We see a similar richness in the alphabet. At its most basic level, it is a set of symbols for speech sounds. With its letters, you can write out the words of your language. Not just words but whole sentences, paragraphs, and things like dictionaries and thesauruses. And beyond them, anything from books of geometrical theorems, accounts, textbooks, novels, and what many take to be the word of God. And writing leads to systems of symbols—mathematics, music, and dance notation. All give rise to further complexity: From math notation came mathematical axioms and from them, theorems and branches of mathematical knowledge such as geometry, topology, linear algebra, and differential calculus. As with atoms, smaller things construct more complex ones giving rise to further ones. The possibilities of what the mind's software might consist of, and do, are vast and still only partly explored.

By manipulating symbols, the brain derives the power to carry out operations it otherwise could not. For instance, letters give each spoken phoneme a symbol. Learning this allows us to sound out words from their spelling as well as the reverse, figuring out how to spell a word from only its spoken pronunciation. Writing replaces speech, allowing it to do what it could not do by itself, that is, exist over time, be edited, and be reproduced in millions of copies.

Symbols do something else as well. Consider our hands and artifacts. Tools and instruments do not change our hands. We learn how to use them so we can, say, play the violin, but this does not change hand anatomy. Our hands remain much the same; our tendons and bones do not rebuild themselves, stretch, or lengthen to be better adapted for playing violins however much violinists might practice. But symbols change the brain; indeed, they do so more powerfully than do our genes. Here is the power of what Carl Sagan called "extrasomatic knowledge" and our "bargain with nature."

Consider the alphabet again. Learning letters changes how we hear. Most of us can judge that three sounds exist in the word "cat," or that if we swap the initial phonemes in the two names "Chuck" and "Berry" we get "Buck Cherry." But we are not born with these skills. Preschool children and illiterates find it hard to segment speech in terms of the phonemes needed to do these tasks. Rather than being innate, the ease with which we manipulate sounds comes from our having learned the alphabet.[1] Nonalphabetic reading is not enough. Chinese logographic readers find these tasks hard unless they have learned the Chinese equivalent of the alphabet, *pinyin*.

Not only is phoneme awareness changed when someone learns an alphabet, but people's skill in picking up speech errors also increases. Phonetic symbols allow the brain to focus closely on what people actually say, instead of inferring it.[2] This has nothing to do with reading or writing. Somehow merely learning the symbolism for phonemes changes how the brain processes the elementary units of speech, which suggests that symbols can dig deep down into the brain to change how it works. Indeed, functional imaging shows the literate brain to be organized differently from the illiterate one, at least in regard to speech.[3] The corpus callosum that connects the two hemispheres even appears to be thicker (at least in some parts) in individuals who can read.[4]

Violin playing is another example. While practice does not change violinists' hands, it does change their brains. Violinists have a larger corpus callosum, the link between the two sides of the brain, than nonviolinists, at least for the front part, which is concerned, among other things, with hand dexterity. Significantly, this finding is related to the experience of playing when young; this link is not larger in violinists who start after the age of 7.[5] Violinists also have two to three times the area of cortex devoted to their left fingers that nonviolinists have, again provided they started when young (in this case before 12).[6] The brains of keyboard players likewise expand (in this case, the area of the right primary motor cortex dealing with fingers), an expansion linked to increased dexterity in the left hand.[7]

These variations suggest a new avenue for studying the mind. Instead of viewing symbols merely as propagated arbitrary associations, we can now see them as active shapers of the very substrate by which we act, think, and feel. Symbols, working together with our prefrontal cortex and neural plasticity, transform our minds and the nature of our consciousness. Mindware is therefore so much more than what we might imagine merely from its symbols. Our symbols are brain and mind changers. This raises the question of how far modern symbols and mindware have changed the nature of what it is to be human.

The question needs to be asked. The history of the last 40,000 years has witnessed a total transformation in the human inventory of symbols. The first increase came with the Upper Paleolithic. As hunter-gatherers, our ancestors employed symbols for elaborate ceremonies and rituals that could bind people into tribes. But this was only a start. Humans discovered agriculture, and in doing so created new opportunities and needs for symbols. Gradually, the stock of symbols expanded. Much has happened in the last 5000 years, especially recently. The Greek alphabet has existed only for roughly the last 3000 years. Computer programming languages have been around only a few decades. In the last chapter, we explore what lies ahead.

The Rise of Abstract Ideas

Do you doubt the mindware revolution? Then look at ideas and concepts. Some represent concrete, visualizable things, like "cabbage." Others represent abstract concepts, like "debate." But has this always been the case? Take numbers. We have both concrete and abstract ones. An abstract number like "two" can apply to anything. But we also have concrete words that apply only to specific things of which we can easily imagine examples, such as a duet, a couple, or twins. While concrete numbers exist, they make up only a small part of our language compared to abstract ones.

However, this was not always so. Abstract counting words representing, say, the concepts of five and six, did not exist until 5000 years ago.[8] Indeed, no words for numbers, abstract or concrete, as high as five or six seem to have existed. Most "primitive" societies count "one, two, many," and when they do so, they use only concrete numbers. That is, there is one word for one person, another for two people, another for three, and so on. Another series may exist for various numbers of canoes and yet others for various numbers of long objects, flat objects, round objects, and measures. But they lack terms for twoness and threeness.[9]

Like abstract numbers, abstract nouns and verbs appear to be recent. In Homer, nine concrete verbs exist to describe sight, but none of them describes the abstract notion of sight as a function. They refer to concrete aspects of vision: a particular look in one's eyes, or a gesture that incites terror; or looking about carefully; or the feelings experienced in the act of seeing. As the Classical Greek scholar Bruno Snell notes, "There was no one verb to refer to the function of sight as such." Or rather, there was not until after Homer. Abstraction, at least in the Greek language, was started by the Classical Greeks.[10]

Abstract and concrete meanings depend on different aspects of the brain. For instance, consider this former naval officer's attempt to define words:[11]

Debate: Discussion between people, open discussion between groups
Malice: To show bad will against somebody
Deceive: To let people down—give them the wrong ideas and wrong impression
Caution: To be careful how you do something

Nothing amiss there. It comes as a shock, though, when he starts to define common concrete words:

Cabbage: Used for eating, material it's usually made from an animal
Tobacco: One of your foods you eat
Ink: Food—you put on top of food you are eating—a liquid
Frog: An animal—not trained

The officer has had a viral infection, herpes simplex encephalitis, that has injured his brain and with it his understanding of the meaning of words—but only concrete ones. Losing abstract meanings would be understandable, as they are learned later in life and are, in general, harder to understand.[12] But to lose concrete meanings while keeping abstract ones is odd. It suggests that two processes exist, one for concrete meanings and another for abstract ones.

One clue to what is going on lies in the close link that abstract nouns have with verbs. Etymologically, many abstract nouns derive directly from verbs; "knowledge," for instance, comes from "to know." Historically, at least according to Bruno Snell, the first abstract nouns, those of the Classical Greeks, were invented from verbs. "*Nous* is the image-making mind, but it is also the act of image-making mind . . . the act of image-making, and finally it is the individual image, the thought. . . . *Gnome* is the understanding mind, it is the act of understanding, and the particular result of the understanding, i.e., the knowledge gained."[13] (The Gnostics, a competing religion to early Christianity, get their name from this word, as does the word *diagnosis*.) So abstract words such as *image* or *knowledge* are linked, at least historically, to actions—the act of image making and the act of knowing. Abstract nouns are like verbs in that they designate functions and operations, whereas concrete nouns relate to objects and their identification.

Patterns of brain injuries, evoked potentials, and gamma show that the brain stores its representations of actions separately from those of objects. Words linked to actions tend to activate the motor cortex, while words representing objects activate the visual cortex.[14] Thus, for example, verbs describing actions done with the feet, like kicking, activate the foot area and those related to the face, such as speaking, the face area.[15] Nouns for things recognized by what we do with them, such as tools, also activate part of the motor cortex, while nouns for things that are identified by what they look like, such as animals, activate the visual cortex.[16] PET studies likewise find that different parts of the brain light up when people come up with words for colors than when they do this for actions.[17] Different areas (on different sides) of the prefrontal cortex activate when people focus on a mix of abstract verbs and nouns, like *knowledge, pretending,* or *decision,* than when focusing on a comparable mix of body part verbs and concrete nouns, such as *hand, eyes, foot,* and *muscle.*[18] Hence, the part of the brain linking verbs with operations could have, in historical times, become upgraded and reused, to do something novel—process abstract nouns.

The unanswered question is, did this significantly change us? Did it alter our mindware? Does using a single abstract noun for sight instead of half a dozen concrete ones create people that reflect and experience the world in new and different ways? How did it affect the part of mind we call our inner voice? It would seem, on the face of it, that our inner selves would depend much on the availability of words and meanings. If Homer used concrete nouns and the later Classical Greeks used abstract ones, did this development change the Greek mind and the Greek self? It is not an idle question. After all, the rise of the Classical Greeks is an abiding mystery. What they thought and how they lived were utterly different from the thinking and lives of any urban people that came before them. Did the world of Socrates, Euclid, Hippocrates, democracy, and the Olympics come into existence though purely cultural causes? Or was there a deeper mind-changing process at work in the Greek brain?[19]

Mind Upgrades

Let us return to our hands. Often, the best way to learn how to use something is to see someone using it. Thus, a great part of the usefulness of our hands comes from our ability to copy. Indeed, we are born to instinctively map what we see others do in terms of our own body, including our hands.[20] A recent discovery in science is that neurons in the motor cortex—*mirror neurons*—are activated when we see actions done by others. Their activation threshold for making a particular motor movement is reduced by the mere fact of seeing it performed.[21] This suggests our motor cortex is wired to link the hands of others with our own, minimizing the problems in imitating their skills.

Imitation is, however, not the whole story of learning. Practice is as important. We need not only to copy but to refine what we duplicate. And, of course, even copying and practice sometimes are not enough. If we are to learn to play the piano, for instance, we need another's teaching to focus and guide us. As a student, we see how our teacher plays, then we attempt to imitate it. At first we don't play well, but our teacher, following our efforts and using his or her own playing experience, can focus our attention on where we are going wrong. Simple imitative abilities can thus, with coaching, allow us to acquire complex skills.

The prefrontal cortex has skills of its own that need to be picked up— not dexterity skills but thinking and organizing ones. As noted in Chapter 4, the Russian psychologist Vygotsky discovered that children organize what they do by talking to themselves. But this private speech is not acquired in isolation. It comes in large part from the verbal guidance picked up from older children and adults. Children are given external instructions, and these get worked into their private speech, and then into inner speech. They can see the goals and cues organizing another person and use them to organize their own thoughts and behavior.[22] You can hear such copied instructions in the private speech of children as they guide themselves. The more supportive the external commands given to them—explicit when needed and withheld when not—the more quickly children internalize them into private and inner speech.[23] Indeed, the quality of the attachments children experience with their parents and caretakers may also, through such internalization, shape their minds. Think back to the research that showed that good attachments early in life lead to good later ones, and also the research where children delayed gratification for an extra marshmallow. They are linked: Children with secure attachments show a better ability to delay gratification.[24] This suggests that the supportive nature of good attachments boosts the ability of the prefrontal cortices to use inner cues, directly enhancing the inner organization of private speech, and so of self-control.

But that is only the start. External verbalization of tasks is important for breaking down a problem into subproblems. The brain needs to cut down tasks to match them with its skills. It may not always be apparent from the surface form of a problem, however, what skills are needed to solve it. Children must be shown how to figure this out so they can do it on their own later. They have prefrontal working memory sketchpads on which to do this, but they first need to learn how to use them. From instructions learned from outside, they gain that knowledge.

External example and instruction given by adults are not the only ways children learn such skills. By making their own contribution to a shared task, they learn to formulate verbal instructions that coordinate and communicate. Making the cues needed to solve problems is a challenge that helps them organize their working memory. And there is often no better way of doing that than having to communicate with others while working together. In private, we may approach a problem in a way that allows such inner organization. But if we are forced to work with others, we must be explicit about how a problem is to be solved. It may be more work, but it allows the brain to learn better skills for mentally managing itself.[25]

We tend to view thinking as a solitary activity. But the above suggests that—at least in the young—it is important that we articulate our thoughts to other people. Young minds need to "beef up" the quality of the inner cues that will later organize their thoughts and feelings. They do this by picking them up from others and being forced to invent and use them in order to communicate. That, of course, is where secure attachments are critical.

Parents who are hostile or indifferent do not allow such communication and therefore such practice. In this way, the external reality of children readily gets internalized. It shapes them in a variety of ways.

Let us look again at how delay-weak children differ from delay-strong ones. Parents described their delay-weak children with phrases such as

Tends to go to pieces under stress, becomes rattled and disorganized
Is shy and reserved, makes social contacts slowly
Is stubborn
Teases other children
Reverts to more immature behavior when under stress

Children who could delay well, however, were described with comments such as

Is verbally fluent, can express ideas well in language
Uses and responds to reason
Is attentive and able to concentrate
Is playful, thinks ahead
Is resourceful in initiating activities
Is curious, exploring, eager to learn, open to new experiences[26]

Children able to acquire skills using internal symbolism can manage and inhibit their impulses and arousal better than other children. Thus, they can control their frustrations, anxieties, and uncertainties. Instead of being victims of their emotions—rattled, shy, teasing, and immature under stress—they let emotions add to their experience of life. That is why they are attentive, thoughtful, resourceful, and curious. And they are likely to grow up to be parents capable of giving their own children secure attachments and so pass on these prefrontal skills.

If we change our viewpoint, we can see that something else happens when people learn from others. Take the example of hand skills. Since they pass between people and across generations, they in effect reproduce themselves. They propagate by replication through being imitated and taught. Your skill in tying shoelaces was passed on to you. You saw others doing it as a child, you imitated them, perhaps unsuccessfully, until someone showed you how. Though we do not see it in this way, your attempt to learn shoelace-tying skills caused them to be passed on from someone else to you. No doubt in the future someone will learn the skill from you and continue its spread. And this is not true just for us. Chimps' ability to hammer nuts is propagated through copying and teaching, thus hopping from one generation to the next.

Richard Dawkins called such a unit of imitation a *meme*, which is short for a mind gene.[27] The ability to tie a knot or hammer nuts is a replicating manual skill. Like a gene, it spreads through duplication. Also like a gene, it has variations. Some are duds, so they do not get passed on; others are effective and thus become the ones we learn. Dawkins' focus is on the replication of the mind's contents, an example of which is manual skills. He therefore looks at the contents of the mind in terms of the spread and survival of individual memes. Our interest here is neurological—how does the brain enrich itself with what it copies and learns from others? The most important and transformative of what Dawkins calls memes we call mindware, the software of the brain. Mindware is what the prefrontal lobe picks up and passes on to other prefrontal lobes through symbols and its use of them.

Mindware Evolution

Mindware is like software in that it can change even though its hardware remains the same. You can upgrade your computer just by changing its software. Look at the history of word processors, for example. The writers of the first word-processing software had to figure out how to make their programs user friendly. The first word processors used hard-to-learn keystrokes and lacked aids like help screens. From the problems people had with these early programs, programmers learned how to make them easier. Thus, each year better software gets written. DOS, a text-based operating system for the PC, went through more than 10 versions between 1981 and 1993. Each one was an improvement on the earlier operating software, though the chips they used need not have changed to make this improvement possible.

Has the human equivalent of software—mindware—been upgraded in a similar way? Are we acquiring better mindware than our ancestors? Could it be that while the bodies of humans have not altered in important ways for the last 100,000 years, our minds—through neural plasticity and what Carl Sagan called "extrasomatic knowledge"—have? Maybe the sense of "who I am," though intimate to us, is rather recent. Perhaps a few thousand years ago, or further back, people's brains lacked that sense of "I" we now take for granted as the core of our individual identity. Is it possible that our ancestors' experience was as different from ours as a text-based DOS operating system is from a graphics-based one, like Windows?

Only a few psychologists and philosophers have attacked this problem. One has been Julian Jaynes, who in the 1970s proposed the radical idea that only modern people have consciousness.[28] Daniel Dennett, the philosopher of the mind, has suggested that the mind's status as a "virtual machine"[29] cries out for the development of a "software archeology." What is striking, however, is not so much their ideas as the fact that Jaynes, Dennett, and others have ventured to suggest what would have been unimaginable a generation ago. Such ideas, eminently plausible in the wake of the discoveries of neural plasticity, are likely to become an increasingly important part of science. The latest findings on the working of our brains cannot help but change the foundations upon which we have based the way we look at our prehistory. The past is not what it used to be.

Human evolution did not fix our brain's information processing but instead created reprogrammable neural circuits that could evolve new kinds of intelligence. That tells us something profound, and not just about ourselves. If evolution could not have hardwired in our brains the intelligence to program a Mars probe or engineer a rocket pad, then it could not have done so elsewhere in the universe. "Extrasomatic knowledge" was necessary to develop not only our minds but any kind of minds. Extraterrestrial beings—if they exist—might have arisen from very different pasts—boron- or carbon-based life, two legs or six—but we can infer that their intelligence must be based on plastic processes that could be upgraded, like ours, with acquired "software." And here is another thought: They would have faced the same technological problems as we did and so their development would have converged toward much the same software as ours did. After all, the constraints on technology are the same across the universe, and therefore they must push in the same direction—on intelligence. This would lead it to mold its "software" evolution in much the same ways. The very source of our uniqueness among animals—our ability to replace genetic with extrasomatic inheritance—could mean that we are kin with all intelligent beings in the universe.

Neanderthals and the Moon

We are a bright species. We have gone into space and walked on the moon. Yet you would never have guessed that if you traveled back to between 100,000 and 40,000 years ago. At that time our ancestors and Neanderthals coexisted. Neanderthals were like us but physically stronger, with large bones and teeth, protruding brows and face, and hardly a chin. Perhaps what we lacked in brawn we made up for in brains. But for most of our history, our species was not bright enough to act very differently from Neanderthals, let alone be more successful than they were. Only around 40,000 to 32,000 years ago, in Western Asia and Europe, did Neanderthal people disappear, to be replaced by our species.

Why did we coexist with Neanderthals for 60,000 years—a far longer case of hominids living side by side than any other in human history? And why did we eventually win out? Brains alone cannot provide the answer, as Neanderthals may in fact have had the larger ones. Perhaps they lacked the long vocal chamber needed for speech. Equal certainty exists among those who study the base of their skulls that they did and that they did not.[30] If they did lack one, then this could be the explanation, but maybe not, since even without a voice box, gestures can communicate, as can be seen among the deaf. Indeed, hunters find advantages in using sign language (speech sounds would warn off potential prey), and not just while hunting but in everyday life. Anthropologists find that hunter-gatherers use sophisticated sign languages to complement their speech.[31] Sign language might even have other advantages—evidence even suggests that it is easier to learn than speech: Deaf children start to pick up signs earlier than hearing ones learn to speak.[32] So "spoken speech" is not in all ways superior to "signed speech." It is not something that can explain our replacement of the Neanderthals.

The reason we—anatomically modern humans—won out lies, we suspect, not in being brighter or better able to speak but in our very physical frailty and our resulting need to exploit our minds. Neanderthals, stronger than us, did not need to take this route. They could survive with their physical strength rather than tapping into the potentials of their brains. An analogy is with countries: The richest ones, such as Switzerland, Finland, Singapore, and Japan, are not blessed with, but rather lack natural resources. Without them, they have been forced to use their brains to innovate, providing products and services ranging from mobile phones to diplomacy.

Brains rather than brawn allowed our ancestors to edge out the Neanderthals. That might appear an ad hoc conclusion until you realize that 40,000 years ago humans had hardly begun to tap into their brains. Being wimps was not a disadvantage in the long run, because it led to a focus on how better to use their brains.

Such an idea neatly explains why anatomically modern people kept company with Neanderthals for more than half of the last 100,000 years and then jumped ahead. Our ancestors might have had a better reason to exploit the potential in their heads, but they still had to discover how. A supercomputer is not much use if you can only program it to be a desk calculator. The mindware that made us needed to be learned somehow. Computers have software because of the work of programmers. But who or what wrote our mindware programming? Nothing—it had to be learned through trial and error. And that took time. Even if the earliest people were better communicators of mindware than their Neanderthal neighbors, they still had to first learn what they were to communicate. Human history is the history of the human brain learning to extend itself and discovering how to do new things. The development of mindware might have gone on for 100,000 years before bearing fruit.

Learning

The word "learning" rarely goes with history. But all we are is the product of learning, both in our personal past and in our species. Without the learning of past people, we would not live in the societies we live in today. Without our own personal learning of mental skills, we would not be the people we are. These are trivial observations, and yet there is something fundamental at the heart of both present and past learning. Learning, either as an individual or as a species, is a much more difficult task than you might think. To understand the history of our species' learning of mindware, it is best that we look first at that of children.

Why do children not stay children? Most psychologists explain it in terms of growing up through stages. But this raises the question, how do they pass from one stage into another? Suppose you cannot do something; how do you learn to do it? How can something in the mind come from nothing? A child's mind in many ways is a developmental blank page with all its neural plasticity. What happens to fill it with abilities? Surprisingly, few psychologists try to answer that question. When they seek to explain how children change, they use terms like *assimilation* and *accommodation*. But such terms evade more than they elucidate.

Here is a partial solution: It is easiest to learn something, paradoxically, if you can already do it, at least a little. Getting that little is often the hardest part.

Consider the problem of learning how to speak. Children learn by doing; so, not surprisingly, the best way of learning to speak is to talk with others. Only by doing so can children learn new words and develop their skills in forming sentences. But how do children get onto the first rung of the language ladder? How can they start speaking with others if that skill presumes they already have the ability to speak?

The problem affects learning to read. The best way to learn new words is to pick them up while reading books. But how do we even start on that road if we cannot stretch our young minds to read even a little? Further, though the words we read are phonetic, we do not normally recognize them through sounding them out. Visual identification is quick—it takes about a tenth of a second—and we do it automatically, without mental effort. But if we identify words by sight, how do we get to learn the link between a word's visual pattern and its meaning for the tens of thousands of words in our reading vocabulary? The problem is not confined just to children but is also found, as we will see, in computers. Indeed, we have already met one variety of it in the problem faced by neural networks seeking to train themselves. The problem has a name, and a story.

Baron Karl von Münchhausen was a famous liar. He once boasted that, trapped and sinking into a swamp, he lifted himself up by his bootstraps and so was able to carry himself to safety.[33] The story is fiction, but the problem is real. It happens every time you turn on a computer. Designers of computers call it the *bootstrapping problem*. For a computer to work, it needs an operating system that lets it load programs into memory. But the operating system is itself a program. How is it loaded when no operating system exists in its memory to load it? The best way to start a computer exists only when, paradoxically, it has already started. So somehow the computer needs to pick itself up using its own "bootstraps."

In the case of computers, the problem is sidestepped by having the computer store a small *bootstrap routine* in nonvolatile memory. This allows it to read the full operating system into memory, so pulling itself up by its own "bootstraps." In the case of children, they "boot" themselves into language by a number of tricks. You do not need to know how to speak to "talk" with your playmates. It is enough to be able to repeat bits of overheard or remembered conversation or even single words.[34] Surprisingly, given neglect of this fact by scientists, people have remarkable skill at repeating words they overhear but do not understand.[35] Listen to a foreigner in a café speaking in a language that you do not know but that uses much the same phonemes as your own (such as French or German for an English speaker). It is easy to repeat their words verbatim without understanding a single one of them. Even mentally retarded and brain-damaged people can do it. Repeating overheard words is, of course, not proper speech. But it gives a child the pronunciation of words they can try to use later in conversations and so find out what they mean. It is a trick that can successfully get a child on the road to speaking with others.

Children boot themselves into reading by using the information stored in how a word is spelled to figure out how it sounds. We may not sound out words as adults, but as children we do. Children sounding out words can use that information to quickly identify what they are. Thus they can train their sight recognition of words and overcome a problem that otherwise would prevent their quick learning of a sight-reading vocabulary.[36] (There are other means, but they are not as speedy.)

The Booting of Mindware

Old people sometimes complain that young people today have it easy. And they are right, but in a sense they do not intend.

When we are young, we are surrounded by people who are willing to talk to us and teach us. We take that situation for granted. Unlike us, the earliest people had nothing to pass on to their children. Somehow they had to discover all the things we learn and, as importantly, that things existed to be learned. Our evolution up to 100,000 years ago left our brains dumb. We lacked even the knowledge that it was possible to do new and unimagined things. Booting thus applies not only to children's learning skills but to the start of what paleoanthropologists call the "human career."

People need a reason for doing things, but it will often follow, not precede, their doing them. Why invent a bow and arrow when a person can accurately throw a killing stick or a boomerang (a killing stick that returns)?[37] A killing stick is a good way of killing game; certainly it had better results than the first attempts people would have made with bows and arrows. Moreover, there is no advantage in inventing only bows and arrows. Before they can be useful, additional skills must be learned, such as how to find poisons for the arrow tips and the best strategies for hunting with them. But these abilities and knowledge can only follow, not precede, the invention of bows and arrows in the first place.

This is a booting problem. Many inventions make sense only after they have come into use with other activities that gradually make them superior to their alternatives. But this will not happen unless the invention already exists. So how do things get going in the first place? !Kung hunter-gatherers poison their arrow tips with an extract from beetle larvae found 3 feet underground near certain trees that might be as much as half a day's walk away![38] How did they learn about such poisons? They would not be useful without bows and arrows, and in turn, bows and arrows would not be better killing tools than throwing sticks if not tipped with poison. So why discover them? The native Australians did not. (Though it can't be because of the competing existence of boomerangs, since these were first made outside Australia. Several were found in Tutankhamen's tomb, and another, dated to 25,000 BC, in a cave in Poland.)[39] If knowledge and skills get better and better, how can innovations arise, when the first efforts will be inferior to those already in existence?

Indeed, this need for inventions to work together explains, we suggest, a puzzle in paleoanthropology. Stone tools such as blades, scrapers, and choppers, associated with the rise of superhunting in the Paleolithic, 40,000 years ago, first occur 60,000 to 50,000 years ago or even earlier, but die out. They are part of pre–Upper Paleolithic (PUP) technologies.[40] In 1995, a 400,000-year-old wooden hunting spear was found among butchered horse remains,[41] suggesting skilled organized hunting had arisen but then failed to take off. Why all this pre–Upper-Paleolithic technology that went nowhere? Probably these tools were not that great an improvement unless combined with other changes on what was already in existence, and so they sparked no revolution.

All technologies face the booting problem at some point. Consider the wheel, surely a useful invention, yet never used in half the world—the Americas. This is not because early Americans did not know of the wheel. Toltecs and Aztecs in ancient Mexico used small-wheeled effigies (they were previously thought to be toys). But the gap between knowing about something and knowing it is useful can be enormous. In the case of the wheel, the rugged terrain of Mexico has been cited as a problem, along with the lack of an animal suited to pull a cart. The largest animal the ancient Mexicans could have used as a draft animal was a dog. All the familiar pack and draft animals, such as horses (the ones the Indians rode came from the Spaniards), donkeys, camels, and oxen, came later from the Old World. But this is a poor explanation; there were other people in the New World with draft animals (llamas) and roads, namely, the Incas. And it does not explain the lack of water wheels, pulleys, winches, millstones, and the potter's wheel.[42] The space in which ideas get invented is filled with more gaps than we realize. If they lacked pack animals, why didn't the early Mexicans use wheels for wheelbarrows, a convenient invention that can be used any-where? Wheelbarrows are convenient, but you cannot know that until you have a working one and experience in using them. You often cannot have one without the other, so there is no reason to think twice about the wheel. Hidden gullies like this block invention and make impossible what, look-ing back, appears to be so easy. Hindsight is "blindsight" when it makes the past look like an inevitable precursor to the present.

Everyday money is another example of this. For a long time people had been using bars of metal as a kind of currency; they had also been using seals to show ownership and provenance. But it was only in the late seventh cen-tury BC that they came together to make coins.[43] Or take pottery. The first pots were made by the Jomon people in Japan only about 12,000 years ago, but kilns were firing ceramic "Venus" figures in Czech and Slovakian areas as far back as 26,000 years ago.[44] Somehow the technology to make fire-hard-ened clay pots had been invented but not the idea, or perhaps the circum-stances were not right to reinforce the invention. Why did the Jomon hunter-gatherers take the step to pot-making but not those in Czech and Slovakian areas? We simply do not know. Or to take another example, metal working was highly developed in the Americas. The Northwest Coast Indians were skilled at working naturally occurring nuggets of copper into valued status items called *coppers*. The Incas were likewise skillful goldsmiths. As with clay-firing skills, metal working had been discovered without being put to practical use. They made no knives, nails, saws, plow blades, or other tools and weapons. Again, did they lack the idea, or perhaps the need? We do not know.

The problem, we suggest, is that what is critical for making an invention is to know what you cannot yet know: its future usefulness. What we take to be inventions are really those things that happened by chance in those rare circumstances that could allow them to take off. That is, until recently. One of our species' greatest inventions came along only a century or so ago. It was the idea that there were things that would be useful if we could only discover them. People saw a need and tried to fill it. Without the idea of invention (and patents that make such inventions worthwhile), there was no motivation to go out and seek new technologies.

The rise of human technologies therefore was a series of accidents that led to them only in hindsight. One surmises that the first bows and arrows arose by a process of improvisations that only by looking back appears to be one of discovery. The rise of Middle Eastern agriculture seems to have been like this, more a series of ad hoc improvisations to cope with a changing climate at the end of the last Ice Age than a discovery that was to change the surface of the Earth.[45] The history of technological innovation until the nineteenth century and professional inventors like Thomas Edison was one not of invention but of blind serendipity.

Social Change

It was not only physical technology that advanced in this way. We discussed how our species discovered and extended family bonds. These were sufficient for people living in the fluid world of simple hunter-gatherer bands. After the rise of agriculture our species had to learn how to make socially structured and settled communities. There were early developments at the beginning of the Upper Paleolithic that prepared the way.

We see this in hunting. The first human-apes probably ate the meat of large animals only when they were able to scavenge kills made by big cats, such as lions and saber-toothed tigers. They had to scrounge the remains of buffalo, bison, mammoths, elephants, and woolly rhinoceroses. If people hunted, it was only for smaller animals like antelopes and deer. As one scientist, Richard Klein, notes, they "were probably very ineffective hunters."[46] This is because they were not yet socially organized. That changed with the start of the Upper Paleolithic, 40,000 years ago. In part, what made the difference was new killing technologies—bows and arrows, and corrals (enclosures, natural ones or ones they built, for killing herds of animals). But it was mainly due to a new ability to organize and communally hunt. To kill a herd of large animals such as buffalo or bison in a corral, hunters must work as a group under a leader.[47] Some have to build a corral or trap; others locate appropriate animals; and still others maneuver and drive them into the corral or trap, where yet another group lies in wait to make the kill. The butchery and division of the meat also requires agreements that are easy to arrange when people live under a chief in subjugation (as they did later in agrarian societies) but are far from easy when people live independent lives for most of the year and come together only seasonally for communal hunts. Such organizing had to be facilitated by a system of beliefs that could bring people together temporarily to elect a leader, hunt, and then dissolve back into independent family units. About 40,000 years ago, people discovered how.

Modern hunter-gatherers on the Pacific Northwest coast of North America illustrate what might have happened. These complex hunter-gatherers live in a wildlife-rich part of the world, so they can afford to make labor-intensive and elaborate objects—canoes 60 ft long, 6 ft across, and capable of carrying 60 people; cedarwood houses 40 ft wide, up to 150 ft long, and 14 ft high. And, significantly, they construct symbolic artifacts, such as dance masks, works of art, and totem poles used in potlatches (gift-bestowing parties), dances, ceremonies, and rituals. These totems and rituals are important, since they act to bind people together into a complex social entity.

Simple hunter-gatherers don't have the resources or the need to create the symbolic artifacts that support a complex social structure. Their symbols are few and simple. In the right circumstances, however—as we can see with the Northwest Coast Indians—hunter-gatherers will evolve the complex social organizations that let them make salmon weirs, hunt whales, and fear tribal war.

Complex hunter-gatherers express themselves in hard, heavy, and durable materials, such as stone, amber, and ivory. Dig in the ground and you find them and see the evidence of their symbols. The first imperishable symbols appear in Russian burials perhaps as far back as 40,000 years ago.[48] Their society was complex, with ranked social lives based around ceremonies and rituals. But we should not infer that earlier people lacked symbols, only the complex way of life that led them to invest their time in making ones that could outlast them.

The new use of symbols to cement complex social affiliations was as important as the bow and arrow. Nothing upgraded human life as much as large-scale cooperation. The first moves in this direction led us to harvest animals that were previously too big or difficult to hunt—in Russia, mammoths and woolly rhinos; on the northwest coast of America, salmon and whales. Using ceremonies, people discovered how to allow large numbers of people to meet together for seasonal hunts and select leaders. Mega-animal hunting led to houses made of mammoth tusks and bones, and to a spate of big animal extinctions. It also led to symbolic representation; mammoths and woolly rhinos no longer exist, except in the painted images left by their killers.[49]

When looking back at our past we should remember that technology and organizational know-how had to be acquired first. The first people faced the hardest task of all: to learn things while not knowing there was anything to learn. Nothing gave them the idea that there was more to life than what they already did and knew. Evolution did not evolve our brains and then give us a list of things we might do with them. Countless people living in countless prehistoric communities had to find out how to do new things the hard way, through a trial-and-error process drawn out over many tens of thousands of years.

It is as if evolution provided our subspecies, *Homo sapiens sapiens*, with a supercomputer but no instruction manuals or programs or even a hint of its potential. Suppose hunter-gatherers received a supercomputer in a box. They might take it out of its box and think it was a table. A few thousand years later their descendants might find out by chance how to use it as a grocery calculator.

Here we meet a profound thing we take for granted. Our brain has vast potential, but this can be realized only if a way can first be found to unlock it. That has taken a long time. Somehow, mindware had to be created. And then it took millennia upon millennia for its potential to be manifested. Musical notation and writing, for example, had to be discovered and then improved over time. Then it took many more centuries for composers to move from Baroque to Classical to Romantic to Modern. The same goes for a thousand and one things that give us the experience we refer to as "ourselves."

Imagine looking at carbon, oxygen, hydrogen, nitrogen, and other atoms before the rise of life: It would be hard to recognize that they had the potential to form animals and plants. That potential was recognizable only when catalysts arose that could build bonds between these atoms and thus make the biomolecules (including enzymes—organic catalysts—and DNA) that create the biomolecular systems we call life that can maintain and reproduce themselves. The same happened with the prefrontal cortex and our minds. The prefrontal cortex is a neurocatalyst that gives new functions to the brain. Without it, the potential of neural plasticity that arose in earlier brains went unused. But early processes lead to other, more powerful processes. Just as biological catalysts require a complex manufacturing process, with RNA splicing, refolding, chaperones, phosphorylation, and other biochemical changes, so the prefrontal cortex has, over the millennia, come to work in equally complex ways, with its replicative code not in DNA but in symbols and the teaching each generation gives to the next. The power that symbols give to the prefrontal cortex has, as with biological catalysts, become more subtle and intricate. And like living cells, brains are adept at maintaining and reproducing the human systems of which they are part.[50]

We are now in a position to see the outlines of how evolution produced the miracle of your existence where FF stands for fission-fusion:

$$NP + PC + FF \text{ ape skills} + \text{symbols} + 10^9 \text{ hours} = \text{your mind}$$

This revised version of the formula developed earlier in this book says that neural plasticity plus prefrontal cortex plus fission-fusion ape skills plus symbols plus time equals your human thinking and feeling. This formula brings to awareness the overlooked importance and power of neural plasticity. It highlights the prefrontal cortex not just as the rational part of our brains but also as a conductor coordinating and integrating itself and the workings of all the brain's other areas—the maestro of the human journey from ape to you. Thus, we suggest, we need to grasp that a second evolution has overtaken our brains: that of our minds. Mindware has dramatically upgraded our species. We are each, individually, a part of this event. One can only wonder at the scale of change that has happened in the last 40,000 years, made possible by neural plasticity, our prefrontal cortex, and our superskill in improving our mindware.

Like the big bang that made our universe and all the atoms that make up our world, something dramatic expanded our minds. The exploitation of our species' potential has been accelerating fast in the last billion hours. Can we say anything about its future? Where are we going in our odyssey?

18

Third-Millennium Brain

To see the future, we need to look at the past. Time—over a hundred millennia, that "billion hours"—was needed to evolve your mind. First, the prefrontal cortex found symbols—developed the mindware—for bonding across place and time. Later, it reworked them to create art, writing, abstract ideas, and the notations of music and mathematics. In doing so, it created something entirely new—human experience. With its increased flexibility, the human brain, aided by its front part, learned to realize itself in a near-infinite number of ways. But this did not happen easily—it required tens of thousands of years before even its simplest possibilities were explored. The human brain has thus made a long odyssey of learning to acquire the potentials we take for granted as modern people. In the past 12,000 years, the human brain has been on an accelerated spree of self-education. People learned how to farm, to live in cities, to read and write, and to believe in one God. Especially within our own lifetimes there has been an ever-quickening pace of invention and discovery. More recently, we have learned how to work nine to five, vote, play computer games, read about science, and surf the World Wide Web. The list of innovations and changes is endless. At each period of history, people found they could use or exploit their brains in new ways. In each, the human brain used its cells to make fresh skills. And it found better ways to pass on what it had learned. In writing and computer memories, in movies and the long-distance perception of phones and television, elements in Earth's crust that previously were not used by life have been refashioned by human technical intelligence to store the past, summon it up again, and sense, over increasingly great distances, the cosmic environment of which we are a part. From hunter-gatherers who never saw beyond a social group larger than perhaps 500 people, our brains now live in a world of billions.

Not all this was progress. Indeed, if we could go back and stop the first humans from using their hidden potential, we might be tempted. The lives of hunter-gatherers were in many ways better than those of the early farmers. Hunter-gatherers were healthier than those who came after them. Their teeth, for a start, were not rotted by cereal carbohydrates, and with no single staple food they avoided the nutritional deficiencies that came with farming.[1] With low density of population, they had few contagious diseases. Moreover, not living with domesticated animals, they did not pick up their infections. But farmers did—horribly. It is believed that measles crossed over to humans from the rinderpest of cattle or the distemper of dogs with which farmers lived. Smallpox likewise came from cattle (or possibly from monkeys), influenza from pigs or chickens, and the common cold from horses— the list goes on.[2] Famine did not severely affect hunter-gatherers, as they could move or shift to unaffected foods in drought years. The unlucky farmers, in contrast, might store food for a bad year but could perish if their crops failed to grow for several years in a row. Before there were mechanical aids to sow, weed, harvest, and mill produce, farming and crop processing were punishingly hard labor. The bones of the first farmers have been found to be deformed by the posture they habitually used to grind food; evidence suggests that it wore out their knee joints and lower backs.[3] Compared to what people had previously known, farming must have been an awful life. But diseases and hard labor were not the only—or the worst—thing that came to impoverish us.

Hunter-gatherers lived together largely as equals with independent minds. To survive, each hunter-gatherer had to be an active learner, sharp and alert to the world in which they lived. Farmers, in contrast, needed only to copy what worked for their ancestors. Cross-cultural research finds that the minds of modern subsistence farmers, unlike those of hunter-gatherers, are not individualistic but conformist.[4] Thus, with farming, people learned how to be nasty—very nasty—to each other. With the rise of farming also came gross inequality. For most humans, life became one of hard want. But not for all. Over the many poor, a privileged few led easy lives as members of ruling elites (nearly always given this privilege by religion). Brains started to hold ideologies and beliefs that blocked the potential of the human mind. They often taught themselves that they were nothing but dirt at the feet of their gods. Instead of maintaining mental independence, the human brain crippled itself with submission and conformity. Worse, not only did brains tie themselves down, but those with physical and cultural muscle trod on the potential of those who lacked power. Women became "the weaker sex," their brains "inferior." Something as incidental to the potential of our brains as our skin color became a measure of what the brain could do. Black people with potential no different from those of white ones came to be seen by them as fit only for slavery or second-class citizenship. Here our attitudes as much as our technology divide us from other animals. We put down others—and do worse. Only human brains enslave, rule, and ridicule. We are the only species to kill others because we think we are better—a superior race or possessors of "the true religion." It is a terrible story in the annals of history. But however much we dislike it, it is one of which all our lives are the products.

Yet we stand in awe at what has happened, both good and bad. And so should you. It is your story.

The Coming Brain Age

This book is a kind of biography—not of a person but of this extraordinary species we are. It tells the story of our brain's growing capacity to exploit its potentials. And what a story it is. Nothing at the start of our journey in Africa could have predicted the kind of creature we were to become millions of years later. Biographies usually recount lives that have ended or largely finished, but the story of the human brain could go on for many more millions of years.

If we have changed so radically in the past, why not in the future? Our species, after all, is a young one. Anatomically modern people date back probably only 100,000 or perhaps (the figures cannot be known for sure) 200,000 years. The average age of vertebrate animals in the fossil record is a few million years. If we don't destroy ourselves, we should last at least that long.

If our brains have shifted greatly from what they once were, it is reasonable to suggest that they might shift again. We do not know what unknown potentials might still lie hidden in them. Someone in the Upper Paleolithic could never have imagined the development of farming, cities, or politics. So is the future impossibly hard for us to imagine. But we can make intelligent guesses. How things have changed in the past gives us clues to what might happen next. We suggest that in the future, our brain's ability to learn may become autocatalytic as it begins to focus increasingly on changing its ability to learn.

We are discovering faster and faster how to use our superbiocomputer, but we lack one last group of symbols that would let us understand and change the use of our minds. We have, as it were, taken our superbiocomputer out of its box and started to program it, but we are still searching for its manuals. But slowly, and at an accelerating rate, we are writing the first pages of one. We are discovering the hidden abilities of our brain. It is knowledge that in future generations will increasingly come to change us.

Each generation of the human species "programs" its young human minds by creating opportunities—"gifted environments"—in which potential can develop. In doing this, each generation in effect writes the limits of how far the mindware of the next generation can go. This has been happening since the earliest days of our species, but ancient people were not aware that this is what they were doing. In ignorance, they gave growing minds only a limited opportunity to explore what their brains were capable of. Indeed, Nicolas Blurton Jones and Melvin Konner note of hunter-gatherer bands such as the !Kung, "There is almost no direct teaching." Young people acquire skills by watching and listening to other people and then trying them out for themselves. A common phrase is, "You teach yourself."[5] Only recently have humans tried to deliberately educate young minds.

We are now entering a new situation. We now know that our brains are not fixed but have many unexplored potentials that we can—and will— unlock. The future of our odyssey lies in our skill at making better gifted environments to enable their discovery. We bet that knowing this may turn out to be one of the biggest things our species will ever learn.

From the invention of the wheel to virtual reality, the history of human innovation has been that it extends our bodies. The brain is next. As we pass into the third millennium, we are at the crest of a great wave of high technology. High-tech brain scanners—PET, fMRI, and MEG—are now exposing the working brain to scientists. But this technology is almost certainly only the start. We will soon break free from the technological limitations that prevent us from looking at the workings of our own brains in our everyday lives.

Current brain scanners have a problem. They are room fillers with million-dollar price tags, which limits them to research and clinical uses. Some, such as PET scanners, need specially delivered radioactive isotopes (the cyclotrons that make them must be within 2 hours' driving time). The need to limit radiation exposure precludes repeated scanning and puts any research done with PET under the control of ethics committees. Other kinds of scanners, such as fMRI, use intense sources of energy flicked around in a strong magnetic field. The scanning is done in heavy donut-shaped machines. Indeed, in the case of fMRI and MEG, it has to be done in special "clean rooms" shielded by copper wiring against electrical interference.

Since most scanners do not take "snapshots," people's heads need to be carefully immobilized in plastic foam head supports so their movements don't blur the image. It is the stuff of specialist technicians. This all, needless to say, limits the widespread use of brain scanners.

But, as more scanners are built, their prices will fall. And, with time, experience in using them will give us the know-how to make brain scanning simpler. Even so, it is difficult to imagine miniaturization's making the room-sized scanners wearable. But the exotic technology of one generation easily becomes the everyday technology of the next. Consider personal computers: Like scanners, the earliest computers were kept in special dust-free "clean rooms." They were hideously expensive and large. They needed to be run and maintained by special technicians in research institutes. But computers shrank in a generation from room fillers to handheld personal digital assistants.

Few people foresaw this. Back then, the physical principles of early tubes and transistors and their manufacture seemed to put a lower limit on size. And there were other practical problems. How would punch cards or punched tape fit into a pocket-sized computer? Where would the market for them be? As in the start of the computer era, we can see that the physical principles upon which today's brain scanners are based will prevent their miniaturization. PET and fMRI will never be cheap or small, just as tube-based computers could never be cheap or small. We need new technologies to see into our minds—and they are coming.

We may think our brains are locked out of our reach in our skulls as if in a cranial safe, but they are not. Do you remember, as a child, shining the light of a flashlight through your hands and hoping to see your bones in the transmitted red glow? The skull and the membranes surrounding the brain are similarly semitranslucent to red light. Indeed, on a sunny day, you could read this book inside your skull.[6] As children we could never see our bones in the red glow, but scientists can. For a start they can use light just beyond red, the *near infrared*, to which bones and flesh are more transparent.[7] With *transillumination imaging* they can see human brains—though so far only in children—with less than 1-cm resolution.[8] A different technique is *near-infrared spectroscopy* (NIRS). This lets us see the brain working not by shining light through the skull but by reflecting light off the brain. When a part of the brain works, more blood flows through it, and, as important, more oxygenated blood (which has a different shade). So as more blood rushes to the cortex, its color changes slightly. Several research groups are shining light at the brain in this way to see how it works.[9] This is letting researchers watch the prefrontal cortex turn on as it thinks.

These two techniques will make brain scanning more accessible. They may only ever let us see broad changes over the cortex, but that is enough to allow us to monitor the activity of the prefrontal lobe. And, of course, with refinement they could become more sensitive. While no substitute for million-dollar scanning machines, they can give readings of the brain's activity many times a second. Thus, unlike most other tools for seeing the brain, they can see it in action. And they can see blood changes not seen by other scanners. PET and fMRI can see only volume changes in blood in the cortex, but NIRS can see changes in the ratio of deoxygenated to oxygenated blood even when this is not accompanied by an increase in overall blood volume. This may turn out to be important. Preliminary work with NIRS suggests that older people think without any change in the local volume of blood getting to their brains.[10] It is a small point but a hint of the extra subtlety available for seeing our brains in action. NIRS's real advantage, however, is not subtlety but safety. The approval of an ethics committee is needed to give someone radioactive substances, as required for PET, but not for shining light through someone's head. And it can be done repeatedly. There is no need to worry about stray magnetism or electrical signals (which easily upset MRI and MEG machines). Equally importantly, it could be cheap, portable, and robust. Having your brain scanned will eventually become as routine as having your temperature or blood pressure taken.

Where such scanners will go and how long we will have to wait for them is hard to know. The first MRI image of a living animal was made in 1974, but it was another 15 years before MRI could be used to picture the brain's activity. Progress takes time, but the potential is there for it to happen. Perhaps technological difficulties not foreseeable at this stage will limit their takeoff. But if they fail, others can take their place. What about ultrasound? Familiar for picturing the unborn, its potential for taking images of the brain is now being explored.[11] Blood flow in the brain changes its electrical impedance, something that can be detected with electrodes and used to map its activity.[12] Then there are T-rays, terahertz (million millions of cycles per second) radiation that is found in the spectrum between microwave radiation and infrared light. Not as familiar as those sorts of radiation, T-rays have until recently been hard to create. But they are safe, and they can now be produced by recently developed emitters. These open up the possibility of revolutionary devices, from new means of checking for explosives at airports to new types of scanners for breast cancer, and maybe for brain activity.

The oldest technique for looking in the brain—measuring its electrical activity—is getting a new lease on life. Since the late 1920s, starting with the work of Hans Berger, scientists have measured brain activity by the electrical potentials they could detect on the scalp. But the technique was always limited. Picking up such potentials requires careful and time-consuming fixing of electrodes to the scalp. The activity detected was hard to see in terms of what was happening in the brain within. Worse, the skull distorts the passage of electrical potentials. But ways of overcoming these problems are now being developed. One is to detect magnetic activity in the brain—MEG; magnetic fields are not affected as they pass to the surface. Another is to use brain scans to measure the thickness of the skull and then correct for the distortions it causes. New "dry" methods now exist that replace the messy and inconvenient process of applying electrodes with gels and silver pastes.[13] And computer graphics have made it much easier to interpret the results.

Such technology will literally change how we see ourselves. Who we are lies in our prefrontal cortex and what it does. Developing the ability to peep at it as we live our everyday lives may very well launch us on the next phase in the evolution of human intelligence. It will certainly alter our notion of self-knowledge. When ancient Greeks posted the phrase "know thyself" at the Delphic Oracle over 2000 years ago, the concept was there, but not the tools. Now the tools are coming.

Every science starts off stumbling about to get the tools that will let it take off. We have seen this happen with physics and biology. Soon psychologists will get the tools to look anew at everything that makes up human life. Cheap scanners will be as revolutionary to them as the cheap microscope was to biologists. They will have "mindscopes" to study the links between personality, social psychology, and our prefrontal cortices. With them, neurology and psychology may merge into a new third-millennium brain science.

And the excitement will not be confined to the halls of academia. Criminologists and law enforcement officers will be able to check on the prefrontal cortices of psychopaths. There may be a new generation of lie detectors (telling the truth is likely to display different patterns of prefrontal cortex activity than telling lies). Therapists and psychiatrists will be able to monitor the prefrontal cortices of their clients as they work over their problems and follow their treatments. Military and aviation authorities might be interested, since soldiers and those supervising or flying aircraft need to work without tiring mentally. Most air crashes now are due to "human error" largely attributable to fatigue. There may be a means, if we can monitor the prefrontal cortex, to keep it working even after long periods without sleep, or to spot when its concentration begins to flag.

Scanning could become a key part of everyday know-how in the next century. The president may have to give bulletins not only on his or her health but also on the sharpness of the workings of his prefrontal cortex. We might monitor ourselves and thus know how to push ourselves to our limits but no further. We might do it before making big decisions, to check how much our dorsolateral cortex is in control. Children might play with it. Tired of their computer games, they might look at what parts of their brain would light up as they talked, moved, read, did arithmetic, or imagined things. "Hey, I bet I can make my prefrontal cortex light up more than yours." Does this sound frivolous? Playing computer games that let you rip off a victim's head and leave the spinal cord hanging out would have sounded like a rather bad use of computer technology 20 years ago. A catalytic conversion of the potential of brains experimenting with scanners could happen fast. Truth is stranger than fiction and in this case, we submit, stranger even than science fiction. If the prefrontal cortex is a foreign land of which we now know only the early reports of cranial pioneers, the future will give us detailed and in-depth tours. If the alphabet changes how we hear phonemes, how might a new awareness of our brain's activity while we think and feel change how we develop? What would happen to the learning of a skill if we could look and monitor how it activated our brains? Suppose children wanting two marshmallows instead of one could see the activation of their prefrontal cortices. Would they catch on better to the trick of organizing their brain to delay gratification? Would those less good at delaying discover internal ways to improve? What if children were given games in which winning depended on how well they activated their prefrontal cortices with inner cues? The promise of biofeedback may finally be realized: Seeing our brains, we might learn how to increase activation in selected areas. It might change what our brains are capable of.

After all, a major limit on learning is our ability to bootstrap ourselves into new skills. One of the things a young brain needs to pick up is how to upgrade sensory and motor skills to develop new ones. What if a teacher were able to correct children's attempts, say at solving math problems, not in terms of what they scribbled on paper but on how they used their brains? It might be that the human brain is limited at present by its inability to focus the best part of the brain for each skill required. Freed from this limit, perhaps even the dimmest might be able to outdo Einstein.

We must be honest: We do not know—we have the only the barest hints of how we can exercise the brain. There was a time not so long ago when people used only a few bars and a mat to exercise. Then machines were developed that could work specific muscles. But these were crude. Muscles need different resistances to work against at different stages of their contraction. So variable resistance weight machines were invented. But even these do not fully exercise muscles, since their fibers fatigue at different times. New computer-controlled machines are on the way that will test a muscle's strength as it tires and change the machine's resistance for the optimal customized work-out.[14]

Will the exercise of our mental abilities go along the same path? Technology is coming into place to make it possible. Maybe it will offer us nothing. After all, not everyone exercises with an exercise machine. But it could equally well become as much a part of tomorrow's school life as the school bell.

The Twenty-First-Century Mind

We can guess what technologies might be around the corner, but what of a hundred or more years in the future? Who a century ago could have predicted the rise of jumbo jets and microchips? Theoretically, if we had Lucite skulls filled with diodes, fiber optics, and detectors, we could picture directly and unhindered the changing patterns of blood activation made by our brains. We could see better what drugs, foods, and even pheromones do to us, enhancing the possibilities for health, longevity, and general happiness. Maybe, generations from now, babies will have their growing skulls "circumcised" and replaced with transparent ones wired with optical fiber detectors to monitor the activity of neurons even deep in their brains. We cannot predict what technological avenues might open up or turn out to be blocked. But the outlines of a startling new brain revolution are coming into view.

If these thoughts seem too far out, consider something equally wild that has already happened to our species. Imagine not knowing what you look like. We take it for granted, yet mirrors, videos, and photographs all depend on technology. The vast majority of human brains that have ever lived did so before the invention of ways to see themselves. That means that except for the occasional reflection in a calm pool of water, people in the past had no way of knowing what they looked like. Open your photo album. It is so easy, yet before the middle of the nineteenth century no one except the very rich could see what they looked like when they were young. Today, children grow up with embarrassing videos of themselves as babies, naked and babbling in the bath. We take this self-awareness of our physical appearance as normal and natural. But it is a very recent phenomenon.

Now imagine a life filled with snapshots and videos of minds at work. Computer software lets you display in a corner of the screen a gauge of how much its processor is being used. Imagine in that corner a gauge not of the computer's processor power but of yours—a PET- or MRI-like scan of your brain, alive and changing as you think. Imagine looking at a watchlike thing on your wrist and checking the state of your prefrontal cortex as you might the time. "This is an important decision—let's check to see if my brain's OK before making it." Or "Oh my God, look at my visual cortices. I don't think I can have another drink tonight." Imagine getting a video report card on the development of your child's prefrontal cortex over the last year. Such familiarity with our brain's activity may be as everyday and natural to us tomorrow as our familiarity with our faces is now.

There will be new technologies—*braintech*—built around our ability to see into our brains. For example, to stretch your imagination as to what might be coming, look at reading. Reading is grossly inefficient. If you cannot recognize a word, you have to skip it or break off and look it up in a dictionary. If you need to skip or check too many words, a text becomes unreadable. Wouldn't it be nice if texts could help us read them? Imagine if, when you came to a difficult-to-read word, the text anticipated your problem and sounded it out for you, or even swapped it with one it guessed you could read more easily. Instead of being silent and passive, the text would actively read along with you. You would carry on smoothly, inputting through your ear what your eye could not read. Such "smart texts" would help children learn to read by softening out their initial difficulties. And adults would find them useful for reading technical books or books in another language.

A book that reads along with its reader could easily be made now. A scanner monitoring our brain could spot the words we found hard. When we read, our eye movements expose what is going on in our brains. Scientists measuring reading eye movements find that our eyes linger over unfamiliar words and make more eye movement regressions. This is also the case when we have problems comprehending a sentence.[15] Our pupils even dilate slightly when we find words hard.[16] The technology for spotting eye movements and dilation exists already; virtual reality goggles track eye movement to stabilize the image a person sees in them. Maybe the next development in computers will be virtual texts, sound and visual images and writing actively paced with neural monitoring of their comprehension.[17]

This may be only the beginning. Could we directly interfere with the brain's workings? It is a fear that people have had before, but with regard to our bodies. Mary Wollstonecraft Shelley wrote *Frankenstein* in 1818. Around her marched the new technologies of the nineteenth century. The first steamboat was built in 1802; in 1829 the "Rocket" pulled a coach at the then incredible speed of 30 mph. Shelley feared the body would be the next thing to become mechanized. But however much engineering has advanced, from the steam engine to the microchip, Frankenstein monsters have never arisen. The reason is not hard to see. Nothing can control the body better than the body itself, but we can use technological aids. Engineers don't change our bodies, they just give us things to control. They do not get us traveling faster by fusing axles and wheels to our pelvises. What technology does is give us wheeled vehicles that we control through pedals and steering wheels. Likewise, technology does not change us so that we can cut things with tendons hooked to sharpened steel fingers; rather, it gives us scissors. Technology does not fuse body and machine but enhances our existing abilities with new dimensions: not new eyes but telescopes, microscopes, and televisions; not new ears but Walkmans, CD players, radios, and mobile phones.

Braintech will enhance the brain's abilities in a similar way. The brain is already too good at controlling itself. Its problem is that it is blind to itself and limited in how it uses its abilities and learns new ones. The kind of technologies we anticipate, such as smart texts, are not mind-Frankensteins but the sort of trivial thing that once mass-produced by SONY or another consumer goods manufacturer on the Pacific rim, we tend to take for granted. Smart texts will be to our senses and minds what television is to our eyes and the telephone is to our voice—portable extensions of our natural abilities.

We do tolerate more intrusive technologies, however, when we are ill. If your joints wear out, high-tech replacements are fitted by surgeons. If your kidneys fail, you can be offered dialysis. When your teeth rot and are pulled out, your dentist measures you for dentures. Here technology crosses the boundary of what is natural and puts in its place something unnatural. But who objects to artificial hips, artificial kidneys, and artificial teeth? While we doubt most people will want braintech to directly interfere with their uninjured brains, it is easy to see that many may wish to use such technology for injured ones.

Ever since the eighteenth century, people have wondered if magnetism or electricity affects our minds. Indeed, hypnosis was discovered by Franz Anton Mesmer while playing with magnetics and "animal magnetism." He was wrong; the power of his "animal magnetism" was that of his own personality and powers of persuasion. Yet in a broader sense he was right. Contemporary researchers are in full swing activating and inhibiting small areas of the cerebral cortex with small electronic and magnetic pulses on the surface of the scalp in a technique called *transcranial magnetic stimulation* (TMS). With it they can delay eye movements, or impair your visual func-

tions or your scanning of visual memory or your recall of things from verbal short-term memory. They can inhibit or excite thumb movements or
stop you from counting out numbers or cause you to make errors in doing
so.[18] Magnetic pulses applied over your forehead can keep your prefrontal
cortex from working[19] or, strangely enough, can actually enhance its functioning for at least some kinds of reasoning.[20] They can pace alpha oscillations if applied at their frequency.[21] A new form of antidepressant treatment
has been discovered using TMS,[22] and there are hints it might one day be a
treatment for Parkinson's disease.[23] And most intriguingly, TMS can affect
the immune system. Magnetic fields applied over the right hemisphere
when people do mental tasks depress CD8+ T cell function, and so suppress
immune responses, but applied over the left one, they increase them.[24]
Similar findings are found for CD4+ T cells, the ones that induce an
immune response and are lost in AIDS, though with more variable effects
for the right side of the brain.

   These applications are not the "sledgehammer" electrical shocks once
used to treat depression. Rather, they are delicate tweezers that can tweak
what the brain does. At present they are research tools used to reversibly
"lesion" and stimulate the brain. But it is possible that magnetic stimulation
of the brain might lead to new ways to exercise the brain, enhance its functions, and change how we think.

   Where will it all lead? Neuro-hell or neuro-heaven? Some might fear
braintech as a new means of brain control. But brain control is already with
us and has rather a lot of methods available to it for changing people's
minds: knives, imprisonment, ideology, taxes—the list is fairly long. We tend
to doubt that brain technology will add much. There are, and will remain,
more direct ways of inflicting irritation and pain. But braintech will offer
what we lack at present—ways to heal and to expand the use of our brains.
That may make us freer, more fulfilled people. When it will arrive we
cannot know, but it will be along sometime—probably well within our own
lifetimes. We could be the first generation to start exploring braintech and
making it part of our everyday lives.

   The brain is the last frontier faced by technology. Braintech will change
what it means to exist as a human being. But how? We can get a taste of our
future with virtual reality. Virtual reality is a brain technology that can
extend our senses into a world simultaneously real and unreal, a world generated by a computer. The first layer of its infrastructure has been built, and
it will continue to evolve with better connections and faster computer chips.
What exotic and ecstatic futures might lie in wait for us in such a virtual
world? Perhaps we will develop new esthetics that are now beyond us in the

way that the music of Mozart, Beethoven, or Ravel would have been unimaginable to medieval monks who sang and heard only plainsong. Human experience in the future is likely to make similar or even bigger jumps. They have happened in the past; why rule them out for the future?

We can thus modify one more time the formula running throughout this book:

$$NP + PC + FF \text{ ape skills} + \text{mindware} + \text{braintech} = \text{future minds}$$

Faintly but clearly over the horizon, another wave is building in the continuing evolution of human intelligence.

Brain Rights

As we begin to read the manual of our brains, our views of who we are will change. We may identify ourselves with the face we see in the mirror, but science suggests a truer identification would be with our brains. This has profound consequences. Under our skulls, no brain is illegitimate to our species—"half-caste," "mixed race," "mulatto," or in some other way a "not so good" example of humanity. We all share the same-colored brains, so subcranially no one is British, Irish, Arab, Jew, black or white. Race and "purity" of descent do not exist as the basis for any human's inner sense of self-worth or identity; in our brains we are kin. To deny this is the neurological equivalent of adding two and two and coming up with five. Mistaking the box for the contents, as such claims are, is quite simply, nonsense.

We can go further. All humans seek a self-identity, or a story as to who they are. So we have found our stories in myths, believing in special origins or superiority over others. We cling to these myths as if nothing else could supply us with our identities. Turn on the news and you can see their price. Humans worldwide are starving or being killed by war and terrorism. And behind every war or terrorist outrage are people holding to a false belief: that a few behaviors or cultural differences make them separate from or better than other people. When we perceive our enemies as being unlike ourselves, we fail to empathize with them. Identity myths induce a kind of moral insanity, an acquired psychopathy that keeps ordinary people from identifying with the suffering of other members of their species. Instead of offering help after massacres and bombings, they joyfully celebrate in the streets, filled with a sense of their "success." We could do something very different.

We could, instead, orient ourselves to our brains—their vulnerability, richness, history, and giftedness. See through your skull, and you will find a complex world of neurons, axons, and synapses. We are their story. People of different nationalities and cultures have brains of the same origin, history, and makeup. We are members of the same species. The neurosciences are now exploring this common ground of brain behind our surface differences. They offer no myth, but rather an exciting story in which all of us can find who we are. Evolutionarily preserved in the living museums of our skulls lie our brains. And in our brains lies the true account of our origins.

It is a human quest that will never end. This book fits many of the pieces together, but what we know is seen through the limits of today's technology. But the neurosciences in the coming years will be getting newer tools to give us a sharper and clearer focus. Some of the elements put together in this book may well change; no doubt surprises are in store. But this odyssey to discover ourselves in our brains will not mislead us with a past that never was. Those PET and MRI scans are not apparitions, but fact. We no longer need myths to understand the origins of who we are. Our brains may hold vast mysteries, and some details may always escape us, but their story is a true one—at least the nearest we will get. Here, in your skull, are the origins of your thoughts, feelings and sense of existence and being alive. It is to this knowledge we should look when trying to understand our identities, not poorly established ancient tales. We should approach the study of our brains not as a part of a new religion but in respect, wonder, and gratitude. It offers us what our species so greatly needs: knowledge that does not divide. What is true of our brains is also true of yours, and those of others. In finding our identity in our brains lies our future.

Perhaps, if we saw more truly who we are in this way, we would be a more tolerant species. A crazy idea? But wait. Scientists have already discovered something about the brain that has removed one kind of intolerance. In the nineteenth century, neurologists such as Paul Broca realized that left-handed people differed from right-handed ones in the side of their brain controlling their hands.[25] That shifted society's attitude toward left-handers. Once the left-handed were considered to be evil people. They were called "sinister"—from "left" in Latin—or worse. The right hand was the good hand; the right side was the one on which the righteous would sit next to God. There was the right thing to do and the divine right of kings. The English words *correct* and *rectitude* derive from the Latin word *rectus*, for "straight" or "right." One of our grandfathers (Dorion's), born left-handed, suffered physical "discipline" until he learned to use the right (pun intended) hand. There was a cruel intolerance toward 1 in 10 people, because they were left-handed. They were forced to use their right hands like the majority. Children at school, for instance, had their left hands tied to their chairs and were punished if they attempted to write with them. Today, there would be an outcry if teachers tried to force left-handers to do this. We have radically changed our idea of what is normal. In large part this is due to the impact of knowing that the brains of left-handers are organized differently from those of right-handers.[76] Brain science helped changed the norm, which used to be seen as that of the majority but which is now seen as that of the individual. Why would anyone try to interfere with what is natural to the hardware of a person's mind?

Today, our intolerance of another minority, those attracted to their own sex, is moving in a similar direction. If what the majority do is defined as natural, homosexuals are not natural. But what if a person's sexual orientation is wired into his or her brain? In that case, what the majority does is irrelevant to the naturalness of homosexuality. Instead of intolerance and prejudice, we will see each person as having his or her own natural sexuality, gay or straight or any of the many shades in between. Knowledge about our brains will powerfully shape the framework in which we see what is natural to each person.

In the future, we suggest, brain science will change ethics still further. Our treatment of fetuses, babies, and children is careless and needs revision. We know that children have basic food and health needs, but their brains also need a sensitive physical and social world. We each need our own gifted environment. Moreover, the growing brains of humans are more vulnerable than those of other apes. Our brains, after all, are born with that extra year of immaturity outside the womb. Yet no system of ethics takes account of this. Evidence mentioned in Chapter 8 suggests that early attachments are essential for proper brain development. That is worrying. Infant monkeys deprived of their mothers and other social contacts are permanently brain damaged. And their brains are mostly formed *before* birth. What of humans, so much of whose brain development takes place *after* they are born? Do they not need their mother's and social contact even more?

Not that damage does not occur in the womb as well. Alcohol and cocaine—particularly when taken together — are the thalidomides of the unborn mind. More children in America are born retarded each year due to their mothers' alcoholism—somewhere near 7000—than are born with Down's syndrome (4400).[27] Researchers had hoped that such children with fetal alcohol syndrome (FAS) would grow out of their early problems. Alcohol-affected children are also facially different, and that does become reduced with age. Disturbingly, however, their cognitive and social problems tend to worsen.[28] We often hear about civil rights, ethnic rights, national rights, human rights, animal rights; but in the future we may need to add a new one: brain rights. We will see that not only do people need protection, but so do our neurons.

Each *Homo sapiens* brain is highly vulnerable during its early development, and this gives each, we argue, presently ignored entitlements. Each of our brains should be guaranteed the right to grow unhandicapped and supplied with the best possible nurture and support. Maybe one day there will be a bill of rights for brains stating that every brain needs and should be given optimal physical development, a mother-bond when young, and later in life a gifted environment sensitive to its uniqueness. Such rights might one day be held to be equally as important as those human rights that protect our freedoms when we are adults.

Tomorrow's Humanity

Given everything discussed above, we foresee that the mind has a future, as it has had a past. The odyssey of what Carl Sagan called "extrasomatic knowledge" will go on, but minds will have entered a new era of humanity: the Brain Age. We tend to see ourselves as advanced and those who came before us as primitive, especially in ages before industry, iron, cities, or farming. Thousands of years from now, maybe, people will look back upon us all as primitives, living in the pre-Brain Age.

Whatever the people of the Brain Age ahead of us may be like, we hope that they will look back upon us not belittlingly, but with a sense of gratitude for the steps we are taking now. After all, we are who we are only because of the learning steps made by others before us, from hunter-gatherers and the first farmers to the nineteenth-century engineers of the industrial revolution. In those countless forgotten and uncelebrated lives lie our own origins. Those to come will be who they are only because of what we are doing and learning now. Let us hope they look back at us and have reason to be thankful.

They might have another feeling—wonder. Millions of transistors packed on a computer chip shine with colors as they diffract light shined upon them. They are gemlike, computational sapphires etched on silicon. Roughly 150 billion humans with brains like ours have existed.[29] Each of those 150 billion brains is or was a biocomputer gem woven out of billions of neurons, axons, and synapses. Though each brain may have told different stories as to who it was, each glittered with consciousness and experienced life. Each tried, in its way, to learn and explore what it is to be a human. Although that experience will change, as it has in the past, some things will not. For 100,000 years—perhaps even much longer—brains have felt the same pains and joys that we do. How our emotions tag experience may have changed, but the emotions themselves have not. Your great-great-grand-child, honeymooning on Titan, the largest moon of Saturn, would likely feel the same sorts of bliss in adventure and euphoria that your hunter-gatherer ancestors felt biting into fermented fruit beneath the shade of a long-gone tree. Our symbols, alliances, and technologies may have changed, but the basic story of emotional, bipedally supported brains struggling to learn from each other and the universe around them has not.

There is an equality in that.

Notes

Chapter 1 The Cosmic Mirror

1. Number of galaxies: NASA estimates 125 billion using the Hubble space telescope. See, for example, http://abcnews.go.com/sections/scitech/DailyNews/galaxies990107.html.
2. Date of the Big Bang: Carl Sagan in 1977 gave it as 15 billion years ago; the present estimated range is 9-16 billion years ago; some people pick 14 billion. However, here we follow the recent review which suggests the best guess is 12.5 billion years ago: (Sneden, 2001).
3. Origin of atoms: (Krauss, 2001). Neutron star collisions and gold, the work of the physicist Stephan Rosswog, *New Scientist*, April 7, 2001.
4. Earth's violent early history: (Bernstein, Sandford, and Allamandola, 1999; Kerr, 1999).
5. First photosynthesis: (Marais, 2000); dangers of oxygen: (Reiter, 2000).
6. Evidence of first eukaryotes: (Knoll, 1999; Kerr, 1999).
7. Early life: (Margulis, 1982).
8. Genes for head development: (Galliot and Miller, 2000).
9. Origins of vertebrates: (Holland and Chen, 2001; Zimmer, 2000).
10. Embryologists question the Haeckelian idea that all embryos are similar—nonetheless it needs a specialist to see the differences: (Richardson, Hanken, Gooneratne, et al., 1997).
11. Neural crest cells: (Gilbert, 2000).
12. Origins of bone: (Zimmer, 2000).
13. Origins of immune system: (Laird, De Tomaso, Cooper, and Weissman, 2000).
14. Mammalian teeth: (Vaughan, Ryan, and Czaplewski, 2000).
15. Origins of primates: (Conroy, 1990).
16. Origins of color vision: (Dominy and Lucas, 2001).

17. It is not clear exactly how, but changes to the abdominal muscles occurred with the rise of apes. It may have had to do with holding one's guts in while vertical and throwing the stomach forward to swing. Breathing, which usually involves the diaphragm, can also involve these muscles.
18. Human faces: (Young, 1993). The human eye: (Kobayashi and Kohshima, 2001).
19. Chins: (Schwartz and Tattersall, 2000).
20. As few as three genes controlling skin color: (Sturm, Box, and Ramsay, 1998)
21. UV, folic acid, and skin color: (Jablonski and Chaplin, 2000).
22. Theories on skin color: (Robins, 1991).
23. Aristophanes's joke on shaving and effeminacy: (Thesmophoriazusae, 130 ff).

Chapter 2 Up from Dragons

1. Wolf and echidna blood temperatures: (Guinness Book of Records, old editions).
2. Between 1 and 10 billion lizards in Italy: (Avery, 1978: 245).
3. Comparative mouse and lizard energy needs at various air temperatures: (Vaughan, 1986: 446).
4. The exceptions occur mostly on a few tropical islands.
5. Effects of sleep deprivation on speech: (Harrison Horne, and Rothwell, 2000).
6. REM sleep deprivation: (Endo, Horne, and Rothwell, 1998).
7. REM was reported not to occur in one mammal—the echidna—but recent research finds it does: (Nicol, Andersen, Phillips, and Berger, 2000).
8. REM, various details: (Antrobus, 1994).
9. Dreams in slow-wave sleep: (Cavallero, Cicogna, Natale, et al., 1992).
10. Breathing during REM sleep: (Kubin, Davis, and Pack, 1998).
11. Loss of REM paralysis and attacking of bed partners: (Olson, Boeve, and Silber, 2000).
12. Sleep terrors, sleepwalking, and slow-wave sleep: (Schenck, Pareja, Patterson, and Mahowald, 1998).
13. For a case of REM linking to learning: (De Koninck, Christ, Hebert, and Rinfret, 1990).
14. Disturbance of REM sleep and loss of learning: (Plihal and Born, 1997; 1999).
15. Reptile wakefulness and sleep folded into mammalian sleep: (Nicolau, Akaarir, Gamundi, et al., 2000).
16. Reptile awakeness and slow-wave sleep: (Nicolau, Akaarir, Gamundi, et al., 2000; González, 1999).
17. K-complexes and cognitive processes: (Niiyama, Fushimi, Sekine, and Hishikawa, 1995).

18. Rial quote on reptile brain and sleepwalking: (Nicolau, Akaarir, Gamundi, et al., 2000: 397).

19. Sleep and consolidation of memories and skills: (Smith, 1995; Karni, Tanne, Rubenstein, et al., 1995; Wagner, Gais, and Born, 2001).

20 Some argue that consolidation does not link to sleep: (Siegel, 2001).

21. Consolidation, slow-wave and REM sleep, and different kinds of memory and learning: (Plihal and Born, 1997; 1999).

22. Different processes in memory areas during memory consolidation: (Hasselmo, 1999).

23. Sleep and reliving experiences of the previous day: (Wilson and McNaughton, 1994; Dave and Margoliash, 2000).

24. Some 50 percent of dreams relate to events from the previous day: (Harlow and Roll, 1992).

25. Tetris and dreams: (Stickgold, Malia, Maguire, et al., 2000).

26. Learning French and dreams: (De Koninck, Christ, Hebert, and Rinfret, 1990).

27. Frontal areas deactivated during REM: (Maquet, Peters, Aerts, et al., 1996).

28. Disconnection between frontal and posterior areas during REM: (Perez-Garci, del-Rio-Portilla, Guevara, et al., 2001).

29. Eye movements and protection of the cornea from suffocation: (Maurice, 1998).

30 There are problems with Maurice's theory, not the least of which is why it is not only the eyes that twitch but other parts of the body as well.

31. Eye movements in REM and dream imagery richness: (Hong et al., 1997).

32. Eye movements and dreams in slow-wave sleep: (Conduit, Bruck, and Coleman, 1997).

33. Eye movements and dream imagery: (Conduit, Bruck, and Coleman, 1997).

34. Ponto-geniculo-occipital potentials, REM, and gamma synchronization: (Amzica and Steriade, 1996).

35. Triune model of the brain: (MacLean, 1990).

36. DNA dates for last shared ancestor: (Ruvolo, Pan, Zehr, et al., 1994).

37. Stone hearths: (Klein, 1989: 313).

38. First bows and arrows: (Klein, 1983: 43).

39. First lamps: (de Beaune and White, 1993).

Chapter 3 Neurons Unlimited

1. Neural plasticity: general review (Buonomano and Merzenich, 1998; Grafman and Christen, 1999; Yuste and Sur, 1999; Kujala, Alho, and Näätänen, 2000; Merzenich, Recanzone, Jenkins, et al., 1988); for hearing (King and Moore, 1991; Rauschecker, 1995); for vision (Kaas, Krubitzer, Chino, et al., 1990); for reorganization up to 10–14 mm (Pons, Garraghty,

Ommaya, et al., 1991; Pons, 1994; Yang, Gallen, Ramachandran, et al., 1994).

2. Multiple brain maps: (Felleman and Van Essen, 1991; Merzenich and Kaas, 1980: 32).

3. Brodmann and the Vogts: (Gorman and Unützer, 1993).

4. Observation of 150–200 connectivity areas: (Roland, 1993: 131–133).

5. Brains as variable as faces: (G. Gray, cited in Whitaker and Selnes, 1976: 844).

6. Anterior commissure: (Demeter, Ringo, and Dorty, 1988)—it is the smaller of the two links between our two cerebral hemispheres (the other is the corpus callosum). Massa intermedia: (Yamamoto, Rhoton, and Peace, 1981: 337)—it projects into the third ventricle and often links the surfaces of the two thalami.

7. Primary visual cortex, threefold variability: (Whitaker and Selnes, 1976: 846).

8. Twofold variation in amygdala and hippocampus volume: (Giedd, Vaituzis, Hamburger, et al., 1996: fig. 2).

9. Cerebral cortex varying twofold: (Giedd, Snell, Lange, et al., 1996: fig. 4; see also the review in Skoyles, 1999a).

10. For example, this situation does not arise in the monocular area of the lateral geniculate that projects to the visual cortex that receives input from the visual field seen in one eye—the nasal area: (Zeki, 1993: 223).

11. Braille and the mapping of the right index finger tip: (Pascual-Leone and Torres, 1993; note the researchers excluded Braille readers who used both index fingers). Shrinking of maps for other fingers: (Pascual-Leone, Wassermann, Sadato, and Hallett, 1993b).

12. Amputees and brain plasticity: (Halligan, Marshall, Wade, et al., 1993; Ramachandran, Rogers-Ramachandran, and Stewart, 1992). MRI and face maps moving into hand areas 3.5 cm: (Yang, Gallen, Ramachandran, et al., 1994; Yang, Gallen, Schwartz, et al., 1994).

13. Brain tumors and shifts of motor maps up to 4.2 cm: (Seitz, Huang, Knorr, et al., 1995).

14. People walking after spinal injuries: (Wernig, Nanassy, Müller, 2000; Wernig, Müller, Nanassy, and Cagol, 1995).

15. Shifting chest nerves to regain arm control: (Mano, Nakamuro, Tamura, et al., 1995).

16. Neural plasticity of erotic areas after penis removal: (Wilkin and Kaplan, 1982).

17. Orgasms following sex change surgery: (Wilkin and Kaplan, 1982: 210).

18. Syndactyly and finger maps: (Mogilner, Grossman, Ribary, et al., 1993).

19. Phantom limb pain and cerebral reorganization: (Flur, Elbert, Knecht, et al., 1995).

20. Chronic back pain and neural plasticity: (Flor, Braun, Elbert, and Birbaumer, 1997).

21. Back pain and exercise: (Frost, Lamb, Klaber, Moffett, Fairbank, and Moser, 1998).

22. Repetitive strain injuries: (Byl, Merzenich, and Jenkins, 1996).

23. Tinnitus, phantom sounds, and auditory map reorganization: (Mühlnickel, Elbert, Taub, and Flor, 1998).

24. Auditory cortex growing as "visual cortex": (von Melchner, Pallas, and Sur, 2000; Sharma, Angelucci, and Sur, 2000; Gao and Pallas, 1999; Roe, Pallas, Kwon, and Sur, 1991; Sur, Pallas, and Roe, 1990). Also found in surgically treated hamsters: (Frost, 1999).

25. The blind mole rat and auditory input redirected by evolution into the visual cortex: (Doron and Wollberg, 1994).

26. Developmental specification of cortex by connections: (O'Leary, Schlaggar, and Tuttle, 1994).

27. Visual cortex of the blind remaining intact: (Breitenseher, Uhl, Wimberger, et al., 1998).

28. Activated primary visual cortex in people born blind: (Veraart, De Volder, Wanet-Defalque, et al., 1990; De Volder, Bol, Blin, et al., 1997).

29. Sound processing, the visual cortex, and those born blind: (Leclerc, Saint-Amour, Lavoie, Lassonde, and Lepore, 2000; Weeks, Horwitz, Aziz-Sultan, Tian, et al., 2000; Alho, Kujala, Paavilainen, et al., 1993; Kujala, Alho, Paavilainen, et al., 1992: on page 469 they state "the blind might use. . . even occipital brain areas in sound localisation"). Note, though, some scientists suggest changes are limited to the secondary visual areas.

30. Tactile imagination and the visual cortex: (Uhl, Kretschmer, Lindinger, et al., 1994); Braille and the visual cortex: (Hamilton and Pascual-Leone, 1998; Sadato, Pascual-Leone, Grafman, et al., 1998; Pascual-Leone, Hamilton, Tormos, Keenan, and Catalá, 1999; Uhl, Franzen, Lindinger, Lang, and Deecke, 1991; Sadato, Pascual-Leone, Grafman et al., 1996; Cohen, Ceilnik, Pascual-Leone, et al., 1997).

31. Strokes destroying visual cortex destroy Braille reading: (Hamilton, Keenan, Catalá, and Pascual-Leone, 2000).

32. Blind cats using visual areas for better hearing: (Korte and Rauschecker, 1993).

33. Blind people hearing speech twice as well as those with sight: (Niemeyer and Starlinger, 1981).

34. Blind people having faster reaction times: (Kujala, Lehtoksoski, Alho, et al., 1997).

35. Echolocation and humans: (Kellog, 1962).

36. Blind people better able to locate sound: (Lessard, Paré, Lepore, and Lassonde, 1998).

37. Burns Taylor, the blind kid on a bike: (McCarty and Worchel, 1954).

38. Active auditory cortex in those born deaf: (Catalán-Ahumada, Deggouj, De Volder, et al., 1993).

39. Deaf cats "seeing" with their auditory cortex: (Rebillard, Carlier, Rebillard, and Pujol, 1977).

40. Deaf "hearing" sign language in auditory cortex: (Nishimura, Hashikawa, Doi, et al., 1999).

41. Brain reorganization in those born deaf: (Neville and Lawson, 1987).

42. Neurons making up only one-fortieth of cortex: (Haug, Kühl, Mecke, et al., 1984: 364, tables 3 and 4).

43. Ratio of glia to neurons—range: (Blinkov and Glezer, 1968); 1 to 9 ratio with neurons: (Travis, 1994); but some suggest roughly equal numbers: (Haug, Kühl, Sass, and Wasner, 1984: table 5 and p. 372, where they put it at 55 to 50).

44. Fifteen billion neurons: (Haug, Kühl, Mecke, et al., 1984).

45. 350 miles of capillaries: (Blinkov and Glezer, 1968: 267).

46. No easy generalizations on inputs and outputs of cerebral cortex: (Jones, 1985: 100).

47. Cell migration and dispersal: (Tan and Breen, 1993; Walsh and Cepko, 1992).

48. Cortex layers, input and output development: (McConnell, 1992; Shatz, 1992).

49. Variation in our bodies, reviews: (Darwin, 1871/1981: 39–40; Williams, 1967).

50. Threefold variation in number of foveal cones: (Curcio, Sloan, Packer, et al., 1987).

51. Four retinal photopigments: (Jameson, Highnote, and Wasserman, 2001).

52. Sinuses having 20-fold variation: (Williams, 1967: 26–27).

53. Blood temperature variation: (Mackowiak, Wasserman, and Levine, 1992); note each person's temperature varies—lowest at 6 am, highest at 6 pm—and even this varies, with some changing only 0.05°C (0.1°F), others up to 1.3°C (2.4°F).

54. Variations in muscles: (Anson, 1962: 179, 250, and 207). Variations in sinuses and stomachs: (Williams, 1967: 13–31).

55. Man with seven fingers (three fingers replacing an omitted thumb): (Dwight, 1892: 478).

56. Not 36 bodies, as cited by Charles Darwin: (1871: 39).

57. The catalog of human variation: (Bergman, Thompson, and Afifi, 1984).

58. Salamander embryo transplants and adapting brain: (Twitty, 1966: 25).

59. Laying down of cortex circuits by connections: (O'Leary, Schlaggar, and Tuttle, 1994).

60. Primary visual cortex circuits shaped by inputs: if eye input is removed early on, the primary visual cortex takes the form of secondary visual cortex: (Dehay, Giroud, Berland, et al., 1996).

61. Late blindness changing brains: (Alho, Kujala, Lehtokoski, et al., 1995).

62. Aging and neural plasticity: (McIntosh, Sekuler, Penpeci, et al., 1999).
63. Neural plasticity and capillaries: (Black, Sirevaag, and Greenough, 1987).
64. New neurons: (Kempermann, Kuhn, and Gage, 1998).
65. Neural plasticity and axonal sprouting: (Darian-Smith and Gilbert, 1994).
66. Three to 6 percent changes in thickness of cortex (Renner and Rosenzweig, 1987: 14-16).
67. Professional keyboard players having thicker cerebral cortex: (Amunts, Schlaug, Jäncke, et al., 1997).
68. Greatest flexibility in first 10 years: (Chugani, 1994).
69. Twins and PET scans: (Clark, Klonoff, Tyhurst, et al., 1988); twins and EEG: (Lykken, McGue, Tellegen, and Bouchard, 1992: 1567; Lykken, Tellegen, and Thorkelson, 1974; Polish and Burns, 1987).
70. Flexibility of growing brain: Some nonuniformities exist in the neural plate birth zone of the cortex (for an example involving the primary visual cortex, see Dehay, Giroud, Berland, et al., 1993). However, no evidence exists that such nonuniformities affect the cortex's later flexibility or ability to adapt: (Roe, Pallas, Kwon, and Sur, 1991; Sur, Pallas, and Roe, 1990).
71. Silent strokes: (Chodosh, Foulkes, Kase, et al., 1988).
72. Size and lack of symptoms in brain tumors: (Olivero, Lister, and Elwood, 1995).
73. One side of the brain compensating for atrophy in the other half: (Skoyles, 1999a ; Griffith and Davidson, 1966; Smith, 1974; Smith and Sugar, 1975; Wigan, 1844/1985).
74. Many small brain injuries versus a single big brain injury: (Feeney and Baron, 1986).
75. Brain damage and excitotoxins: (Lynch and Dawson, 1994).
76. Distant effects of brain injuries: (Feeney and Baron, 1986; Højer-Pedersen and Petersen, 1989; Metter, Riege, Hanson, et al., 1984; Kosslyn, Daly, McPeek, et al., 1993a).
77. Loss of two-thirds of somatosensory cortex without impairment: (Roland, 1993: 116).
78. After strokes, finger maps invading face maps: (Weiller, Ramsay, Wise, et al., 1993: 187).
79. Human evolution packing our brains with nonsensory and nonmotor cortex: (Deacon, 1990a).
80. Primary visual cortex activation and speed of picturing letters: (Kosslyn, Thompson, Kim, et al., 1996).
81. Mental imagery and the visual cortex: (Farah, Péronnet, Gonon, and Giard, 1988; Kosslyn, Alpert, Thompson, et al., 1993b; Le Bihan, Turner, Zeffiro, et al., 1993; Kosslyn, Thompson, Kim, and Alpert, 1995; Miyashita, 1995).
82. Imagined running and increased heart rate: (Decety, Jeannerod, Germain, and Pastene, 1991).
83. Motor sequencing and the supplementary motor area: (Roland, Larsen, Lassen, and Skinhøj, 1980; Rao, Binder, Bandettini, et al., 1993).

84. Tapping phantom fingers: (Ersland, Rosen, Lundervold, et al., 1997).
85. Knowing a grapefruit is bigger than an orange: (Goldenberg, Podreka, Steiner, et al., 1989).
86. Chinese abacus operators: (Stigler, 1984); abacus activating primary visual cortex (Pütz, Miyauchi, Sasaki, et al., 1996).
87. Paul Wittgenstein: See the letter by his student Erna Otten and comment by Oliver Sacks: (Otten and Sacks, 1992).
88. Judging picture of left and right hands: (Parsons, Fox, Downs, et al., 1995).
89. Slight movements of tongue and lips during inner thought: (Sokolov, 1972: 157–167). Note it is an overflow phenomenon: People with paralyzed speech apparatus have unaffected memory.
90. Broca's area and inner speech: (Hinke, Hu, Stillman, et al., 1993; McGuire, Silbersweig, Murray et al., 1996; also see Luria, 1973: 319–320); also possibly Wernicke's area: (Ingvar, 1993).
91. Imagining and hearing song and the superior temporal cortex: (Zatorre, Halpern, Perry, et al., 1996).
92. Memory buffering of overheard and articulated speech: (Baddeley and Salamé, 1986; Baddeley, Thomson, and Buchanan, 1975; Salamé and Baddeley, 1982).
93. Speech buffers in thought: (Sokolov, 1972).
94. An individual with at least partial visual inner speech: (Campbell and Butterworth, 1985)
95. The case of someone after a stroke "thinking in images": (Levine, Calvanio, and Popovics, 1982).
96. Children learning use of speech memory: (Cowan, Cartwright, Winterowd, and Sherk, 1987; Gathercole, Adams, and Hitch, 1994).
97. Abacus experts' special visual-spatial buffers: (Hatano and Osawa, 1983).
98. Imaginary and real exercise strength gains: (Yue and Cole, 1992).
99. People immobilized in casts losing less strength if they do imaginary exercises: (Pascual-Leone and Blanco, 1994).

Chapter 4 Superbrain

1. Dr. Egas Moniz, Nobel Prizes, and prefrontal lobotomies: (Valenstein, 1986, 1990); Nobel Prize citation quote: (Valenstein, 1990: 539).
2. Psychologist's comment on Carolyn: (Valenstein, 1986: 245).
3. IQ and prefrontal lobe removal: (Shallice, 1988: 329). While prefrontal cortex injuries may not affect clinical IQ tests (Wechsler Adult Intelligence Scale), they do affect IQ measures of fluid intelligence g: (Duncan, Burgess, and Emslie, 1995).
4. Man with IQ over 150 after lobotomy: (Hebb, 1939: 79–80, case IV).
5. The size of the human frontal lobe compared to the rest of the brain is no larger than might be expected for an ape brain of similar size: (Semendeferi,

Damasio, Frank, and Van Hoesen, 1997). However, the frontal lobe contains areas in addition to the prefrontal cortex; when this is taken into account, the frontal lobe is proportionately larger: (Semendeferi, Armstrong, Schleicher, et al., 2001).

6. Relative area of prefrontal cortex in different animals: (Fuster, 1980: 5, citing Brodmann; Deacon, 1990b).

7. Prefrontal cortex and metabolic activity when blindfolded and earplugged: (Roland, 1984, 1993: 472).

8. Meditation and prefrontal cortex: (Anand, Chhina, and Singh, 1961; Banquet, 1973: 146; Kasamatsu and Hirai, 1966).

9. Moments before sleep and prefrontal cortex: (Santamaria and Chiappa, 1987: 348; Markand, 1990: 171).

10. Prefrontal cortex having 16 times as many synapses (spines): (Elston, 2000).

11. The prefrontal cortex and pleasures of alcohol: (Tiihonen, Kuikka, Hakola, et al., 1994b).

12. Right frontal cortex and orgasm: (Tiihonen, Kuikka, Kupila, et al., 1994a).

13. Prefrontal cortex functions: A wide literature exists (for example, Fuster, 1989, 1997; Levin, Elsenberg, and Benton, 1991; Passingham, 1993; Shimamura, 2000). The ideas in this chapter (apart from the musical metaphor) are derived from Tim Shallice's supervisory attention system (SAS) model: (Shallice, 1988; Shallice and Burgess, 1991).

14. Prefrontal cortex and integration: (Prabhakaran, Narayanan, Zhao, and Gabrieli, 2000).

15. Prefrontal cortex and multitasking: (Burgess, Veitch, De Lacy Costello, and Shallice, 2000).

16. The prefrontal cortex and internal cues: (Fuster, 1980; Passingham, 1985, 1993; Goldman-Rakic, Bates, and Chafee, 1992; Godefroy and Rousseaux, 1997).

17. "Orchestra without conductor": existence noted by Bruno Walter (Bamberger, 1965: 154–155). One also existed in Moscow, called "Persimfans": (Galkin, 1988: 520). Orpheus, a conductorless orchestra, also exists at present in New York City.

18. Increased need for conductors after Beethoven: (Galkin, 1988: 7 and elsewhere).

19. Dates of first conducting: (Robertson, 1974).

20. Prefrontal cortex and task changing: (Dove, Pollmann, Schubert, et al., 2000).

21. Prefrontal cortex and executive function: (Adcock, Constable, Gore, and Goldman-Rakic, 2000; Bunge, Klingberg, Jacobsen, and Gabrieli, 2000; Fuster, 2000; Owen, Schneider, and Duncan, 2000; Paulesu, Frith, and Frackowiak, 1993; Petrides, Alivisatos, Meyer, and Evans, 1993; Smith, and Jonides, 1999).

22. Sustained attention and prefrontal cortex: (Knight, 1991; Wilkins, Shallice, and McCarthy, 1987).

23. Novelty and the prefrontal cortex: (Daffner, Mesulam, Scinto, et al., 2000; Knight, 1984; 1994; Desmedt and Debecker, 1979; Halgren, Baudena, Clarke, et al., 1995).

24. Women shaving their faces and other odd people and lack of surprise in a woman with prefrontal cortex damage: (Brazzelli, Colombo, Della Sala, and Spinnler, 1994: 38).

25. Imagining and perceiving music: (Zatorre, Halpern, Perry, et al., 1996).

26. Working memory and our prefrontal cortex: (Goldman-Rakic, 1994, 1996).

27. Prefrontal cortex, delay, and animal research: (Fuster, 1989; Stamm, 1987; Constantinidis, Franowicz, and Goldman-Rakic, 2001).

28. Prefrontal cortex preparing, waiting, and evaluating: (Basile, Rogers, Bourbon, and Papanicolaou, 1994; Burgess, Quayle, and Frith, 2001; Pochon, Levy, Poline, et al., 2001).

29. Temporary stopping of the prefrontal cortex with magnetic stimulation: (Pascual-Leone and Hallett, 1994).

30. Delay of gratification and later academic success: (Mischel, Shoda, and Peake, 1988; Mischel, Shoda, and Rodriguez, 1989).

31. Brain scans show delay upon lighting up prefrontal and visual cortices: (Goldberg, Randolph, Gold, Berman, and Weinberger, 1994).

32. Prefrontal cortex injuries, imitation, and external cuing of actions: (Lhermitte, Pillon, and Serdaru, 1986: first quotation, 328; Lhermitte, 1986: second quotation, 338).

33. Prefrontal cortex and imagination: (Schupp, Lutzenberger, Birbaumer, Miltner, and Braun, 1994).

34. A 6-hour New York–to–Los Angeles car journey and 90 percent of Americans men: (Eslinger, Grattan, Damasio, and Damasio, 1992: 765).

35. Estimation and prefrontal cortex: (Shallice and Evans, 1978).

36. Prefrontal cortex's encoding and decoding role in memory: (Allan, Dolan, Fletcher, and Rugg, 2000; Fletcher, Shallice, and Dolan, 2000; Johnson, Saykin, Flashman, et al., 2001; Kirchhoff, Wagner, Mari, and Stern, 2000; Konishi, Wheeler, Donaldson, and Buckner, 2000; Lee, Robbins, Pickard, and Owen, 2000; Opitz, Mecklinger, and Friederici, 2000; Ranganath, Johnson, and D'Esposito, 2000; Rugg, Fletcher, Frith, et al., 1996; Shallice, Fletcher, Frith, et al., 1994; Tulving, E., Kapur, S., Craik, et al., 1994).

37. Prefrontal cortex, recollection, and memory management: (Metcalfe, 1993: 14–17; Della Sala, Laiacona, Spinnler, and Trivelli, 1993). Prefrontal injuries affecting free but not externally cued recall: (Jetter, Poser, Freeman, et al., 1986).

38. Prefrontal cortex injury and confabulation: (Rapcsak, Polster, Glisky, and Comer, 1996).

39. Wisconsin Card Sorting Test and prefrontal cortex: (Drewe, 1974).

40. Prefrontal cortex, functional (magnetoencephalography) imaging and Wisconsin Card Sorting Test: (Wang, Kakigi, and Hoshiyama, 2001).

41. Prefrontal cortex and grouping ideas: (Freedman, Riesenhuber, Poggio, and Miller, 2001).
42. Prefrontal cortex and learning: (see Chapters 6 and 14).
43. Prefrontal cortex and planning: (Burgess, Veitch, De Lacy Costello, and Shallice, 2000; Goel and Grafman, 2000; Goldberg and Podell, 2000; Koechlin, Corrado, Pietrini, and Grafman, 2000; Lepage and Richer, 2000).
44. Prefrontal cortex and cognitive flexibility: (Eslinger and Grattan, 1993).
45. Supervisory attention: (Shallice, 1988: 332–352; Shallice and Burgess, 1991).
46. Speechless inner "image" speech: (Levine, Calvanio, and Popovics, 1982). One university student's sketchpad is likely to be visual (her visual buffer is better than her verbal one): (Campbell and Butterworth, 1985).
47. Deaf signers and short-term memory: (Shand, 1992).
48. Prefrontal cortex and our mental sketchpad: (Petrides, Alivisatos, Meyer, and Evans, 1993; Postle, Berger, and D'Esposito, 1999; Jahanshahi, Dirnberger, Fuller, and Frith, 2000). Supporting the nonprefrontal location of short-term buffers (as opposed to an executive controlling them) is the lack of prefrontal cortex activation in nondemanding short memory tasks: (Paulesu, Frith, and Frackowiak, 1993).
49. The inner thoughts of the deaf being sign-based: (Jackson, 1958: 126–127).
50. Thinking in signs activating the same areas as thinking verbally: (McGuire, Robertson, Thacker, et al., 1997).
51. Inner arm "voice" of deaf: (Max, 1937).
52. Helen Keller and "finger memory": (Keller, 1903: 293).
53. Vygotsky and self-organizing use of speech: (Vygotsky, 1962: 29–36, 86–88, and 225–236; see also Luria, 1980: 308–309; Frauenglass and Diaz, 1985; Beck, 1994).
54. Prefrontal cortex and recall, monitoring, and manipulation of tones: (Zatorre, Halpern, Perry, et al., 1996). Most brain activity coherences during musical imagery occur over left prefrontal cortex: (Petsche, Richter, von Stein, et al., 1993: 142).

Chapter 5 Mind-Engine

1. Electrode studies and tuning/modulating: (Desimone, 1996; Desimone, Miller, and Chelazzi, 1994: 86; Miller and Desimone, 1994: 522 and note 13; Miller, Erickson, and Desimone, 1996). PET studies: (Roland, 1993: 139, 225–726).
2. Two-part activation of visual cortex on attending to a word's letters: (Posner, 1994: 7401–7402).
3. Prefrontal and anterior cingulate organization (200 ms) of processing in the Wernicke's area (700 ms): (Snyder, Abdullaev, Posner, and Raichle, 1995).

4. Visual cortex, lateral geniculate nucleus, and links between them: (Llinás and Paré, 1991: 525.) What might be going on between the cortex and the lateral geniculate nuclei? Llinás and Paré (1991) suggest a reverberatory circuit responsible for our gamma brain waves; Edelman suggests a reentrant loop. Note that visual input can reach the secondary visual cortex through some other minor links. But, though minor, these might be sufficient to explain the phenomenon of "blind sight": (Cowey and Stoerig, 1991).

5. Lateral geniculate nuclei active for 320 ms: (Dinse and Krüger, 1994: table 1).

6. Neurons in the superior temporal lobe recognizing shapes within 20 ms: (Rolls, and Tovee, 1994).

7. Oscillations lasting 40 to 100 seconds: (Albrecht, Royl, and Kaneoke, 1998).

8. 1000-Hz brainwaves: (Klostermann, Funk, Vesper, et al., 2000).

9. Deep sleep brain rhythms: (Steriade, Contreras, Dossi, and Núñez, 1993).

10. Alpha rhythms: (Markand, 1990).

11. Mu brain activity: (Arroyo, Lesser, Gordon, et al., 1993a).

12. Tau, or hearing alpha: (Tesche and Hari, 1993).

13. Alpha: A committee for the International Federation of Societies for Electroencephalography and Clinical Neurophysiology defines alpha as oscillations between 7 and 13 Hz linked to vision. This definition is too narrow. Mu (motor alpha) ranges from 8 to 22 Hz (Tesche and Hari, 1993: figs. 2c and 2d) but is functionally akin to alpha. Beta: This frequency range (14–40 Hz) overlaps functionally with both alpha and gamma, especially over the motor and sensorimotor cortex in the upper mu range. Frequency, however, is only one aspect of oscillations: Similar frequencies may reflect different processes. For instance, alpha shares much of its frequency range with the spindles of sleep, yet these are generated by quite different processes: (Steriade, Contreras, Dossi, and Núñez, 1993). The same observation applies to the overlap between alpha and gamma, which are known to be separately generated: (Tesche and Hari, 1993).

14. Prefrontal cortex active when one is blindfolded and earplugged: (Mazziotta, Phelps, Carson, and Kuhl, 1982; Roland, 1984, 1993: 472).

15. Frontal alpha and drifting into sleep: (Markand, 1990: 171; Santamaria and Chiappa, 1987: 348; but note Hari, 1993: 1046–1047, explains it in terms of a shift between visual and auditory alpha).

16. Meditation and alpha, including yoga: (Anand, Chhina, and Singh, 1961); TM: (Banquet, 1973); zen: (Kasamatsu and Hirai, 1966); qi gong: (Zhang, Li, and He, 1988).

17. Meditators with alpha rhythms when their eyes are open: (Anand, Chhina, and Singh, 1961; Banquet, 1973: 146; Kasamatsu and Hirai, 1966).

18. Visualizers and alpha: (Slatter, 1960).

19. Gamma synchronization, general: (Basar and Bullock, 1992; Bressler, 1990; König and Engel, 1995; König, Engel, and Singer, 1995; Kulli and Koch, 1991; Llinás and Paré, 1991; Ribary, Ioannides, Singh, et al., 1991; Singer, 1993, 1994; von der Malsburg, 1995).
20. Alpha and gamma oscillations: (Bressler, 1990).
21. Gamma, arousal, and the reticular formation: (Munk, Roelfsema, König, et al., 1996).
22. Problems detecting gamma activity: (Rockstroh, Elbert, Birbaumer, and Lutzenberge,1983: 105; Spydell and Sheer, 1982; Spydell, Ford, and Sheer, 1979).
23. The binding problem and its solution by synchrony and gamma: (Engel and Singer, 2001; Gray, 1999; Reynolds and Desimone, 1999; Singer, 1999; von der Malsburg, 1999; Wolfe and Cave, 1999); for criticisms: (Shadlen and Movshon, 1999).
24. The binding problem in memory: (Cohen and Eichenbaum, 1993: 286–287) and thinking: (Shashri and Ajjanagadde, 1993).
25. Optical imaging of synchronization: (Arieli, Shoham, Hildesheim, and Grinvald, 1995).
26. The sweep of gamma oscillations: (Llinás and Paré, 1991: 531; Ribary et al., 1992).
27. Chattering cells: (Gray and McCormick, 1996).
28. Thalamus relay neurons, changes in processing, and synchronization: (Sillito, Jones, Gerstein, and West, 1994); 40-Hz responses to visual bars in humans: (Lutzenberger, Pulvermüller, Elbert, and Birbaumer, 1995).
29. Synchronization and scene segmentation: (Engel, König, Kreiter, et al., 1992; Engel, König, and Singer, 1991; von der Malsburg, 1995).
30. Strabismic amblyopia, synchronization, and binding: (Roelfsema, König, Engel, et al., 1994).
31. Resetting of gamma oscillations: (Llinás and Ribary, 1993; Ribary et al., 1992).
32. Gamma and temporal binding: (Joliot, Ribary, and Llinás, 1994).
33. Gamma, binding, and optical illusions: (Tallon, Bertrand, Bouchet, and Pernier, 1995).
34. Evoked responses and gamma: (Hari, 1993: 1057; Pantev, Makeig, Hoke, et al., 1991).
35. Gamma and hearing: (Hari, 1993; Pantev, Makeig, Hoke, et al., 1991; Ribary et al., 1992; Tesche and Hari, 1993; Tiitinen, Sinkkonen, Reinikainen, et al., 1993).
36. Gamma synchronization between the two hemispheres: (Engel, König, Kreiter, and Singer, 1991).
37. Gamma linking activity in human cortex 9 cm apart: (Desmedt and Tomberg, 1994).

38. Gamma and preparation to move: (Kristeva-Feige, Feige, Makeig, et al., 1993; Murphy and Fetz, 1992; Pfurtscheller, Neuper, and Kalcher, 1993; Sanes and Donoghue, 1993). Synchronized coherence in the premotor cortex: (Vaadia, Haalman, Abeles, et al., 1995).

39. Synchronized coherence across the brain: (Bressler, Coppola, and Nakamura, 1993; Desmedt and Tomberg, 1994). Synchronized coherence in the prefrontal cortex: (Aertsen, Vaadia, Abeles, et al., 1991).

40. Gamma and fast and slow reactors: (Jokeit and Makeig, 1994).

41. Gamma and selective attention: (Tiitinen, Sinkkonen, Reinikainen, et al., 1993). Gamma linking prefrontal cortex and sensory parietal cortex during attention: (Desmedt and Tomberg, 1994).

42. Hippocampal gamma and quotes on possible role in memory: (Bragin, Jandó, Nádasdy, et al., 1995).

43. Music perception, coherence, and high beta activity: (Petsche, 1996; Petsche, Richter, von Stein, et al., 1993; Petsche, von Stein, and Fitz, 1996). Recently, Petsche's group has found a gamma link to music (Bhattacharya and Petsche, 2001).

44. Gamma synchronization and reasoning: (Shashri and Ajjanagadde, 1993).

45. Gamma and problem solving: (Spydell, Ford, and Sheer, 1979; Spydell and Sheer, 1982).

46. Gamma and left hemisphere neural assemblies processing words: (Lutzenberger, Pulvermüller, and Birbaumer, 1994).

47. Gamma and learning disability: (Sheer, 1976).

48. Alzheimer's disease and reduced gamma: (Ribary, Ioannides, Singh, et al., 1991).

49. Schizophrenia and gamma: (Haig, Gordon, De Pascalis, et al., 2000).

50. Gamma binding as a kind of computer clock: (Pöppel, 1989: 227).

51. Subthreshold oscillations in neurons and gamma: (Llinás, Grace, and Yarom, 1991; Nuñez, Amzica and Steriade, 1992; Wang, 1993).

52. Oscillations and synchronization over 2 mm: (König, Engel and Singer, 1995).

53. The intraliminar thalamic nuclei and gamma: (Llinás and Ribary, 1993). Note other brain nuclei have been suggested to play a role in activating gamma, such as the nucleus basalis of Meynert: (Metherate, Cox, and Ashe, 1992).

54. Pulvinar and attention: (Olshausen, Anderson, and Van Essen, 1993; Leberge, 1995; Robinson and Peterson, 1992).

55. Pulvinar and gamma: (Shumikhina and Molotchnikoff, 1995).

56. Generation of gamma by reticular thalamus: (Llinás, Grace, and Yarum, 1992; Pinault and Deschênes, 1992).

57. Reticular thalamus: (Blenner and Yingling, 1993; Mitrofanis and Guillery, 1993; Yingling and Skinner, 1977).

58. Basal ganglia (external segment of the globus pallidus) control over reticular thalamus: (Parent and Hazrati, 1995: 117).

59. The name is such a mouthful we have spared the reader in the body of our text: the *acetylcholine ascending projection of the pedunculopontine tegmental nucleus.*

60. Prefrontal control of brain activity through the reticular formation: (Desmedt and Debecker, 1979: 660–661).

61. Reticular thalamus and the prefrontal cortex: (Blenner and Yingling, 1993; Rockstroh, Elbert, Birbaumer, and Lutzenberger, et al., 1983: 102–104; Yingling and Skinner, 1977).

62. Failure of alpha biofeedback: (Rockstroh, Elbert, Canavan, et al., 1990: 130–131).

63. Biofeedback control of gamma: (Bird, Newton, Sheer, and Ford, 1978).

64. Reticular thalamus and initiation: (Raeva and Lukashev, 1993).

65. Prefrontal cortex and parietal sensory cortex gamma synchronization: (Desmedt and Tomberg, 1994).

66. Slow potentials: (Rockstroh, Elbert, Birbaumer, and Lutzenberger, 1983; Rockstroh, Elbert, Canavan, et al., 1990).

67. Increased neuron activity and slow cortical potentials: (Rockstroh, Elbert, Canavan, et al., 1990: 50–55; Rockstroh, Müller, Cohen, and Elbert, 1992; Sandrew, Stamm, and Rosen, 1977).

68. PET and slow potentials: (Lang, Lang, Podreka, et al., 1988).

69. Slow potentials predicting task accuracy: (Morgan, Wenzl, Lang, et al., 1992).

70. Slow potentials and facilitation of learning and attention: (Sandrew, Stamm, and Rosen, 1977).

71. Reticular thalamus and slow potentials: (Rockstroh, Elbert, Birbaumer, and Lutzenberger, 1983: 103; Yingling and Skinner, 1977: 91–92).

72. Slow potentials and the prefrontal cortex: (Rockstroh, Elbert, Birbaumer, and Lutzenberger, 1983: 168–169; Rockstroh, Elbert, Canavan, et al., 1990: 178–181).

73. Prefrontal cortex modulation of early attentive sensory processing: (Knight, 1991: 142). For a role in the later N100: (Joutsiniemi and Hari, 1989; see also Desmedt and Debecker, 1979 for N120).

74. Delays of 14 ms between auditory and prefrontal cortex: (Zappoli, Zappoli, Versari, et al., 1995).

75. Prefrontal injuries and gating of very early activity in the auditory cortex: (Knight, Scabini, and Woods, 1989).

76. The tuning and recruitment of cortical fields by the prefrontal cortex: (Roland, 1993: 139, 226).

77. The effects of cooling the prefrontal cortex on its modulation of other brain areas: (Goldman-Rakic and Chafee, 1994).

78. Unilateral prefrontal injuries and unilateral cessation of brain activity: (Zappoli, Zappoli, Versari, et al., 1995).

Chapter 6 Neural Revolution

1. Heavy brains in two mentally retarded people: (Markowitsch, 1992: 10; citing work published by Matiegka in 1902). Note the 1995 *Guinness Book of Records* (Matthew, 1995) records one at 5 lb 1.1 oz. Twenty-lb sperm whale brain: *Guinness Book of Records* (although later editions do not continue to carry this fact).

2. Brain and body weights in various animals: (Blinkov and Glezer, 1968: 342–345); in general: (Deacon, 1990b; note that the allometric equations linking body and brain weights imply small animals will have large brain–body ratios).

3. No new components—certain neuron types have been claimed to be unique to us (such as calretinin-containing pyramidal cells in anterior cingulate cortex layer V); however, they are only more common in our species: (Hof, Nimchinsky, Perl, and Erwin, 2001).

4. Human brains three times expected mass: (Passingham, 1973).

5. Re 812-lb chimps: data extrapolated from figure 12 and equation 18 in Martin (1982: 28).

6. Primary sensory and motor cortices same relative size in apes and humans: (Deacon, 1990a).

7. For further reading, see these introductions to neural networks: (Allman, 1989; Cowan and Sharp, 1988; Gurney, 1997; Hinton, 1992; Koch, 1997; McClelland and Rumelhart, 1988; Rumelhart, McClelland, and the PDP Research Group, 1986).

8. Neural network training and the problem of correction feedback for supervised networks: (Skoyles, 1991a, 1991b; also Share, 1995).

9. Error correction and information processing: Networks can be trained in different ways—supervised, unsupervised (such as Boltzmann), and autoassociative. These comments, while focused on supervised training, also apply if it is done unsupervised (Skoyles, 1991b). Notice that latent semantic analysis (Landauer and Dumais, 1997) offers a deep insight into how new information can be picked up in terms of extracted eigenvectors. But space prevents, at least in this book, discussion of this and the prefrontal cortex.

10. Duncan's work on IQ and the prefrontal cortex: (Duncan, Burgess, and Emslie, 1995; Duncan, Emslie, Williams, et al., 1996; Duncan, Johnson, Swales, and Freer, 1997).

11. The late development of the prefrontal cortex: (Delalle, Evers, Kostovic, and Uylings, 1997; Eslinger, Grattan, Damasio, and Damasio, 1992; Heinen, Glocker, Fietzek, et al., 1998; Kostovic, 1990: 234-245; Marosi, Harmony, Sánchez, et al., 1992; Paus, Zijdenbos, Worsley, et al., 1999; Sowell, Thompson, Holmes, et al., 1999; Steen, Ogg, Reddick, and Kingsley, 1997; Thatcher, 1991; Thatcher, Walker, and Giudice, 1987; Woo, Pucak, Kye, et al., 1997).

12. Age 16 and adult levels of prefrontal cortex synaptic density: (Huttenlocker, 1979; Kostovič, 1990: 235). Patricia Goldman-Rakic has questioned

this, using work on rhesus monkeys and the assumption that human and monkey brains develop similarly. They do not. Rhesus monkey brains are born to expand 70 percent, but human ones 350 percent.

13. Adolescent prefrontal cortex refining links within itself: (Hudspeth and Pribram, 1990). Evidence suggesting late volume changes during adolescence linked to white matter maturation: (Jernigan, Trauner, Hesselink, and Tallal, 1991).

14. Refinement of links around puberty: (Woo, Pucak, Kye, et al., 1997).

15. Kostović's quote: (Kostović, 1990: p. 223).

16. Enriched learning environment and greater prefrontal activation: (Raleigh, McGuire, Melega, Cherry, Huang, and Phelps, 1996).

17. Prefrontal working memory development: (Luciana and Nelson, 1998; Cycowicz, 2000; Sowell, Delis, Stiles, and Jernigan, 2001).

18. Prefrontal maturation and delay and selective attention: (Casey, Trainor, Orendi, et al., 1997; Casey, Giedd, and Thomas, 2000; Espy, Kaufmann, McDiarmid, and Glisky, 1999).

19. Prefrontal cortex regulation of cognitive development: (Case, 1992; Stuss, 1992).

20. Two visual systems and streams of processing: (Livingstone and Hubel, 1987; Ungerleider and Mishkin, 1982).

21. Hearing better while seeing speakers: (Obler, Nicholas, Albert, and Woodward, 1985).

22. Blindness and learning to speak: (Mills, 1987).

23. Ga-ga, ba-ba, and da-da: (McGurk and MacDonald, 1976).

24. McGurk effect and auditory cortex visual input: (Sams, Aulanko, Hämäläinen, et al., 1991).

25. Ears making sounds to aid intelligibility of speech against noise: (Giraud, Garnier, Micheyl, et al., 1997).

Chapter 7 Machiavellian Neurons

1. The Machiavellian animal mind: (Humphrey, 1976).

2. Morphinelike substances secreted after grooming: (Keverne, Martensz, and Tuite, 1989).

3. Grooming in chimps: (Goodall, 1986: 387–408, Figan and Flo, 402).

4. Monkey grandmothers aiding grandchildren's survival: (Fairbanks, 1988).

5. Melissa and Getty: (Goodall, 1986: 74).

6. Social life of monkeys and apes: (Smuts, 1985; de Waal, 1989; Whiten and Byrne, 1988).

7. Goblin's hero-worship of Figan: (Goodall, 1990: 48 and 120-122). For further reading, see www.upfromdragons.com and these introductions to neural networks: (Allman, 1989; Cowan and Sharp, 1988; Gurney, 1997; Hinton, 1992; Koch, 1997; McClelland and Rumelhart, 1988; Rumelhart, McClelland, and the PDP Research Group, 1986).

8. Figan and alliances: (Goodall, 1986: 65).

9. Whom to share food with and whom not to: (Nishida, Hasegawa, Hayaki, et al., 1992; discussion of manipulative meat sharing strategies in chimps).

10. Satan and Goblin: (Goodall, 1986: 387).

11. Motherhood and reptiles: (MacLean, 1990: 136–137). Dinosaurs and their young: (Paul, 1989: 42–43).

12. Robins and throwing the baby out with the bathwater: (Welty, 1982: 407).

13. The presence of interleukin-10 (an immunomodulating anti-inflammatory cytokine) in human milk: (Garofalo, Chheda, Mei, et al., 1995).

14. Naturally occurring benzodiazepinelike substances in human breast milk: (Dencker, Johansson, and Milsom, 1992).

15. Mammalian mothers regulating internal homeostasis of their infants: (Hofer, 1984; 1987).

16. Touch and the growth of young: (Schanberg and Field, 1987).

17. Mother's touch shaping later behavior: See references in Chapter 8.

18. Similarities among primate cries: (MacLean, 1990: 401). Individuality of human cries: (Gustafson, Green, and Cleland, 1994).

19. Mothering and morphinelike substances: (Bridges and Grimm, 1982; Panksepp, Siviy, and Normansell, 1985).

20. Play in mammals: (Fagen, 1981); in birds: (Fagen, 1981: 202–205).

21. Play in adults: (Fagen, 1981: 438–445).

22. Day-old babies distressed at another's distress: (Sagi and Hoffman, 1976).

23. Chimpanzees clasping each other: (de Waal, 1989: 12); spider monkeys: (MacLean, 1990: 402).

24. Conflict is important in social innovation. The political power of allied groups not only to withstand the harshness of nature but to sacrifice and control their individual members as they battle it out with each other is discussed at length in Bloom (1997).

25. Animals and morality: (Darwin, 1871/1981; de Waal, 1996).

26. Ethics in chimpanzees: (de Waal, 1991).

27. Near-murder of Goblin: (Goodall, 1992).

28. Figan, bananas, and suppression of food calls: (Goodall, 1986: 125).

29. The key role of play in social skill learning: (Van den Berg, Hol, Van Ree, et al., 1999).

30. Social relationships and social cognition: (Cheney, Seyfarth, and Smuts, 1986).

31. Manipulations in primate social life: (De Waal, 1989; Smuts, 1985).

32. Social survival driving brain expansion: (Sawaguchi and Kudo, 1991).

33. Figan versus Goblin: (Goodall, 1986: 68 and 78).

34. Shifting groups (fission-fusion) in chimps: (Goodall, 1986: 146–171); as ecological adaptation to gathering dispersed foods: (Ghiglieri, 1984: 174–175).

35. Goodall quote on chimps and firearms: (Goodall, 1986: 530).
36. Fission-fusion social life: Lions, bottlenose porpoises, eastern gray kangaroos, and spider monkeys also show fission-fusion. Chimps have taken it much further: Their societies are more fragmentary and flexible, with highly organized large group level defenses.
37. Chimps having bigger brains for their body size than gorillas: (Martin, 1982: 28–29 and fig. 12).

Chapter 8 The Troop within Our Heads

1. "A freeman? . . . : Euripides, *Hecabe* (lines 863–869).
2. Milgram's experiment: (Milgram, 1963; 1974; Miller, 1986).
3. Experts not predicting obedience: (Milgram, 1963: 375; 1974: 27–31).
4. Monkeys refusing to electrocute: (Masserman, Wechkin, and Terris, 1964).
5. Research on perceptions of the morality of Milgram's research (Bickman and Zarantonello, 1978; Miller, 1986: 110–115).
6. Comments about the avoidance of social nature research by social psychologists: (Batson, 1990: 336).
7. Second quote on individualism: (Fiske, 1992: 689).
8. Chinese anthropologist's comment: (Hsu, 1971: 34).
9. Wave maneuvers in birds and people: (Potts, 1984). AI computer simulations suggest individual-level processes may be involved.
10. "We" and "they": (Forsyth, 1990: 395; Turner, 1987).
11. The slightness of circumstances needed for prejudice: (Tajfel, 1970; Turner, 1987: 26-27).
12. Social groups and inhumanity; to be fair to women, the negative aspects of group behavior are confined largely to male, not female, ones: (Rodseth, Wrangham, Harrigan, and Smuts, 1991: 231).
13. Menstrual synchronization: (McClintock, 1971). Note there is some evidence that it is synchronized via the olfactory system: (Russell, Switz, and Thompson, 1980).
14. Synchronization of male circadian cortisol rhythms: (Vernikos-Danellis, 1980).
15. Social synchronization of the basic rest-activity cycle: (Meiser-Koll, 1992: 274–276).
16. Importance of social synchronization from another perspective: (Hofer, 1984: 187).
17. Synchronization in social interaction: (Kendon, 1970; Hatfield, Cocioppo, and Rapson, 1994).
18. For a general discussion of the physiological effects of social interaction: (Hofer, 1984). Small and unnoticed touches between people have been shown to have an effect on behavior: (Crusco and Wetzel, 1984).
19. Smiles: (Ekman, Davidson, and Friesen, 1990).

20. Empathy and physiological synchronization: (Levenson and Ruef, 1992). Such synchronization was found with positive emotions and depended upon a person's state of relaxation.

21. Oxytocin and bonding: (Carter and Getz, 1993; Insel and Young, 2000, 2001). Vasopressin and bonding: (Insel, Wang and Ferris, 1994; Pitkow et al., 2001; Uvnas-Moberg, 1997c; 1998; Winslow, Hastings, Carter, et al., 1993). For a book of articles on these chemicals and affiliation: (Carter, Lederhendler, and Kirkpatrick, 1987).

22. Marmosets and oxytocin receptors: (Wang, Moody, Newman, and Insel, 1997).

23. Oxytocin and infant attachment: (Insel, 1993: 231–232).

24. Oxytocin and relaxation: (Uvnas-Moberg, 1997b; 1998).

25. Oxytocin and selective serotonin reuptake inhibitor [SSRI] antidepressant drugs: (Uvnas-Moberg, Bjokstrand, Hillegaart, and Ahlenius, 1999).

26. Endorphins secreted after grooming: (Keverne, Martensz, and Tuite, 1989). Changed opiate receptors in people who commit suicide: (Gabilondo, Meana, and García-Sevilla, 1995).

27. Talk and people: (Dunbar, 1992).

28. Most activities more enjoyable when done in company: (Csikszentmihalyi and Larson, 1984: ch. 9).

29. William James quote on our social nature: (James, 1891: 293–294).

30. Lone sailors and their boats: (Hofer, 1984: 193).

31. Imaginary companions: (Critchley, 1979: 2–3).

32. Social isolation carrying a risk comparable to that of smoking: (House, Landis, and Umberson, 1988: 541). "Social relationships, or the relative lack thereof, constitute a major risk factor for health—rivaling the effects. . . of cigarette smoking, blood pressure, blood lipids, obesity and physical activity." Social isolates twice as likely to die as others: (House, Landis, and Umberson, 1988: 541–542). One study found "a relative risk ratio for mortality of about 2.0 indicating that persons low on the [social network] index were twice as likely to die as persons high on the index" (p. 541). Another study found similar results, with the added observation that "the greatest increase in mortality risk occur[s] in the most socially isolated third" (p. 542).

33. Living alone and recurrent heart attacks: (Case, Moss, Case, et al., 1992).

34. Cancer relapse and social contact: (Spiegel, 1991).

35. Social support and immune systems: (Cacioppo, 1994; Cohen, Kaplan, Cunnick, et al., 1992).

36. Up to 47 percent increase in weight of newborn human babies when handled: (Field, 2001).

37. Rat pups, premature babies, and touch: (Polan and Ward, 1994; Schanberg and Field, 1987).

38. Old Anna and tender loving care: (Montagu, 1978: 78).

39. Socially isolated monkeys and brain damage: The work concerned only subcortical brain structures. The damage was related to neurochemical (substance P, leucine-enkephalin, and tyrosine) changes in the striatum (Martin, Spicer, Lewis, et al., 1991). The researchers did not report on the cerebellum, hippocampus, anterior cingulate cortex, or neocortex. Postnatal increase in brain size in rhesus monkeys versus humans: (Martin, 1982: fig. 15).
40. Chugani findings on Romanian orphans: (Chugani et al., 2001).
41. Adrian Raine's work on violence: (Raine, Brennam, and Mednick, 1994).
42. One should not ignore the many individuals who survive hard upbringings with no apparent ill effects. Such robust individuals will tell us much about how to minimize the effects of disadvantage.
43. Mammalian mother regulating the internal homeostasis of her infants: (Hofer, 1984; 1987; 1996).
44. Breakdown of homeostasis in separated infants: (Hofer, 1996).
45. Working models and attachments: (Bowby, 1969).
46. Infant attachment leading to adult attachment: (Benoit and Parker, 1994; Main, 1990; Shaver and Clark, 1994).
47. Adult attachments with children: (Benoit and Parker, 1994; Main, 1990; Shaver and Clark, 1994).
48. Attachment and religion: (Kirkpatrick, 1992).
49. Social support and immune systems: (Cacioppo, 1994; Cohen, Kaplan, Cunnick, et al., 1992).
50. Impact of loving and caring parents 35 years later: (Russek and Schwartz, 1997).
51. Maternal (social) referencing: (Klinnert, Campos, Emde, and Svejda, 1983; Walden and Ogan, 1988); and visual cliff: (Sorce, Emde, Campos, and Klinnert, 1985). Social referencing in chimps: (Evans and Tomesello, 1986).
52. Social groups and norms: (Turner, 1987).
53. Social reference and conformity: (Feinman, 1992).
54. Myron Hofer on attachment and regulation: (Hofer, 1996: 577; see also Hofer, 1984; 1987).
55. Self-esteem as sociometer of group rejection: (Leary, Tamber, Terdal, and Downs, 1995).

Chapter 9 Our Living Concern

1. Blood pumped into the brain each minute: (Rempp, Brix, Wenz, et al., 1994: table 1 and using a 60:40 ratio suggested by them).
2. Speed of blood through brain: (Saito, Yoshokawa, Nishihara, et al., 1995: fig. 2, control); 1 minute after electroconvulsive therapy, it goes as high as 170 cm/s; in children it can be as high as 240 cm/s: (Pegelow, Wang, Granger, et al., 2001).

3. Anterior cingulate, reviews: (MacLean, 1990; Roland, 1993: 358–360; Vogt, Finch, and Olson, 1992).
4. Our minds' lack of access to their own workings: (Gopnik, 1993; Nisbett and Wilson, 1977).
5. Anterior cingulate active 2 s before actions: (Shima, Aya, Mushiake, et al., 1991).
6. The key paper that started it all: (Stroop, 1935; see also MacLeod, 1991).
7. Anterior cingulate and attention to action (Barch, Braver, Sabb, and Noll, 2000; Carter, Macdonald, Botvinick, et al., 2000; Casey, Thomas, Welsh, et al., 2000; Kiehl, Smith, Hare, and Liddle, 2000; Leung, Skudlarski, Gatenby, et al., 2000; Liddle, Kiehl, and Smith, 2001; MacDonald, Cohen, Stenger, and Carter, 2000; MacLeod and MacDonald, 2000; Pardo, Pardo, Janer, and Raichle, 1990; Peterson, Skudlarski, Gatenby, et al., 1999).
8. Anterior cingulate and saying words beginning with a common letter or making random finger movements: (Frith, Friston, Liddle, and Frackowiak, 1991).
9. Go–no-go tasks and anterior cingulate: (Leimkuhler and Mesulam, 1985; Kawashima, Satoh, Itoh, et al., 1996).
10. Man with the alien hand: (Feinberg, Schindler, Flanagan, and Haber, 1992: 19–20). Gary Goldberg (1985: 579) links this syndrome with the supplementary motor cortex, but more recent work finds a role for the anterior cingulate: (Hashimoto and Tanaka, 1998).
11. Freud's *über ich*, or over-I, was translated into the Latin-based "superego." This word has been tainted by decades of theorizing in the absence of neuroanatomical reality checks or empirical evidence. The notion of a cingulate over-self, by contrast, is based on actual observations of brain function.
12. Social inhibition and the anterior cingulate: (Garrett, Wood, and Keyes, 1994).
13. Giving up seats on trains and Milgram: (Milgram and Sabini, 1978: 37).
14. Minimal effect of punishment on crime: (Paternoster, Saltzman, Waldo, and Chiricos, 1983).
15. Anticipation and the anterior cingulate: (Drevets, Burton, Videen, et al., 1995; Murtha, Chertkow, Beauregard, et al., 1996).
16. Anterior cingulate and ANS: (Vogt, Finch, and Olson, 1992: 438).
17. Pain, anticipation, and the anterior cingulate: (Koyama, Kato, Tanaka, and Mikami, 2001; Sawamoto, Honda, Okada, et al., 2000).
18. Illness increasing pain sensitivity: (Wiertelak, Smith, Furness, et al., 1994).
19. Thalamic pain nuclei and their various cortical projections: (Craig, Bushnell, Zhang, and Blomquist, 1994).
20. Anterior cingulate, pain, and PET: (Casey, Minoshima, Berger, et al., 1994; Derbyshire, Jones, Devani, et al., 1994; Jones, Brown, Friston, et al., 1991). General discussion of pain and anterior cingulate: (Vogt, Sikes, and Vogt, 1993).

21. Pain and anterior cingulate lesions: (Ballantine, Bouckoms, Thomas, and Giriunas, 1987). The researchers did not seek to damage the anterior cingulate cortex but the fibers underneath it linking the surrounding medial-orbital prefrontal cortex and the hippocampus. However, as Brent Vogt and others (1992) note, this could not be done without partially injuring it.

22. Anterior cingulate and opiate receptors (also patient's quote): (Jones et al., 1991a; 1991b).

23. Acupuncture and the anterior cingulate: (Wu, Hsieh, Xiong, et al., 1999).

24. Anterior cingulate and nonrepeating of numbers: (Petrides, Alivisatos, Meyer, and Evans, 1993).

25. Anterior cingulate and the verbal selection of verbs to fit nouns: (Raichle, Fiez, Videen, et al., 1994; Snyder, Abdullaev, Posner, and Raichle, 1995).

26. Cingulate cortex and attention to reality: (Johnson, 1991: 188–189). Patient's quote: (Whitty and Lewin, 1957: 73).

27. Monkey vocalizations, anterior cingulate, and auditory cortex: (Müller-Preuss, Newman, and Jürgens, 1980).

28. Stopping auditory hallucinations by keeping the mouth open: (Bick and Kinsbourne, 1987).

29. Stopping hallucinations by repeatedly humming a single note: (Green and Kinsbourne, 1990).

30. Anterior cingulate interneurons reduction in schizophrenia: (Benes, McSparren, Bird, et al., 1991). Functional reduction: (Tamminga, Thaker, Buchanan, et al., 1992; Seigel, Buchsbaum, Bunney, et al., 1993: 1329).

31. Anterior cingulate and schizophrenic auditory hallucinations: (Cleghorn et al., 1992).

32. Cingulate, mothering, play, and crying (MacLean, 1990; Vogt, Finch, and Olson, 1992: 438–439).

33. Vasopressin receptors in anterior cingulate: (Insel, Wang, and Ferris, 1994: tables 2 and 3).

34. Jaak Panksepp on attachment and opioids: (Nelson and Panksepp, 1998; Panksepp, Siviy, and Normansell, 1985; Panksepp, Nelson, and Siviy, 1994).

35. Hypothalamus as chief controller of the autonomic systems: (Kojima, Ogomori, Mori, et al., 1996; Swanson, 2000).

36. Hypothalamus and male sexual orientation: (LeVay, 1991).

37. Oxytocin made in hypothalamus: (Carter, 1992); effects elsewhere in brain: (Insel and Shapiro, 1992).

38. Hypothalamus activation and imagining music: (Kato, Zhu, Strupp, et al., 1996).

39. Amygdala: Some anatomists question the identification of the many areas making up the amygdala as constituting a single brain area: (Swanson and Petrovich, 1998).

40. Right amygdala larger than left: (Giedd, Vaituzis, Hamburger, et al., 1996).

41. Fear and the right amygdala: (Coleman-Mesches and McGaugh, 1995).

42. The amygdala and perception of human emotions: (Adolphs, Tranel, Damasio, and Damasio, 1994).

43. The amygdala and *Playboy*: (Bauer, 1982: 704).

44. The amygdala and facial expressions and other social stimuli: (Brothers, Ring, and Kling, 1990; Brothers, 1992; Young, Aggleton, Hellawell, et al., 1995). Strange versus family faces: (Seeck, Mainwaring, Ives, et al., 1993).

45. A 27-year-old woman with amgydalotomy quote: (Jacobson, 1986: 441–442).

46. Amygdala and emotional associations: (Sarter and Markowitsch, 1985).

47. Rats, sex, and the amygdala: (Everitt, Cador, and Robbins, 1989).

48. Peirce on indexes: (Peirce, 1932: 143, 160–162; quote, p. 161).

49. Amygdala and learning from experience: (Gaffan, Murray, and Fabre-Thorpe, 1993).

50. Amygdala and perception: (Paradiso et al., 1999; Tabert et al., 2001; Taylor, Liberzon, and Koeppe, 2000).

51. Anterior cingulate and laughter: (Arroyo, Lesser, Gordon, et al., 1993b).

52. Anterior cingulate and initiation: (Cohen, Kaplan, Zuffante, et al., 1999; Devinsky, Morrell, and Vogt, 1995).

53. Regional brain activity and voluntary and spontaneous smiling: (Ekman and Davidson, 1993).

54. Making expressions, making feelings: (Levenson, Ekman, and Friesen, 1990; Schiff and Lamon, 1989).

55. The insula's mapping of our bodies: (Schneider, Friedman, and Mishkin, 1993).

56. Pain and the insula: (Casey, Minoshima, Berger, et al., 1994; Coghill, Talbot, Evens, et al., 1994).

57. Washington University research group on the insula/anterior cingulate: (Raichle, Fiez, Videen, et al., 1994).

58. Prefrontal cortex links to amygdala: (Amaral, Price, Pitkänen, and Carmichael, 1992: 48–53; Fuster, 1989: 15, 21; Morecraft, Geula, and Mesulam, 1992); to hypothalamus: (Fuster, 1989: 21; Kievit and Kuypers, 1975; Morecraft, Geula, and Mesulam, 1992; Nauta, 1971; Ongur, An, and Price, 1998).

59. Orbital prefrontal cortex and autonomic nervous system: (Neafsey, 1990). Note that Neafsey defines the orbital prefrontal cortex to include the insula.

60. Prefrontal cortex working with anterior cingulate for nonemotional tasks: (Paus, Petrides, Evans, and Meyer, 1993); for an emotional one: (Derbyshire, 1994).

61. Anticipation of electric shocks and orbital cortex: (Drevets, Burton, Videen, et al., 1995).

62. Rats and prefrontal extinction of fears: (Morgan, Romanski, and LeDoux, 1993).

63. Prefrontal cortex and conditioned response extinguishment: (Molchan, Sunderland, McIntosh, Herscovitch, and Schreurs, 1994: 8126).

64. Frontal lobes and unlearning fear: (Barinaga, 1992, reporting the work of Michael Davis and Christian Grillon).

65. War-related posttraumatic stress disorder, anterior cingulate, and orbital prefrontal cortex: (Bremner, Staib, Kaloupek, et al., 1999).

66. Pain, orbital prefrontal cortex, and atypical facial pain: (Derbyshire, Jones, Devani, et al., 1994).

Chapter 10 Doing the Right Thing

1. Seven-year-old girl with acquired psychopathy: (Eslinger, Gratten, Damasio, and Damasio, 1992).

2. Orbital prefrontal cortex injuries' greater effects in the young: (Eslinger, Gratten, Damasio, and Damasio, 1992; Price, Daffner, Stowe, and Mesulam, 1990).

3. Lack of prefrontal cortex, lack of startle response: (Damasio, Tranel, and Damasio, 1990).

4. Links between the prefrontal cortex and the brain's sex areas (the hypothalamus, septum, and preoptic area): (Fuster, 1989: 21, 32).

5. Orgasm and the prefrontal cortex: (Tiihonen, Kuikka, Kupila, et al., 1994).

6. Research on people born psychopaths and prefrontal cortex: (Raine, 1993: ch. 6; Raine and Venables, 1992: 303–304; Raine, Buchsbaum, Stanley, et al., 1994).

7. Defective smell labeling, psychopaths, and orbital cortex: (Lapierre, Braun, and Hodgins, 1995); and those with damage orbital cortex: (Maloy, Bihrle, Duffy, and Cimino, 1993).

8. Violent crimes and brain damage: (Raine, 1993: 110–113).

9. PET study of murderers: (Raine, 1993: 146–150; Raine, Buchsbaum, Stanley, et al., 1994).

10. Indeed, brain scans are now firmly entrenched in the courtrooms: (Motluk, 1997).

11. Prefrontal cortex blood flow reduction and sleep loss: (Thomas, Sing, Belenky, et al., 1993: 17 percent dorsolateral, 24 percent orbital; they found also decreases of 18 percent in the cerebellum and thalamus).

12. Effects of sleep loss: (Horne, 1993).

13. Prefrontal cortex and choosing options: (Damasio, Tranel, and Damasio, 1991; Damasio, 1994).

14. Card games, orbital cortex and concern about the future: (Bechara, Tranel, Damasio, and Damasio, 1996).

15. Self-induced sadness and orbital prefrontal cortex: (Pardo, Pardo, and Raichle, 1993).

16. Emotions for sustaining reciprocal cooperation in friendship: (Trivers, 1971: 48–54). Evolutionary link between the orbital cortex, reciproca-

tion emotions, and other deficits in psychopaths: (Rose and Moore, 1993).

17. Prefrontal cortex and social referencing: (Kalin, 1993).
18. Face processing and the orbital prefrontal cortex: (Horwitz, Maisog, Kirschner, et al., 1993); in 2-month-old babies: (Tzourio, De Schonen, Pietrzyk, et al., 1993).
19. Prefrontal cortex and attachment: (Schore, 1994; 1996).
20. Quote on belongingness: (Markus and Kitayama, 1991: 226).
21. The subtle entrapment of Milgram's experiment: (Ross and Nisbett, 1991: 53–58).
22. Telefraud: (Coyne, 1991).
23. Twenty-two prefrontal cortical areas in macaque orbital cortex: (Carmichael and Price, 1994).
24. Conflict in the brain: (Levy and Trevarthen, 1976).
25. Activation decreases in brain areas during tasks: (Kawashima, O'Sullivan, and Roland, 1993; Seitz and Roland, 1992; Shulman, Fiez, Corbetta, and Buckner, 1997).
26. Left and right of empirical and decontextualized reasoning: (Deglin and Kinsbourne, 1996).
27. Lateralization of memory retrieval and encoding: (Shallice, Fletcher, Frith, et al., 1994; Tulving, Kapur, Craik, et al., 1994).
28. Immune function and right and left prefrontal cortices: (Kang, Davidson, Coe, et al., 1991); for similar difference when the two sides of the brain are externally activated: (Amassian, Henry, Durkin, et al., 1994).
29. Two sides of the prefrontal cortex and different emotions: (Grafman, Vance, Weingartner, et al., 1986; Tomarken, Davidson, Wheeler, and Doss, 1992; Davidson, 1992).
30. Right prefrontal negative and left prefrontal positive connection to emotions: (Davidson, 1992; Tomarken, Davidson, Wheeler, and Doss, 1992).
31. Right prefrontal activation in infants and separation: (Davidson and Fox, 1989; Fox, Bell, and Jones, 1992).
32. Left prefrontal side of the brain and looking on the positive side of things: (Davidson and Fox, 1989; Davidson, 1992; Tomarken and Davidson, 1994).
33. Right prefrontal cortex and orgasm: (Tiihonen, Kuikka, Kupila, et al., 1994). Note the researchers found that Brodmann's area 10 became activated, which is usually taken to be part of the "nonemotional" dorsolateral prefrontal cortex.
34. The prefrontal cortex and pleasures of alcohol: (Tiihonen, Kuikka, Hakola, et al., 1994); right hemisphere activation in flotation: (Raab and Gruzelier, 1994).
35. Hypnosis, hypnotizability, and right-side activation: (Gruzelier, 1998; Gruzelier, Allison, and Conway, 1984; Gruzelier and Warren, 1993; MacLeod-Morgan and Lack, 1982; Spiegel, 1992).

36. The two sides of the brain inhibiting each other: (Levy and Trevarthen, 1976; Kinsbourne, 1974).

Chapter 11 Where Memories Are Made

1. The man-with-two-wives dialogue: (Alexander, Stuss, and Benson, 1979); also Stuss (1991: 71–73), which gives a slightly different version. The condition is clinically known as Capgras syndrome, or reduplicative paramnesia. It can be found not only following brain damage but also in schizophrenics. (Interestingly, the areas of the brain thought to be involved in schizophrenia—as in Capgras syndrome—are the prefrontal and temporal lobes.)
2. The case of R. K.: (Staton, Brumback, and Wilson, 1982).
3. The woman with the strange cat: (Reid, Young, and Hellawell, 1993).
4. John Locke linking identity with memory: (Locke, 1689/1975, Bk II, ch. XXVII, sections 9–10).
5. For a similar analysis of Capgras syndrome: (Hirstein and Ramachandran, 1997).
6. Files and headers in memory: (Morton, Hammersley, and Bekerian, 1985).
7. Smell and memory: (Engen, 1991).
8. Proust and madeleine dipped in tea: (Proust, 1982: 48–51).
9. Right side of the brain and identification: Loss of voice identification—phonagnosia (Kreiman and Van Lancker, 1988); of face identification—prosopagnosia (Landis, Regard, Bliestle, and Kleihues, 1988).
10. Identification separate from familiarity, with Capgras syndrome being due to the former's failing to check against the latter: (Ellis and Young, 1990).
11. The man with two homes: (Kapur, Turner, and King, 1988).
12. Hippocampus and hanging: (Medalia, Merriam, and Ehrenreich, 1991). Interestingly, the blood supply to the brain is cut off by only a 36-lb pull on a noose—much less than someone's weight.
13. Hippocampus, blood vessels, and epilepsy: (MacLean, 1990: 416). The relationship of the hippocampus to epilepsy is more complex: For instance, the hippocampus is vulnerable to kindling, something that can underlie some of the causes and effects of epileptic fits: (Majkowski, 1988). The excitotoxins mentioned in Chapter 3 also have a role: (Ronne-Engström, Hillered, Flink, et al., 1992).
14. Fourteen million hippocampal neurons: (Brown and Cassell, 1980).
15. Hippocampus: Strictly "hippocampus area" which includes some smaller bits of the brain nearby related to input and output, the parahippocampus, and the entorhinal and perirhinal cortices.
16. The hippocampus as a cognitive map: (Eichenbaum and Cohen, 1988; O'Keefe and Nadel, 1978; O'Keefe and Speakman, 1987; Wilson and McNaughton, 1993). The description of "bugging" the rat's own map derives from the work of Michael Recce of University College London.

17. Hippocampus and smell: (Vanderwolf, 1992).

18. Autoassociative memory and the hippocampus: (Rolls and O'Mara, 1993). Note that autoassociative networks exist chiefly in the CA3 region; networks in CA1 seem to compare activations in these.

19. Hippocampus place cells and clues: (O'Keefe and Speakman, 1987).

20. Storage of long-term and short-term memories outside the hippocampus: (Sobotka and Ringo, 1993). However there is selective activation of the hippocampus during long-term but not short-term memory recall: (Grasby, Frith, Friston, et al., 1993; see also Nadel, Samsonovich, Ryan, and Moscovitch, 2000).

21. Hippocampus and the binding of different information: (Cohen, Ryan, Hunt, et al., 1999).

22. Hippocampus and memory networks: Karl Pribram has called the hippocampus the "black hole" of the neurosciences: (Pribram, 1986: 329). Perhaps this is because brain scientists tend to see only the trees (of neurons) but not the forest (what they collectively do). Most theorizing on the hippocampus's role in memory limits itself to *long-term potentiation* (LTP) and *consolidation*. But LTP can account only for information storage, not for memory. For memory to exist, information storage is needed to organize other information storage. LTP is a building block of such a process but is not the process itself. This is why it can have a role in temporary memory storage (Rawlins, 1985; Squire, 1992) and also in the updating of long-term memories stored elsewhere (Squire, 1992).

23. Hippocampus and memory mapping: (Cohen and Eichenbaum, 1993: 288; Teyler and DiScenna, 1987).

24. Simonides and the Method of Loci: (Cicero, *De oratore*, II, lxxxvi, 351–354).

25. Pon memory limit being not storage but recall: (Ericsson, 1985).

26. Activation of parahippocampus during learning of positions of objects in virtual reality maze: (Aguirre, Detre, Alsop, and D'Esposito, 1996).

27. Hippocampus and object-location memory, exploring maps, virtual and real places: (Maguire, Burgess, Donnett, et al., 1998; Mellet, Briscogne, Tzourio-Mazoyer, et al., 2000; Owen, Milner, Petrides, and Evans, 1996).

28. Taxi drivers recalling routes around London: (Maguire, Frackowiak, and Frith, 1997).

29. Spatial memory and "super" memory: (Ericsson, 1985).

30. Prefrontal cortex, recall, and memory: (Della Sala, Laiacona, Spinnler, and Trivelli, 1993; Shallice, Fletcher, Frith, et al., 1994; Tulving, Markowitsch, Kapur, et al., 1994).

31. Hippocampus linked with prefrontal cortex: (Goldman-Rakic, Selemon, and Schwartz, 1984).

32. Prefrontal cortex injuries and recall problems: (Della Sala, Laiacona, Spinnler, and Trivelli, 1993; Incisa Della Rocchetta and Milner, 1993; Jetter, Poser, Freeman, and Markowitsch, et al., 1986).

33. Files stored around cortex: (Damasio and Damasio, 1994).
34. Polar parts of temporal lobe and recall of personal details: (Markowitsch et al., 1993).
35. Interference in word-pair learning and prefrontal cortex: (Uhl, Podreka, and Deecke, 1994).
36. Prefrontal cortex injuries and source memory: (Janowsky, Shimamura, and Squire, 1989).
37. Prefrontal cortex and order memory: (Kesner, Hopkins, and Fineman, 1994).
38. Consolidation: (Alvarez and Squire, 1994; Buzsáki, 1996; McClelland, McNaughton, and O'Reilly, 1995).
39. Hippocampus and prepermanent links: (Alvarez and Squire, 1994).
40. The hippocampus and novelty detection: (Tulving, Markowisch, Kapur, et al., 1994).
41. Failure of meta-awareness as to the lip pressures used in trumpet playing: (Davies, Kenny, and Barbenel, 1989).
42. Person without a map of the world: (Fisher, 1982).
43. Constitution Avenue: (Bisiach and Luzzatti, 1978—note the Italian authors asked their patient about the Piazza del Duomo in Milan). The problems of those with hemineglect are more complex than simply ignoring things on one side. Evidence exists that they can use memories and information on the neglected side; their problems lie in activating the engrams that would allow them to be used: (Meador, Loring, Bowers, and Heilman, 1987).
44. Unilateral neglect of left turns: (Bisiach, Brouchon, Poncet, and Rusconi, 1993).
45. Egocentric and allocentric hippocampus neurons: (Tamura, Ono, Fukuda, and Nakamura, 1990).
46. Locations of egocentric and allocentric maps: (Kesner, Farnsworth, and DiMattia, 1989; Semmes, Weinstein, Ghent, and Teuber, 1963; Stein, 1992).
47. Posterior cingulate cortex quote: (Vogt, Finch, and Olson, 1992: 441; see also Olson, Musil, and Goldberg, 1993).
48. Posterior cingulate and encoding memories: (Fletcher, Frith, Grasby, et al., 1995; McDonald, Crosson, Valenstein, and Bowers, 2001).
49. Posterior cingulate and sense of familiarity: (Shah, Marshall, Zafiris, et al., 2001).

Chapter 12 What Are We?

1. Minsky on consciousness: (Minsky, 1987: 151).
2. Dennett on consciousness: (Dennett, 1987: 160).
3. The brain and its relationship with its body: (Cotterill, 1998).
4. Phantoms at the dentist's: (Patrick Wall, personal communication); similar experiences can follow spinal anesthetic block during labor and even a

blood pressure cuff on the arm: (Melzack, 1992: 91, 95)—and limbs that "go to sleep."

5. Amputee not realizing operation has taken place: (Simmel, 1956: 640).
6. Phantom limbs, general accounts: (Habel, 1956; Melzack, 1990; 1992; Mitchell, 1872/1965; Riddich, 1941; Simmel, 1956).
7. "Telescoping" of phantoms: (Simmel, 1956: 643).
8. Mu and phantom limbs: (Gastaut, Naquet, and Gastaut, 1965).
9. Soldier's phantom hand grasping bomb: (Riddoch, 1941: 203).
10. Rings and watches on phantoms: (Habel, 1956: 632).
11. Phantom breasts: (Aglioti, Cortese, and Franchini, 1994; Melzack, 1990: 89).
12. Phantom breasts, noses, and penises: (Riddoch, 1941, 207). Phantom tongue: (Hanowell and Kennedy, 1979). Weir Mitchell on erect penis: (Mitchell, 1872/1965: 350, see also Melzack, 1990: 89; and Fisher, 1999).
13. Orgasms in paraplegics: (Money, 1960).
14. Phantoms after accidents: (Ettlin, Seiler, and Kaeser, 1980).
15. The Vehe Amassian illusion: (Amassian, Cracco, and Maccabee, 1989).
16. Shortened phantoms lengthening into artificial limbs: (Riddoch, 1941: 199–200; Simmel, 1956: 644; Mitchell, 1872/1965: 352).
17. Nielsen hand embodiment illusion: (Nielsen, 1963; also see Ramachandran and Blakeslee, 1999).
18. Nielsen illusion and phantom arms: (Ramachandran and Rogers-Ramachandran, 1996; Ramachandran and Blakeslee, 1999).
19. Phantom breast nipple and tip of nose remaining: (Riddoch, 1941: 207). Since people can have phantoms for bladders, wombs, and rectums, other factors are involved in forming phantoms: (Melzack, 1990).
20. Motor and somatosensory cortices: (Evarts, 1987).
21. Right primary motor cortex and right hand discrimination of length: (Kawashima, Roland, and O'Sullivan, 1994).
22. Premotor cortex and ongoing control of movement: (Flament, Onstott, Fu, and Ebner, 1993; Grazino, Yap, and Gross, 1994).
23. Premotor cortex activated by seeing movements: (Rizzolatti Fadiga, Galese, and Fogassi, 1996).
24. Supplementary motor cortex: (Goldberg, 1985; Tanji and Shima, 1994).
25. Corticospinal projection from somatosensory area: (Galea and Darian-Smith, 1994).
26. Cooling of motor followed by somatosensory cortex takeover: (Sasaki and Gemba, 1984).
27. Nonexistence of "muscle cortex": (Schieber, 1990).
28. Motor cortex maps: (Schieber and Hibbard, 1993).
29. Alice in Wonderland syndrome and Charles Dodgson's migraines: (Rolak, 1991); the alternative explanation that it was due to his temporal lobe epilepsy: (LePlante, 1993).
30. Jonathan Swift: (Laplante, 1993: 69).

31. Origins in the womb of genital map's being next to foot map in brain: (Farah, 1998).
32. Orgasms in phantom feet and foot fetish quotes: (Ramachandran, 1993: 10417).
33. The man with three heads, etc.: (Weinstein, 1954).
34. Man with fingers and a leg in his mother's suitcase: (Halligan, Marshall, and Wade, 1995: 178).
35. Woman with extra hand: (Weinstein, Kahn, Malitz, and Rozanski, 1954: 47).
36. Goethe quote and autoscopy: (Lhermitte, 1951: 474).
37. "Leg from the dissecting room" quotes: (Sacks, 1984: 50–52).
38. Quote from woman whose limbs were not her own: (Nielsen, 1938: 555).
39. Music and human emotional expression: (Clynes, 1977).
40. MacLean, subjectivity, and the limbic system: (MacLean, 1990: 578)
41. Prefrontal cortex and readiness potentials: (Singh and Knight, 1990).
42. Slow potentials in movement and various kinds of nonmotor cognition: (Brunia and Damen, 1988; Rockstroh, Elbert, Birbaumer, and Lutzenberger, 1983; Rockstroh, Elbert, Canavan, et al., 1990).
43. Movement related potentials before voluntary relaxation: (Terada, Ikeda, Negamine, and Shibasaki, 1995).
44. Tics lacking readiness potentials: (Fahn, 1993: 13).
45. Readiness potentials and consciousness: (Libet, 1985).
46. Willed action, the prefrontal lobe, and PET: (Frith, Friston, Liddle, and Frackowiak, 1991; Jahanshahi, Jenkins, Brown, et al., 1995).
47. Gamma and consciousness: (Sauve, 1999).
48. Gamma and focusing on touch: (Desmedt and Tomberg, 1994); and motor preparation: (Kristeva-Feige, Feige, Makeig, et al., 1993; Murphy and Fetz, 1992; Sanes and Donoghue, 1993).
49. Gamma, anesthesia, and consciousness: (Kulli and Koch, 1991; quote, 6).
50. Gamma respones and anesthesia: (Kulli and Koch, 1991; Plourde, 1993; Schwender, Madler, Klasing, et al., 1994).
51. Llinás and Paré quote on consciousness and gamma: (Llinás and Paré, 1991: 531).
52. Francis Crick, ideas on a link between consciousness and gamma: (Crick and Koch, 1990; Koch and Crick, 1994).
53. Prefrontal cortex and metabolic activity when blindfolded and earplugged: (Roland, 1984; 1993: 472).
54. Prefrontal cortex and consciousness: (Dehaene and Naccache, 2001; Jack and Shallice, 2001).
55. Dennett on narrative center: (Dennett, 1992).
56. Watson's mobiles over babies' cots: (Watson and Ramey, 1972).
57. Stress of stressful noises reduced by control: (Glass, Singer, and Friedman, 1969).

58. Control turning a frightening toy into an enjoyable one: (Gunnar-Vongnechten, 1972).
59. Mastery motivation in children: (Yarrow, McQuiston, MacTurk, et al., 1983).
60. Distress when children lose control of things: (Lewis, Sullivan, Ramsay, and Alessandri, 1992).
61. Elbow room, see discussion by Dennett: (1984).
62. Psychological research on seeking freedom: (Brehm and Brehm, 1981).
63. Transcendence and gamma: (Banquet, 1973: 146: "this fast activity appeared as beta periods at 20 and 40 c/sec"); two earlier French scientists, Das and Gastaut, have been claimed by Banquet (1973: 146) and also Sheer (1976: 77) to have made similar findings in 1957.
64. Anaxagoras claiming the Sun was a burning stone and the moon had hills: (Barnes, 1987: 237).

Chapter 13 Of Human Bonding

1. The uniqueness of the human species as a social primate: (Rodseth, Wrangham, Harrigan, and Smuts, 1991).
2. Unique duration of human relationships: (Rodseth, Wrangham, Harrigan, and Smuts, 1991).
3. Chimpanzees and cooperation: (Nishida and Hiraiwa-Hasegawa, 1987). Bonobos, compared to chimpanzees, also show less cooperation between males, with females being more strongly bonded: (Kano, 1992: 207).
4. Hamadryas baboons: (Sigg, Stolba, Abegglen, and Dasser, 1982). Note hamadryas baboons have quite a different social lifestyle from the savannah baboons studied by Barbara Smuts and referred to elsewhere.
5. Chimps closest kin to us: (Ruvolo, Pan, Zehr, et al., 1994).
6. Female chimps leaving home: In some places, like Mahale, most do: (Nishida, Takasaki, and Takahata, 1990: 73–74), while few did at Gombe: (Goodall, 1986: 89).
7. Matrilocality quote: (George Peter Murdock cited in Rodseth, Wrangham, Harrigan, and Smuts, 1991: 230). Matrilocality following migration: (Divale, 1984). Note that matrilocal residence is not to be confused with matrilineal descent. Our account ignores exogamy and endogamy, since, in both, daughters leave the family household. Humans show a mix of exogamy and endogamy (as do chimps). Our point is that where exogamy happens, it is on the part of females. One complication is that what counts anthropologically as endogamy is a matter of definition as to the social boundaries of a group—many societies showing endogamy would by different boundary criteria be considered exogamous.
8. !Kung, water holes, and marriage alliances: (Marshall, 1976: 73, 168).
9. Alliances as safety nets: (Gamble, 1983: 204–205).
10. Sex and more intense love: (Tennov, 1979: 78).

11. Up to 10-fold increase in oxytocin in men at orgasm: (Murphy, Checkley, Seckl, and Lightman, 1990).
12. Marmoset and oxytocin: (Wang, Moody, Newman, and Insel, 1997).
13. Monogamy, oxytocin, and prairie voles: (Carter and Getz, 1993; Insel and Shapiro, 1992).
14. Marmosets' oxytocin receptors: (Wang, Moody, Newman, and Insel, 1997).
15. Increased vasopressin receptors and male attachment: (Pitkow, Sharer, Ren, et al., 2001).
16. Male-to-male sex in rhesus monkeys: (Reinhardt, Reinhardt, Bercovitch, and Goy, 1986); in gorillas: (Yamagowa, 1987).
17. Sex reducing tension between two social bonobo bands: (Hashimoto and Furuichi, 1994).
18. Descriptions of sex in bonobos: (De Waal, 1995; De Waal and Lanting, 1997).
19. Female bonding in bonobos: (De Waal, 1995; Kano, 1992: 180–194; Parish, 1994; De Waal and Lanting, 1997).
20. Parental bonding and human evolution: (Lovejoy, 1981; Hamilton, 1984).
21. Extra-pair sex in birds: (Mock and Fuijioka, 1990).
22. Human brains growing a year after birth: (Martin, 1982: 26; Stanley, 1992: 242–243).
23. Obstetrics and human heads: (Rosenberg, 1992).
24. Pelvic outlet increasing 30 percent in area during birth: (Russell, 1969).
25. Helplessness of babies and human evolution: (Stanley, 1992).
26. Hunter-gatherer mothers carrying children nearly 5000 miles: (Lancaster and Lancaster, 1983: 47, citing the work of Richard Lee).
27. Newborn's brain accounting for 60 percent of body's oxygen consumption: (Kuzawa, 1998: table 2).
28. The need for fathers: Hrdy in a recent book questions this: (Hrdy, 1999). However, much evidence goes against her claims about the unimportance of husbands: (Hurtado, 2000).
29. Lack of grief in nonhuman mothers—not entirely absent: (Smuts, 1985: 230; De Waal, 1996: 53–59).

Chapter 14 The Symbolic Brain

1. Masonic rituals: (Knight, 1983: 312–317).
2. Symbols and symbolons: (Phillis, 1987: 204).
3. Drum talk and whistle-speech: (Critchley, 1979: 88–91).
4. Research on blanket-attached children: (Pasman and Weisberg, 1975).
5. Chimpanzees and symbols: (Savage-Rumbaugh, Rumbaugh, Smith, and Lawson, 1980). Children spontaneously inventing language symbols: (Goldin-Meadow and Feldman, 1977).
6. Greeting gestures: (Axtell, 1991: 20–22).

7. The "OK" gesture: (Axtell, 1991: 47–48).

8. Personal identification and smell: (Lord and Kasprzak, 1989).

9. Helen Keller recognizing people by smell: (Keller, 1903: 293).

10. Iconic and symbolic signs: Our approach comes from Peirce: (1932: 134–173).

11. History of the question mark: (Bischoff, 1990: 169–170).

12. Neurons in inferior temporal cortex: (Tanaka, 1993; Tsunoda, Yamane, Nishizaki, and Tanifuji, 2001).

13. Wrong trail: Deacon in his book *Symbolic Species* (1997), while correctly linking the prefrontal cortex with the creation of symbols, does so incorrectly for creating their "semiotic grounding"—symbolic reference—an offshoot of Socrates's problem. Unfortunately, the year in which he published this, Thomas Landauer and Susan Dumais published a paper (1997) that removed the need for such "grounding" by showing that meaning referred to locations mathematically extracted from the context in which they appeared in a many (200)-dimensional eigenvector space, "latent semantic analysis."

14. Unciate fascicle: (Eacott and Gaffan, 1992).

15. Prefrontal cortex and learning of associations: (Asaad, Rainer, and Miller, 1998; Chee, Sriram, Soon, and Lee, 2000; Dimitrov, Granetz, Peterson, et al., 1999; Gomez Beldarrain, Grafman, Pascual-Leone, and Garcia-Monco, 1999; Murray, Bussey, and Wise, 2000; Passingham, 1993: 250–251; Savage, Deckersbach, Heckers, et al., 2001).

16. Prefrontal cortex, learned associations, and people: (Petrides, 1990).

17. The dispersed network involved in processing letters: (Raij, Uutela, and Hari, 2000).

18. The abilities of the prefrontal cortex to form associations across the senses ("temporal integration"): (Fuster, Bodner, and Kroger, 2000).

19. Gamma and the learning of associations: (Miltner, Braun, Arnold, et al., 1999).

20. Popper's World 3: (Popper, 1972; 1976: 180–192).

21. Neural plasticity as the missing link: (Skoyles, 1998; 1999b).

22. Location of speech and written symbols: (Howard, Patterson, Wise, et al., 1992).

23. Japanese logographs: (Sakurai, Momose, Iwata, et al., 1992).

24. Semantic meanings and the posterior superior temporal gyrus area: (Wise, Chollet, Hadar, et al., 1991).

25. Sight-reading of music, localization: (Sergent, Zuck, Terriah, and MacDonald, 1992).

26. Anthropologist quote on ritual: (Radcliffe-Brown: 1965: 145).

27. Tambiah on rituals: (Tambiah, 1968: 202).

28. Emotions and memory: (Cahill, Prins, Weber, and McGaugh, 1994).

Chapter 15 Lucy and Kanzi: Travelers to Humanity

1. Lucy: (Johanson and Edey, 1981); Lucy's hands, chimps, and bipedality: (Hunt, 1994).
2. Lucy's brain: (Hollaway, 1983; Hofman, 1983: 105 and fig. 2).
3. Ramid "root" Lucy: (White, Suwa, and Asfew, 1994). Ramid means "root" in the Afar language spoken in the area where the remains (like Lucy) were found. The name of the Afar language, by the way, is the root of the second part of Lucy's scientific name: *Australopithecus afarensis*. Another species, *Kenyanthropus platyops*, has been discovered to have been walking around Africa 3.5 million years ago: (Leakey, Spoor, Brown, et al., 2001).
4. No ancient chimp bones: (Peterhans et al., 1993).
5. Goodall quotes: Figan and his mother: (1986, 204); about females whimpering for their mothers: (1986: 203).
6. Arousal, hugging, aggression, and reunions: (Goodall, 1986: 332, 344, 366–367).
7. Female chimpanzees moving between bands: (Goodall, 1986: 86–92; Kano, 1992: 92; Nishida, Takasaki, and Takahata, 1990: 73–74).
8. Female genital swelling as passport and protection against male violence: (Goodall, 1986: 483; Hrdy, 1981: 152–153).
9. Attacks on older females: (Goodall, 1986: 92).
10. The murder of Madam Bee: (Goodall, 1986: 511–514, 528–529). Jane Goodall (1986: 524) suggests they killed her to gain her daughters (who were strongly bonded to their mother). We do not question this, though they were not successful in this with Honey Bee. As a group of males, they were better off with her dead (as each would get a better opportunity to mate with her daughters). However, this would change if one of them bonded with her (their personal fitness would be promoted by her aiding their offspring).
11. Figan and Tubi: (Goodall, 1986: 73).
12. Chimp brothers and sisters not knowing each other: (Nishida, 1994: 387).
13. Problems testing chimpanzees' abilities: (de Waal, 1989: 249–250).
14. Breath control, speech, and humans: (MacLarnon and Hewitt, 1999; Vaneechoutte and Skoyles, 1998).
15. Chimps not making sounds without emotions: (Goodall, 1986: 125: "The production of a sound in the absence of the appropriate emotional state seems to be an almost impossible task for a chimpanzee.").
16. Kanzi: (Savage-Rumbaugh, McDonald, Sevcik, et al., 1986; Savage-Rumbaugh and Lewin, 1994; De Waal and Lanting, 1997: 38–41).
17. Kanzi compared with Alia: (Savage-Rumbaugh and Rumbaugh, 1993: 92–99).
18. Kanzi's vocalizations: (Hopkins and Savage-Rumbaugh, 1991).

19. Enriching environment and the prefrontal cortex: (Raleigh, McGuire, Melega, et al., 1996).
20. Chimps and sweets: (Boysen, 1996).
21. Salomé, Sartre, Ricci, Nina, and nut teaching: (Boesch, 1993: 178–177; see also Matsuzawa, 1996).
22. Jane Goodall on Washoe's ability to teach: (Goodall, 1986: 25).
23. Kanzi teaching Tamuli: (De Waal and Lanting, 1997: 159).
24. Effects of foraging demands on mothers on the development of their infants: (Andrews and Rosenblum, 1994).
25. Long-term effect of stress in mothers on their adult children: (Rosenblum, Coplan, Friedman, et al., 1994).
26. Female foraging and social bonds in chimps: (Wrangham, 1986).
27. Differences between bonobo and common chimpanzees: (Kano, 1992: social relations, 206–207; food, 137; see also White, 1988).
28. Foraging and relationships in bonobos: (Kano, 1992: 206–207; White, 1988; 1992).
29. Chimpanzees and hunting: (Goodall, 1986: 267–312, attack on 6-year-old boy: 282; Boesch and Boesch, 1989; Boesch and Boesch-Achermann, 2000).
30. Sharing and cooperation in Taï chimp hunters: (Boesch, 1994; Boesch and Boesch-Achermann, 2000).
31. Taï female chimps forming coalitions like those in bonobos: (Boesch, 1991b : 238–239).
32. Boesch's quote on female status at Taï: (Boesch, 1994: 666).
33. Culture differences among chimp bands: (Boesch, Marchesi, Marchesi, et al., 1994; McGrew, 1992).
34. Dialects of chimp pant-hoots: (Mitani, Hasegawa, Gros-Louis, et al., 1992).
35. Rhesus monkeys learning making up: (de Waal and Johanowicz, 1993).
36. Alarm calls in vervet monkeys: (Seyfarth, Cheney, and Marler, 1980).
37. Thirty-four chimp calls: (Goodall, 1986: 127).
38. Taï chimps signaling among groups: (Boesch, 1991a).
39. Three-part ancient Athenian names: (Hansen, 1991: 96).
40. Bonobos, it should also be noted, communicate the direction in which a party should go by branch-dragging displays: (De Waal and Lanting, 1997: 148) and by deliberately implanting sticks: (Savage-Rumbaugh, Williams, Furuichi, and Kano, 1996).
41. Macho taking over Le Chinois's pant-hoot ID: (Boesch, 1991: 82–83).
42. Gesturing in bonobos: (De Waal and Lanting, 1997: 150–151).
43. Chimps communicating with finger-pointing: (Leavens, Hopkins, and Bard, 1996).
44. Kanzi inventing his own "protogrammar" for putting signs together: (Savage-Rumbaugh and Lewin, 1994: 161–163).

Chapter 16 The Runaway Species

1. Circumstances of Nariokotome boy's burial: (Walker, 1993); and of his discovery: (Leakey and Walker, 1993).
2. Nariokotome boy's estimated height and weight: (Ruff and Walker, 1993).
3. Chimp using hunting skills in interband violence: (Goodall, 1986: 530–531).
4. Java *Homo erectus* at 1.8 million years: (Swisher, Curtis, Jacob, et al., 1994).
5. Importance of gaps in the fossil record: (Martin, 1993, especially fig. 2).
6. Associative mating of like with like: (Thiessen, 1999).
7. Darwin and sexual selection: (Darwin, 1871/1981; Cronin, 1991: 113–249).
8. Darwin quote on sexual selection: (1871/1981: part 1, p. 262).
9. Ibid., part 1, p. 153.
10. Litmus test theory of gaudy plumage: (Hamilton and Zuk, 1982; Zahavi and Zahavi, 1997).
11. Peacocks, elaborateness of trains, size of eye-spots, and increased offspring fitness: (Petrie, 1994).
12. Collared flycatchers and sexual selection: (Sheldon, Merilä, Qvarnström, et al., 1997).
13. Sexual selection and the evolution of the human mind: (Miller, 2000). Our view differs from Miller's in that we view such selection as having happened prior to the formation of parental bonds rather than prior to promiscuous matings.
14. Darwin, op. cit., part 2, pp. 374–375.
15. Beauty and signs of health and fertility: (Etcoff, 1994; Shackelford and Larsen, 1999; Symons, 1995).
16. Darwin, op. cit., part 2, p. 375. On sexual selection and human origins, see also Tanner and Zihlman, 1976.
17. Sexual selection and runaway evolution: (Harvey and Arnold, 1982; Barber, 1995).
18. Man with half a brain and a university diploma: (Griffith and Davidson, 1966).
19. Nico: (Batto, 2000).
20. For information on the calculations, visit our Web site, www.upfrom dragons.com.
21. Limited link between IQ and brain size: (Skoyles, 1999a).
22. Daniel Lyon's brain: (Widler, 1911). Our figure for Nariokotome boy's brain weight comes from his brain volume (909 cc) times the specific gravity of fresh brain tissue (1.09 g/cc).
23. Two percent microcephalics in Clifford Sells' study of 1006 school-aged children: (Sells, 1977).
24. Jay Giedd's MRI study of 104 brains: (Giedd, Snell, Lange, et al., 1996).

25. Edward Sassaman and Ann Zartler finding 39.1 percent of microcephalics were not retarded and 7 percent had average IQs: (Sassaman and Zartler, 1982).
26. Livia Rossi's study of autosomal-dominant microcephalics and the case of C2: (Rossi, Candini, Scarlatti, et al., 1987).
27. For further discussion, see our Web site, www.upfromdragons.com.
28. Changes in hunting skills 40,000 years ago: (Klein, 1983; 1987: 39; 1989: 321–327; Straus, 1987; Zvelebil, 1984).
29. Size of newborn heads: (Martin, 1982: 20–27).
30. Pelvic size and efficient bipedalism: (Day, 1992).
31. Running and our species: (Heinrich, 2001).
32. Ethnographic discussions of running and walking: (Devine, 1985; Nabokov and MacLean, 1980).
33. The hunt for Willie Boy: (Nabokov and MacLean, 1980: 17).
34. Lack of exercise by Americans: (cited in Pate, Pratt, Blair, et al., 1995).
35. Sweat cooling and evolution of running: (Heinrich, 2001; Porter, 1993).
36. Xenophon and running down bustards: (Xenophon: *Anabasis*, I, v).
37. Running and hunting down animals: (Carrier, 1984; Devine, 1985; Heinrich, 2001).
38. Twenty-one-month gestation: (Martin, 1982: 25–27).
39. One possibility we do not have room to discuss is whether big brains were needed to compensate for the handicap of inefficient "mindware."
40. Neanderthal brains bigger than ours: (Stringer and Gamble, 1993: 82).
41. Cases of people with congenital brain agenesis or damage and mental and clinical normalcy: (massive left hemisphere cyst: Ingram, Levin, Guinto, and Eisenberg, 1994; agenesis of right temporal lobe: Kansu and Zacks, 1979; two cases of agenesis of the corpus callosum: Fisher, Ryan, and Dobyns, 1992).

Chapter 17 The Billion-Hour Journey

1. Phoneme segmentation and phonetic symbols: (Morais, Cary, Alegria, and Bertelson, 1979); and Chinese logographic readers: (Read, Zhang, Nie, and Ding, 1986; Huang and Hanley, 1994).
2. Phonetic symbols and speech awareness: (Morais, 1987).
3. Altered brain function in literates: (Castro-Caldas, Petersson, Reis, et al., 1998; Petersson, Reis, Askelof, et al., 2000).
4. Thicker corpus callosum in literates: (Castro-Caldas, Miranda, Carmo, et al., 1999).
5. Violinists having larger anterior corpus callosum: (Schlaug, Jäncke, Huang, et al., 1995).
6. Violinists having two to three times the usual cortical area devoted to the left hand: (Elbert, Pantev, Wienbruch, et al., 1995).

7. Enlarged motor cortex in keyboard players: (Amunts, Schlaug, Jäncke, et al., 1997).
8. No abstract numbers until 5000 years ago: (Schmandt-Besserat, 1992: 197).
9. Concrete and abstract numbers: (Schmandt-Besserat, 1992: 184–187).
10. Verbs for vision in Homer: (Snell, 1953: 1–3). The rise of abstract language with the Classical Greeks: (Snell, 1953: 227–231).
11. Naval officer and concrete words: (Warrington and Shallice, 1984: 842).
12. Survival of abstract concepts but not concrete ones: (Warrington, 1975; Warrington and Shallice, 1984).
13. Snell quote about abstract nouns and verbs: (Snell, 1953: 234).
14. Nouns and verbs located differently in the brain, evidence from lesions: (Daniele, Giustolisi, Silveri, et al., 1994; Damasio and Tranel, 1993); from evoked potentials: (Preissl, Pulvermüller, Lutzenberger, and Birbaumer, 1995); from studies of gamma activity: (Pulvermüller, Preissl, Lutzenberger, and Birbaumer, 1996).
15. Foot and face verbs activating foot and face areas: (Pulvermüller, Harle, and Hummel, 2000).
16. Nouns for tools activating part of motor cortex and those for animals activating the visual cortex: (Martin, Wiggs, Ungerleider, and Haxby, 1996).
17. PET linking words for colors and actions to different areas: (Martin, Haxby, Lalonde, et al., 1995).
18. Prefrontal cortex and mental action verbs and nouns: (Baron-Cohen, Ring, Moriarty, Schmitz, Costa, and Ell, 1994).
19. Other brain change explanations exist for Greek culture: (Jaynes, 1979).
20. Our instinct to link what we see in others with our own bodies: (Fadiga, Fogassi, Pavesi, and Rizzolatti, 1995; Meltzoff and Gopnik, 1993).
21. Threshold for performing a motor activity reduced by seeing it: (Fadiga, Fogassi, Pavesi, and Rizzolatti, 1995).
22. Children imitating goals and organizing cues: (Tomasello, Kruger, and Ratner, 1993).
23. Private speech and the quality of others' commands: (Beck, 1994; Skotko, 1992).
24. Secure attachment and gratification delay: (Jacobsen, 1998).
25. Learning by imitating goals and collaboration: (Tomasello, Kruger, and Ratner, 1993)
26. Delay over sweets, children ratings, and ego resilience: (Mischel, Shoda, and Peake, 1988: 692–695).
27. Memes: (Dawkins, 1989: 192; more general discussion: 188–199, 322–324; 1982: 109–112).
28. Julian Jaynes: (1979—but also see Dennett, 1986). Mind software: (Dennett, 1986; 1991: 190).
29. Dennett and virtual machines: (Dennett, 1991: 209–226).

30. The case against Neanderthal speech: (Lieberman, 1984); against Lieberman's case: (Houghton, 1993); for a discussion: (Lewin, 1993: 169–172).
31. Hunter-gatherers and sign language: (Lévy-Bruhl, 1985: 158–164; Umiker-Sebeok and Sebeok, 1978).
32. Earlier start with signs than speech: (Goodwyn and Acredolo, 1993).
33. Baron von Münchhausen and bootstrapping: (Illingworth, 1986: 42; we have not been able to find traces of this story in the published Baron von Münchhausen adventures).
34. Imitation and language learning: (Fillmore, 1979; Savage-Rumbaugh, McDonald, Sevcik, et al., 1986: 224, suggest this is also important in chimps).
35. For a review arguing phones, enable vocal imitation by functioning as a replicative code: (Skoyles, 1998).
36. Children booting themselves into reading via phonics: (Skoyles, 1988c).
37. Killing stick accuracy: (Bahn, 1987).
38. !Kung and arrow poison: (Marshall, 1976: 149, 152).
39. Boomerangs outside Australia: (Bahn, 1987).
40. PUP technologies: (Stringer and Gamble, 1993: 200–201; also known as pre-Aurignacian: Klein, 1989: 302–303).
41. A 400,000-year-old wooden hunting spear: (Thieme, 1997).
42. Potter's wheel and the Americas: (Oliphant, 1992: 178; Whitehouse and Wilkins, 1986: 106).
43. Origins of coinage: (Whitehouse and Wilkins, 1986: 24); independently invented at the same time by both the Chinese and the Greek-influenced Lydians.
44. The origins of ceramics: (Vandiver, Soffer, Klima, and Svoboda, 1989).
45. Accidental rise of Middle Eastern agriculture: (Henry, 1989).
46. "Ineffective hunters": (Klein, 1987: 39; see also Klein, 1983; Straus, 1987).
47. Killing of buffalo in corrals: (Frison, 1987).
48. Complex hunter-gatherer societies in Russia around 40,000 BC: (White, 1993: 288).
49. Archaeologists on symbols: (Noble and Davidson, 1989); difficulty of survival of early symbolic artifacts: (Marshack, 1976: 140); review of issues: (Lewin, 1993: 161).
50. Humans as a superweb: (Bloom, 2000).

Chapter 18 Third-Millennium Brain

1. Farmers and nutritional deficiencies: (Cohen, 1989: 58–59).
2. Measles and other diseases arising with farmers: (Cohen, 1989: 46–51).
3. First farmers' worn out and deformed joints: (Molleson, 1989; 1994).

4. Hunter-gatherers as individualists versus farmers as conformists: (Berry, 1967; Berry and Annis, 1974; Witkin and Berry, 1975).
5. Lack of education among hunter-gatherers: (Blurton Jones and Konner, 1976).
6. Enough light to read this book inside your skull: (Benaron, Cheong, and Stevenson, 1997).
7. Near-infrared look inside brain: (Benaron and Stevenson, 1993; Berg, Jarlman, and Svanberg, 1993; Corballis, Gratton, Cho, et al., 1994; Creace, 1993: 560; Fabiani, Gratton, Friedman, et al., 1994; Kato, Kamei, Takashima, and Ozaki, 1993).
8. Seeing brain blood flow using light-emitting diodes: (Gratton, Maier, Fabiani, et al., 1994; McCormick, Stewart, Lewis, et al., 1992; Watanabe, Yamashita, Maki, Ito, and Koizumi, 1996); seeing hyperfrontal activity using flashbulbs: (Chance, Zhuang, Unah, et al., 1993; Hoshi and Tamura, 1993a; Villringer, Bötzel, Hock, et al., 1993; and Villringer, Planck, Hock, et al., 1993).
9. Multichannel near-infrared optical imaging: (Hoshi and Tamura, 1993b). Seeing patterns of brain activity using light directly on the brain: (Grinvald, Frostig, Siegel, and Bartfeld, 1991).
10. Older people having changed oxygenated/deoxygenated ratio, constant blood volume: (Hoshi and Tamura, 1993a).
11. Ultrasound and brain scanning: (Motluk, 1997).
12. Electrical impedance and brain imaging: (Tidswell, Gibson, Bayford, and Holder, 2001).
13. New electrocephalographic technology: (1) brainwave visual analyzer (Psychic Lab, New York) reported in *New Scientist*, 6 March 1993; (2) MRI-corrected high-resolution EEG: (Gevins, Le, Martin, et al., 1994).
14. Computerized work-out exercise machines: (Mestel, 1994).
15. Eye movements and reading difficulty: (Rayner and Pollatsek, 1981).
16. Pupil dilation with text difficulty: (Just and Carpenter, 1993).
17. Active texts; for similar ideas: (Negroponte, 1995: 135–136).
18. Delaying of eye movements: (Prior, Bertolasi, Rothwell, et al., 1993); blocking of visual functions: (Amassian, Cracco, Maccabee, et al., 1993; Beckers and Hömberg, 1991); blocking of recall from verbal short-term memory: (Gafman, Pascual-Leone, Alway, et al., 1994); inhibiting or exciting of thumb movement: (Wassermann, Pascual-Leone, Val's-Sole, et al., 1993); arresting of, and inducing of errors in, number counting (Pascual-Leone, Gates, and Dhuna, 1991).
19. Magnetic stimulation inhibiting prefrontal cortex activity: (Pascual-Leone and Hallett, 1994).
20. TMS enhancing prefrontal reasoning: (Boroojerdi, Phipps, Kopylev, et al., 2001).
21. Alpha waves and magnetic fields: (Bell, Marino, and Chesson, 1994).

22. TMS and depression: (George, Wassermann, Williams, et al., 1995).
23. TMS and Parkinson's disease: (Strafella, Paus, Barrett, and Dagher, 2001).
24. TMS and immune function: (Amassian, Henry, Durkin, et al., 1994).
25. Paul Broca, handedness and side of brain: (Harris, 1990: 208).
26. Changed attitudes toward left-handers due to brain science: (Harris, 1990: 208–209).
27. FAS more common than Down's syndrome: (Abel and Sokol, 1987: FAS, 0.18 per 1000; Down's syndrome, 0.12 per 1000); additive effects of alcohol and cocaine on the unborn: (Kurth and Le Quesne, 1993).
28. Alcohol effects on children continuing into adulthood: (Streissguth, 1993).
29. 150 billion people: (Lee and DeVore, 1968: 3).

Bibliography

Abel, E. L., and Sokol, R. J. (1987). Incidence of fetal alcohol syndrome and economic impact of FAS-related anomalies. *Drug and Alcohol Dependence*, 19, 51–70.

Adcock, R. A., Constable, R. T., Gore, J. C., and Goldman-Rakic, P. S. (2000). Functional neuroanatomy of executive processes involved in dual-task performance. *Proceedings of the National Academy of Sciences of the USA*, 97, 3567–3572.

Adolphs, R., Tranel, D., Damasio, H., and Damasio, A. (1994). Impaired recognition of emotion in facial expressions following bilateral damage to the human amygdala. *Nature*, 372, 669–672.

Aertsen, A., Vaadia, E., Abeles, M., Ahissar, E., Bergman, H., Karmon, B., Lavner, Y., Margalit, E., Nelken, I., and Rotter, S. (1991). Neural interactions in the frontal cortex of a behaving monkey. *Journal für Hirnforsche*, 32, 735–743.

Aglioti, S., Cortese, F., and Franchini, C. (1994). Rapid sensory remapping in the adult human brain as inferred from phantom breast perception. *NeuroReport*, 12, 473–476.

Aguirre, G. K., Detre, J. A., Alsop, D. C., and D'Esposito, M. (1996). The parahippocampus subserves topographical learning in man. *Cerebral Cortex*, 6, 823–829.

Aiello, L., and Dunbar, R. (1993). Neocortex size, group size and the evolution of language. *Current Anthropology*, 34, 184–192.

Albrecht, D., Royl, G., and Kaneoke, Y. (1998). Very slow oscillatory activities in lateral geniculate neurons of freely moving and anesthetized rats. *Neuroscience Research*, 32, 209–220.

Alexander, M. P., Stuss, D. T., and Benson, D. F. (1979). Capgras syndrome: A reduplicative phenomenon. *Neurology*, 29, 334–339.

Alho, K., Kujala, T., Paavilainen, P., Summala, H., and Näätänen, R. (1993). Auditory processing in visual brain areas of the early blind: Evidence form event-related potentials. *Electroencephalography and Clinical Neurophysiology*, 86, 418–427.

Allan, K., Dolan, R. J., Fletcher, P. C., and Rugg, M. D. (2000). The role of the right anterior prefrontal cortex in episodic retrieval. *NeuroImage*, 11, 217–227.

Allman, W. F. (1989). *Apprentices of wonders: Inside the neural network revolution.* New York: Bantam Books.

Alvarez, P., and Squire, L. R. (1994). Memory consolidation and the medial temporal lobe: A simple network model. *Proceedings of the National Academy of Sciences of the USA*, 91, 7041–7045.

Amaral, D. G., Price, J. L., Pitkänen, A., and Carmichael, S. T. (1992). Anatomical organization of the primate amygdaloid complex. In J. P. Aggleton, ed., *The amygdala*, pp. 1–66. New York: Wiley-Liss.

Amassian, V. E., Cracco, R., and Maccabee, P. J. (1989). A sense of movement elicited in paralyzed distal arm by focal magnetic coil stimulation of human motor cortex. *Brain Research*, 479, 355–360.

Amassian, V. E., Cracco, R., Maccabee, P. J., Cracco, J. B., Rudell, A. P., and Eberle, L. (1993). Unmasking human visual perception with the magnetic coil and its relationship to hemispheric asymmetry. *Brain Research*, 605, 312–316.

Amassian, V. E., Henry, K., Durkin, H., Chice, S., Cracco, J. B., Somasundaram, M., Hassan, N., Cracco, R. Q., Maccabee, P. J., and Eberle, L. (1994). Human immune functions are differentially affected by left- versus right-sided magnetic stimulation of temporo-parieto-occipital cortex [abstract]. *Neurology*, 44 (suppl. 2), A133.

Amunts, K., Schlaug, G., Jäncke, L., Steinmetz, H., Schleicher, A., Darbringhaus, A., and Ziles, K. (1997). Motor cortex and hand motor skills: Structural compliance in the human brain. *Human Brain Mapping*, 5, 206–225.

Amzica, F., and Steriade, M. (1996). Progressive cortical synchronization of ponto-geniculo-occipital potentials during rapid eye movement sleep. *Neuroscience*, 72, 309–314.

Anand, B. U., Chhina, G. S., and Singh, B. (1961). Some aspects of electroencephalographic studies in Yogis. *Electroencephalography and Clinical Neurophysiology*, 13, 452–456.

Andrews, M. W., and Rosenblum, L. A. (1994). The development of affiliative and agonistic social patterns in differentially reared monkeys. *Child Development*, 65, 1398–1404.

Anson, B. J. (1963). *An atlas of human anatomy*, 2nd ed. Philadelphia: Saunders.

Antrobus, J. S. (1994). Dreaming. *Encyclopedia of Human Behavior*, vol. 2., V. S. Ramachandran, ed. San Diego, Academic.

Arieli, A., Shoham, D., Hildesheim, R., and Grinvald, A. (1995). Coherent spatiotemporal patterns of ongoing activity revealed by real-time optical imaging coupled with single-unit recording in the cat visual cortex. *Journal of Neurophysiology*, 73, 2072–2095.

Arroyo, S., Lesser, R. P., Gordon, B., Uematsu, S., Hart, J., Schwerdt, P., Andreasson, K., and Fisher, R. S. (1993a). Mirth, laughter and gelastic seizures. *Brain*, 116, 757–780.

Arroyo, S., Lesser, R. P., Gordon, B., Uematsu, S., Jackson, D., and Webber, R. (1993b). Functional significance of the mu rhythm of human cortex. *Electroencephalography and Clinical Neurophysiology*, 87, 76–87.

Asaad, W. F., Rainer, G., and Miller, E. K. (1998). Neural activity in the primate prefrontal cortex during associative learning. *Neuron*, 21, 1399–1407.

Avery, R. A. (1978). Thermoregulation, metabolism and social behaviour in Lacertidae. In A. P. Bellairs, and C. B. Cox, eds., *Morphology and biology of reptiles*, London: Academic Press.

Axtell, R. E. (1991). *Gestures: The do's and taboos of body language around the world*. New York: Wiley.

Baddeley, A., and Salamé, P. (1986). The unattended speech effect: Perception or memory? *Journal of Experimental Psychology: Learning, Memory and Cognition*, 12, 525–529.

Baddeley, A., Thomson, N., and Buchanan, M. (1975). Word length and the structure of short-term memory. *Journal of Verbal Learning and Verbal Behavior*, 14, 575–589.

Bahn, P. G. (1987). Return of the Euro-boomerang. *Nature*, 329, 388.

Ballantine, H., Bouckoms, A. J., Thomas, E. K., and Giriunas, I. E. (1987). Treatment of psychiatric illness by stereotactic cingulotomy. *Biological Psychiatry*, 22, 807–817.

Bamberger, C. (1965). *The conductor's art*. New York: Columbia University Press.

Banquet, J. P. (1973). Spectral analysis of the EEG in meditation. *Electroencephalography and Clinical Neurophysiology*, 35, 143–151.

Barber, N. (1995). The evolutionary psychology of physical attractiveness: Sexual selection and human morphology. *Ethology and Sociobiology*, 16, 395–424.

Barch, D. M., Braver, T. S., Sabb, F. W., and Noll, D. C. (2000). Anterior cingulate and the monitoring of response conflict: Evidence from an fMRI study of overt verb generation. *Journal of Cognitive Neuroscience*, 12(2), 298–309.

Barinaga, M. (1992). How scary things get that way. *Science*, 258, 887–888.

Barnes, J. (1987). *Early Greek philosophy*. London: Penguin.

Baron-Cohen, S., Ring, H., Moriarty, J., Schmitz, B., Costa, D., and Ell, P. (1994). Recognition of mental state terms. *British Journal of Psychiatry*, 165, 640–649.

Basar, E., and Bullock, T., eds. (1992). *Induced rhythms in the brain*. Boston: Birkauser.

Basile, L. F., Rogers, R. L., Bourbon, W. T., and Papanicolaou, A. C. (1994). Slow magnetic flux from human frontal cortex. *Electroencephalographic and Clinical Neurophysiology*, 90, 157–165.

Batson, C. D. (1990). How social an animal? *American Psychologist*, 45, 336–346.

Batto, A. M. (2000). *Half a brain is enough? The story of Nico*. Cambridge: Cambridge University Press.

Bauer, R. (1982). Visual hypoemotionality as a symptom of visual-limbic disconnection in man. *Archives of Neurology*, 39, 702–708.

Bechara, A., Tranel, D., Damasio, H., and Damasio, A. R. (1996). Failure to respond autonomically to anticipated future outcomes following damage to prefrontal cortex. *Cerebral Cortex*, 6, 215–225.

Beck, L. E. (1994). Why children talk to themselves. *Scientific American*, 271 (5), 60–65.

Beckers, G., and Hömberg, V. (1991). Impairment of visual perception and visual short term memory scanning by transcranial magnetic stimulation of occipital cortex. *Experimental Brain Research*, 87, 421–432.

Begun, D., and Walker, A. (1993). The endocast. In A. Walker and R. Leakey, eds., *The Nariokotome* Homo erectus *skeleton*, pp. 326–358. Cambridge, Mass.: Harvard University Press.

Beldarrain, M. G., Grafman, J., Pascual-Leone, A., and Garcia-Monco, J. C. (1999). Procedural learning is impaired in patients with prefrontal lesions. *Neurology*, 52, 1853–1860.

Bell, G. B., Marino, A. A., and Chesson, A. L. (1994). Frequency-specific responses in the human brain caused by electromagnetic fields. *Journal of Neurological Sciences*, 123, 26–32.

Benaron, D. A., Cheong, W. F., and Stevenson, D. K. (1997). Tissue optics. *Science*, 276, 2002–2003.

Benaron, D. A., and Stevenson, D. K. (1993). Optical time-of-flight and absorbance imaging of biologic media. *Science*, 259, 1463–1466.

Benes, F. M., McSparren, J., Bird, E. D., SanGiovanni, J. P., and Vincent, S. L. (1991). Deficits in small interneurons in prefrontal and cingulate cortices of schizophrenic and schizoaffective patients. *Archives of General Psychiatry*, 48, 996–1001.

Benoit, D., and Parker, K. C. (1994). Stability and transmission of attachment across three generations. *Child Development*, 65, 1444–1456.

Berg, R., Jarlman, O., and Svanberg, S. (1993). Medical transillumination imaging using short-pulse diode lasers. *Applied Optics*, 32, 574–579.

Bergman, R. A., Thompson, S. A., and Afifi, A. K. (1984). *Catalog of human variation*. Baltimore: Urban and Schwarzenberg.

Bernstein, M. P., Sandford, S. A., and Allamandola, L. J. (1999). Life's far-flung raw materials. *Scientific American*, 281 (1), 42–49.

Berry, J. W. (1967). Independence and conformity in subsistence level societies. *Journal of Personality and Social Psychology*, 7, 415–418.

Berry, J. W., and Annis, R. C. (1974). Ecology, culture and psychological differentiation. *International Journal of Psychology*, 9, 173–193.

Bhattacharya, J., and Petsche, H. (2001). Musicians and the gamma band: a secret affair? *NeuroReport*, 12, 371–374.

Bick, P. A., and Kinsbourne, M. (1987). Auditory hallucinations and subvocal speech in schizophrenic patients. *American Journal of Psychiatry*, 144, 222–225.

Bickman, L., and Zarantonello, M. (1978). The effects of deception and level of obedience on subjects' rating of the Milgram study. *Personality and Social Psychology Bulletin*, 4, 81–85.

Bird, B. L., Newton, F. A., Sheer, D. E., and Ford, M. (1987). Biofeedback training of 40-Hz EEG in humans. *Biofeedback and Self-Regulation*, 3, 1–11.

Bisiach, E., Brouchon, M., Poncet, M., and Rusconi, M. L. (1993). Unilateral neglect in route description. *Neuropsychologia*, 31, 1255–1262.

Bisiach, E., and Luzzatti, C. (1978). Unilateral neglect of representational space. *Cortex*, 14, 129–133.

Black, J. E., Sirevaag, A. M., and Greenough, W. T. (1987). Complex experience promotes capillary formation in young rat visual cortex. *Neuroscience Letters*, 83, 351–355.

Blenner, J. L., and Yingling, C. D. (1993). Modality specificity of evoked potential augmenting/reducing. *Electroencephalography and Clinical Neurophysiology*, 88, 131–142.

Blinkov, S. M., and Glezer, I. I. (1968). *The human brain in figures and tables*. New York: Plenum Press.

Bloom, H. (1997). *The Lucifer principle: A scientific expedition into the forces of history*. New York: Atlantic Monthly Press.

Bloom, H. (2000). *Global brain: The evolution of the mass mind from the Big Bang to the 21st century*. New York: Wiley.

Blurton, N. G., Jones, N., and Konner, M. (1976). !Kung knowledge of animal behavior, In R. Lee and I. DeVore, eds., *Kalahari hunter-gatherer*, pp. 325–348. Cambridge, Mass.: Harvard University Press.

Boesch, C. (1991a). Symbolic communication in wild chimpanzees? *Human Evolution*, 6, 81–90.

Boesch, C. (1991b). The effects of leopard predation on grouping patterns in forest chimpanzees. *Behaviour*, 117, 220–241.

Boesch, C. (1993). Aspects of transmission of tool-use in wild chimpanzees. In K. R. Gibson and T. Ingold, eds., *Tools, language and cognition in human evolution*, pp. 171–183. Cambridge: Cambridge University Press.

Boesch, C. (1994). Cooperative hunting in wild chimpanzees. *Animal Behavior*, 48, 653–667.

Boesch, C., and Boesch, H. (1989). Hunting behavior of wild chimpanzees in the Taï National Park. *American Journal of Physical Anthropology*, 78, 547–573.

Boesch, C., and Boesch-Achermann, H. (2000). *The chimpanzees of the Taï forest*. Oxford: Oxford University Press.

Boesch, C., Marchesi, P., Marchesi, N., Fruth, B., and Joulian, F. (1994). Is nut cracking in wild chimpanzees a cultural behaviour? *Journal of Human Evolution*, 26, 325–338.

Boroojerdi, B., Phipps, M., Kopylev, L., Wharton, C. M., Cohen, L. G., and Grafman, J. (2001). Enhancing analogic reasoning with rTMS over the left prefrontal cortex. *Neurology*, 56, 526–528.

Bowby, J. (1969). *Attachment and loss*, vol. 1: *Attachment*. London: Hogarth Press.

Boysen, S. T. (1996). "More is less": The elicitation of rule-governed resource distribution in chimpanzees. In A. E. Russon, K. A., Bard, and S. T., Parker, eds., *Reaching into thought: The minds of the great apes*, pp. 177–189. Cambridge: Cambridge University Press.

Bragin, A., Jandó, G., Nádasdy, Z., Hetke, J., Wise, K., and Buzsáki, G. (1995). Gamma (40–100 Hz) oscillation in the hippocampus of the behaving rat. *Journal of Neuroscience*, 15, 47–60.

Brazzelli, M., Colombo, N., Della Sala, S., and Spinnler, H. (1994). Spared and impaired cognitive abilities after bilateral frontal damage. *Cortex*, 30, 27–51.

Brehm, S. S., and Brehm, J. W. (1981). *Psychological reactance: A theory of freedom and control*. New York: Academic Press.

Breitenseher, M., Uhl, F., Wimberger, D. P., Deecke, L., Trattnig, S., and Kramer, J. (1998). Morphological dissociation between visual pathways and cortex: MRI of visually-deprived patients with congenital peripheral blindness. *Neuroradiology*, 40, 424–427.

Bremner, J. D., Staib, L. H., Kaloupek, D., Southwick, S. M., Soufer, R., and Charney, D. S. (1999). Neural correlates of exposure to traumatic pictures and sound in Vietnam combat veterans with and without posttraumatic stress disorder: A positron emission tomography study. *Biological Psychiatry*, 45, 806–816.

Bressler, S. L. (1990). The gamma wave: A cortical information carrier? *Trends in Neurosciences*, 13, 161–162.

Bressler, S. L., Coppola, R., and Nakamura, R. (1993). Episodic multiregional cortical coherence at multiple frequencies during visual task performance. *Nature*, 366, 153–156.

Bridges, B. C., and Grimm, C. T. (1982). Reversal of morphine disruption of maternal behaviour by concurrent treatment with the opiate antagonist naloxone. *Science*, 218, 166–168.

Brothers, L. (1992). Perception of social acts in primates: Cognition and neurobiology. *Seminars in the Neurosciences*, 4, 409–414.

Brothers, L., Ring, B., and Kling, A. (1990). Responses of neurons in the macaque amygdala to complex social stimuli. *Behavioral Brain Research*, 41, 199–213.

Brown, M. W., and Cassel, M. D. (1980). Estimates of the number of neurons in the human hippocampus. *Journal of Physiology*, 301, 58P–59P.

Brunia, C. H., and Damen, E. J. (1988). Distribution of slow brain potential related to motor preparation and stimulus anticipation in a time estimation task. *Electroencephalography and Clinical Neurophysiology*, 69, 234–243.

Bunge, S. A., Klingberg, T., Jacobsen, R. B., and Gabrieli, J. D. (2000). A resource model of the neural basis of executive working memory. *Proceedings of the National Academy of Sciences of the USA*, 97, 3573–3578.

Buonomano, D. V., and Merzenich, M. M. (1998). Cortical plasticity: From synapses to maps. *Annual Reviews in Neuroscience*, 21, 149–186.

Burgess, P. W., Quayle, A., and Frith, C. D. (2001). Brain regions involved in prospective memory as determined by positron emission tomography. *Neuropsychologia*, 39, 545–555.

Burgess, P. W., Veitch, E., De Lacy Costello, A., and Shallice, T. (2000). The cognitive and neuroanatomical correlates of multitasking. *Neuropsychologia*, 38, 848–863.

Buzsáki, G. (1996). The hippocampo-neocortical dialogue. *Cerebral Cortex*, 6, 81–92.

Byl, N. N., Merzenich, M. M., and Jenkins, W. M. (1996). A primate genesis model of focal dystonia and repetitive strain injury: I. Learning-induced dedifferentiation of the representation of the hand in the primary somatosensory cortex in adult monkeys. *Neurology*, 47, 508–520.

Cacioppo, J. T. (1994). Social neuroscience: Autonomic, neuroendocrine, and immune responses to stress. *Psychophysiology*, 31, 113–128.

Cahill, L., Prins, B., Weber, M., and McGaugh, I. L. (1994). b-adrenergic activation and memory for emotional events. *Nature*, 371, 702–704.

Campbell, R., and Butterworth, B. (1985). Phonological dyslexia and dysgraphia in a highly literate subject: A development case with associated deficits of phonemic processing and awareness. *Quarterly Journal of Experimental Psychology*, 37A, 435–475.

Carmichael, S. T., and Price, J. L. (1994). Architectonic subdivision of the orbital and medial prefrontal cortex in the macaque monkey. *Journal of Comparative Neurology*, 346, 366–402.

Carrier, D. R. (1984). The energetic paradox of human running and hominid evolution. *Current Anthropology*, 25, 483–495.

Carter, C., and Getz, L. (1993). Monogamy and the prairie vole. *Scientific American*, 268, (6), 70–76.

Carter, C. S. (1992). Hormonal influences on human sexual behavior. In J. B. Becker, S. M. Breedlove, and D. Crews, eds., *Behavioral endocrinology*, pp. 131–142. Cambridge, Mass.: MIT Press.

Carter, C. S., Lederhendler, I. I., and Kirkpatrick, B, eds. (1987). *Integrative neurobiology of affiliation (Annals of the New York Academy of Sciences, 807)*, New York: New York Academy of Sciences.

Carter, C. S., Macdonald, A. M., Botvinick, M., Ross, L. L., Stenger, V. A., Noll, D., and Cohen, J. D. (2000). Parsing executive processes: Strategic vs evaluative functions of the anterior cingulate cortex. *Proceedings of the National Academy of Sciences of the USA*, 97(4), 1944–1948.

Case, R. (1992). The role of the frontal lobes in the regulation of cognitive development. *Brain and Cognition*, 20, 51–773.

Case, R. B., Moss, A. J., Case, N., McDermott, M., and Eberly, S. (1992). Living alone after myocardial infarction. *JAMA*, 267, 515–519.

Casey, B. J., Giedd, J. N., and Thomas, K. M. (2000). Structural and functional brain development and its relation to cognitive development. *Biological Psychology*, 54(1–3), 241–257.

Casey, B. J., Thomas, K. M., Welsh, T. F., Badgaiyan, R. D., Eccard, C. H., Jennings, J. R., and Crone, E. A. (2000). Dissociation of response conflict, attentional selection, and expectancy with functional magnetic resonance imaging. *Proceedings of the National Academy of Sciences of the USA*, 97(15), 8728–8733.

Casey, B. J., Trainor, R. J., Orendi, J. L., Schubert, A. B., Nystriom, L. E., Giedd, J. N., Haxby, J. V., Noll, D. C., Cohen, J. D., Forman, S. D., Dahkl, R. E., and Rapoport, J. L. (1997). A developmental functional MRI study of prefrontal activation during performance of a go-no-go task. *Journal of Cognitive Neuroscience*, 9, 835–887.

Casey, K. L., Minoshima, S., Berger, K. L., Koeppe, R. A., Morrow, T. J., and Fey, K. A. (1994). Positron emission tomographic analysis of cerebral structures activated specifically by repetitive noxious heat stimuli. *Journal of Neurophysiology*, 71, 802–887.

Castro-Caldas, A., Miranda, P. C., Carmo, I., Reis, A., Leote, F., Ribeiro, C., and Ducla-Soares, E. (1999). Influence of learning to read and write on the morphology of the corpus callosum. *European Journal of Neurology*, 6, 23–28.

Castro-Caldas, A., Petersson, K. M., Reis, A., Stone-Elander, S., and Ingvar, M. (1998). The illiterate brain: Learning to read and write during childhood influences the functional organization of the adult brain. *Brain*, 121, 1053–1063.

Catalán-Ahumada, M. Deggouj, N., De Volder, A., Melin, C., Michel, C., and Veraart, C. (1993). High metabolic activity demonstrated by positron emission tomography in human auditory cortex in case of deafness of early onset. *Brain Research*, 623, 287–292.

Cavallero, C., Cicogna, P., Natale, V., Occhionero, M., and Zito, A. (1992). Slow wave sleep dreaming. *Sleep*, 15, 562–566.

Chance, B., Zhuang, Z., Unah, C., Alter, C., and Lipton, L. (1993). Cognition-activated low-frequency modulation of light absorption in human brain. *Proceedings of the National Academy of Sciences of the USA*, 90, 3770–3774.

Chee, M. W., Sriram, N., Soon, C. S., and Lee, K. M. (2000). Dorsolateral prefrontal cortex and the implicit association of concepts and attributes. *NeuroReport*, 11, 135–140.

Cheney, D., Seyfarth, R., and Smuts, B. (1986). Social relationships and social cognition in nonhuman primates. *Science*, 234, 1361–1366.

Chodosh, E. H. Foulkes, M. A., Kase, C. S., Wolf, P. A., Mohr, J. P., Hier, D. B., Price, T. R., and Furtado, J. G. (1988). Silent stroke in the NINCDS stroke data bank. *Neurology*, 33, 1674–1679.

Chugani, H. T. (1994). Development of regional brain glucose metabolism in relation to behavior and plasticity. In G. Dawson, and K. W. Fisher, eds., *Human behavior and the developing brain*, pp 153–173, New York: Guilford Press.

Chugani, H. T., Behen, M. E., Muzik, O., Juhasz, C., Nagy, F., and Chugani, D. C. (2001). Local brain functional activity following early deprivation: A study of postinstitutionalized Romanian orphans. *Neuroimage*, 14, 1290–1301.

Clark, C., Klonoff, H., Tyhurst, J. S., Ruth, T., Adam, M., Rogers, J., Harrop, R., Martin, W., and Pate, B. (1988). Regional cerebral glucose metabolism in identical twins. *Neuropsychologia*, 26, 615–617.

Cleghorn, J. M., Franco, S., Szechtman, B. A., Kaplan, R. D., Szechtman, H., Brown, G. M., Nahmias, C., and Garnett, E. S. (1992). Toward a brain map of auditory hallucinations. *American Journal of Psychiatry*, 149, 1062–1069.

Clynes, M. (1977). *The touch of emotions,* New York: Doubleday Anchor.

Coghill, R. C., Talbot, J. D., Evens, A. C., Meyer, E., Gjedde, A., and Bushnell, M. C. (1994). Distributed processing of pain and vibration by the human brain. *Journal of Neuroscience,* 14, 4095–4108.

Cohen, I. G., Clenik, P., Pascual-Leone, A., Corwell, B., Faiz, L., Dambrosla, J., Honda, M., Sadato, N., Gerfoff, C., Catala, M. D., and Hallett, M. (1997). Functional relevance of cross-modal plasticity in blind humans. *Nature,* 398, 180–183.

Cohen, M. N. (1989). *Health and the rise of civilization.* New Haven, Ct.: Yale University Press.

Cohen, N. J., and Eichenbaum, H. (1993). *Memory, amnesia, and the hippocampal system.* Cambridge, Mass.: MIT Press.

Cohen, N. J., Ryan, J., Hunt, C., Romine, L., Wszalek, T., and Nash, C. (1999). Hippocampal system and declarative (relational) memory: Summarizing the data from functional neuroimaging studies. *Hippocampus,* 9, 83–98.

Cohen, R. A., Kaplan, R. F., Zuffante, P., Moser, D. J., Jenkins, M. A., Salloway, S., and Wilkinson, H. (1999). Alteration of intention and self-initiated action associated with bilateral anterior cingulotomy. *Journal of Neuropsychiatry and Clinical Neuroscience,* 11, 444–453.

Cohen, S., Kaplan, J. R., Cunnick, J. E., Manuck, S. B., and Rabin, B. S. (1992). Chronic social stress, affiliation and cellular immune response in nonhuman primates. *Psychological Science,* 3, 301–304.

Coleman-Mesches, K., and McGaugh, J. L. (1995). Differential involvement of the right and left amygdalae in expression of memory for aversively motivated training. *Brain Research,* 670, 75–81.

Conduit, R., Bruck, D., and Coleman, G. (1997). Induction of visual imagery during NREM sleep. *Sleep,* 29, 949–956.

Conroy, G. C. (1990) *Primate evolution.* New York: W. W. Norton and Co.

Constantinidis, C., Franowicz, M. N., and Goldman-Rakic, P. S. (2001). The sensory nature of mnemonic representation in the primate prefrontal cortex. *Nature Neuroscience,* 4, 311–316.

Cotterill, R. (1998). *Enchanted looms.* Cambridge: Cambridge University Press.

Cowan, J. D., and Sharp, D. H. (1988). Neural nets and artificial intelligence. *Daedalus,* 117, 85–121.

Cowan, N., Cartwright, C., Winterowd, C., and Sherk, M. (1987). An adult model of preschool children's speech memory. *Memory and Cognition*, 15, 511–517.

Cowey, A.,and Stoerig, P. (1991). The neurobiology of blindsight. *Trends in Neurosciences*, 14, 140–145.

Coyne, H. (1991). *Scam*. London: Duckworth.

Craig, A. D., Bushnell, M. C., Zhang, E.-T.,and Blomquist, A. (1994). A thalamic nucleus specific for pain and temperature sensation. *Nature*, 372, 770–773.

Creace, R. P. (1993). Biomedicine in the age of imaging. *Science*, 261, 554–561.

Crick, F., and Koch, C. (1990). Towards a neurobiological theory of consciousness. *Seminars in the Neurosciences*, 2, 263–275.

Critchley, M. (1979). *The divine banquet of the brain*. New York: Raven.

Cronin, H. (1991). *The ant and the peacock*. Cambridge: Cambridge University Press.

Crusco, A., and Wetzel, C. (1984). The Midas touch: The effects of interpersonal touch on restaurant tipping. *Personality and Social Psychology Bulletin*, 10, 512–517.

Csikszentmihalyi, M., and Larson, R. (1984). *Being adolescent*. New York: Basic Books.

Curcio, C. A., Sloan, K. R., Packer, O., Hendrickson, A. E., and Kalina, R. E. (1987). Distribution of cones in human and monkey retina: Individual variability and radial asymmetry. *Science*, 236, 579–582.

Cycowicz, Y. M. (2000). Memory development and event-related brain potentials in children. *Biological Psychology*, 54(1–3), 145–174.

Daffner, K. R., Mesulam, M. M., Scinto, L. F., Acar, D., Faust, R., Chabrerie, A., Kennedy, B., and Holcomb, P. (2000). The central role of the prefrontal cortex in directing attention to novel events. *Brain*, 123, 927–939.

Damasio, A. (1994). *Descartes' error*. New York: Grosset/Putnam.

Damasio, A., and Damasio, H. (1994). Cortical systems for retrieval of concrete knowledge: The convergence zone framework. In C. Koch and J. L. Davis, eds., *Large-scale neuronal theories of the brain*, pp. 61–74. Cambridge, Mass.: MIT Press.

Damasio, A., and Tranel, D. (1993). Nouns and verbs are retrieved with differently distributed neural systems. *Proceedings of the National Academy of Sciences of the USA*, 90, 4957–4960.

Damasio, A., Tranel, D., and Damasio, H. (1990). Individuals with sociopathic behavior caused by frontal damage fail to respond autonomically to social stimuli. *Behavioral Brain Research*, 41, 81–94.

Damasio, A., Tranel, D., and Damasio, H. (1991). Somatic markers and the guidance of behavior: Theory and preliminary testing. In H. Levin., H. Eisenberg, and A. Benton, eds., *Frontal lobe function and dysfunction*, pp. 217–229. New York: Oxford University Press.

Daniele, A., Giustolisis, L., Silveri, C., Colosimo, C., and Gainotti, G. (1994). Evidence for a possible neuroanatomical basis for lexical processing of nouns and verbs. *Neuropsychologia*, 32, 1325–1341.

Darian-Smith, C., and Gilbert, C. D. (1994). Axonal sprouting accompanies functional reorganization in adult cat striate cortex. *Nature*, 368, 737–740.

Darwin, C. (1871/1981). *The descent of Man and selection in relation to sex*. New York: D. Appleton Company.

Dave, A. S., and Margoliash, D. (2000). Song replay during sleep and computational rules for sensorimotor vocal learning. *Science*, 290, 812–816.

Davidson, R. (1992). Emotion and affective style: Hemispheric substrates. *Psychological Science*, 3, 39–43.

Davidson, R. J., and Fox, N. A. (1989). Frontal brain asymmetry predicts infants' response to maternal separation. *Journal of Abnormal Psychology*, 98, 127–131.

Davies, J. B., Kenny, P., and Barbenel, J. (1989). A psychological investigation of the role of mouthpiece force in trumpet performance. *Psychology of Music*, 17, 48–62.

Dawkins, R. (1982). *The extended phenotype*. Oxford: W. H. Freeman.

Dawkins, R. (1989). *The selfish gene*, 2nd ed. Oxford: Oxford University Press.

Day, M. H. (1992). Posture and childbirth. In S. Jones, R. Martin, and D. Pilbeam, eds., *The Cambridge encyclopedia of human evolution*, p. 88. Cambridge: Cambridge University Press.

De Beaune, S. A., and White, R. (1993). Ice age lamps. *Scientific American*, 266(3), 74–79.

De Koninck, J., Christ, G., Hebert, G., and Rinfret, N. (1990). Language learning efficiency, dreams and REM sleep. *Psychiatric Journal of the University of Ottawa*, 15, 91–92.

De Volder, A. G., Bol, A., Blin, J., Robert, A., Arno, P., Grandin, C., Michel, C., and Veraart, C. (1997). Brain energy metabolism in early blind subjects: Neural activity in the visual cortex. *Brain Research*, 750, 235–244.

De Waal, F. B. M. (1989). *Peacemaking among primates*. Cambridge, Mass.: Harvard University Press.

De Waal, F. B. M. (1991). The chimpanzee's sense of social regularity and its relation to the human sense of justice. *American Behavioral Scientist*, 34, 335–349.

De Waal, F. B. M. (1995). Bonobo sex and society. *Scientific American*, March, 58–64.

De Waal, F. B. (1996). *Good natured: The origins of right and wrong in humans and other animals*. Cambridge, Mass.: Harvard University Press.

De Waal, F. B., and Johanowicz, D. L. (1993). Modification of reconciliation behavior through social experience. *Child Development*, 64, 897–908.

De Waal, F. B., and Lanting, F. (1997). *Bonobo: The forgotten ape*. Berkeley, Calif.: University of California Press.

Deacon, T. (1990a). Rethinking mammalian brain evolution. *American Zoologist*, 30, 629–705.

Deacon, T. (1990b). Fallacies of progression in theories of brain-size evolution. *International Journal of Primatology*, 11, 193–236.

Deacon, T. (1997). *The symbolic species, The co-evolution of language and the brain*. New York: Norton.

Decety, J., Jeannerod, M., Germain, M., and Pastene, J. (1991). Vegetative response during imagined movement is proportional to mental effort. *Behavioral Brain Research*, 42, 1–5.

Deglin, V. L., and Kinsbourne, M. (1996). Divergent thinking styles of the hemisphere: How syllogisms are solved during transitory hemisphere suppression. *Brain and Cognition*, 31, 285–307.

Dehaene, S., and Naccache, L. (2001). Towards a cognitive neuroscience of consciousness: Basic evidence and a workspace framework. *Cognition*, 79, 1–37.

Dehay, C., Giroud, P., Berland, M., Killackey, H., and Kennedy, H. (1996). Contribution of thalamic input to the specification of cytoarchitectonic cortical fields in the primate. *Journal of Comparative Neurology*, 367, 70–89.

Dehay, C., Giroud, P., Berland, M., Smart, I., and Kennedy, H. (1993). Modulation of the cell cycle contributes to the parcellation of the primate visual cortex. *Nature*, 366, 464–466.

Delalle, I., Evers, P., Kostovic, I., and Uylings, H. B. (1997). Laminar distribution of neuropeptide Y–immunoreactive neurons in human prefrontal cortex during development. *Journal of Comparative Neurology*, 379(4), 515–522.

Della Sala, S., Laiacona, M., Spinnler, H., and Trivelli, C. (1993). Autobiographic recollection and frontal damage. *Neuropsychologia*, 31, 823–839.

Demeter, S. Ringo, J., and Dorty, R. (1988). Morphometric analysis of the human corpus callosum and anterior commissure. *Human Neurobiology*, 6, 219–226.

Dencker, S. J., Johansson, G., and Milsom, I. (1992). Quantification of naturally occurring benzodiazepine-like substances in human breast milk.*Psychopharmacology*, 107, 67–72.

Dennett, D. (1984). *Elbow space: The varieties of freedom worth wanting*. Oxford: Oxford University Press.

Dennett, D. C. (1986). Julian Jaynes's software archeology. *Canadian Psychology*, 27, 149–154.

Dennett, D. C. (1987). Consciousness. In R. Gregory, ed., *Oxford companion to the mind*, pp. 160–164. Oxford: Oxford University Press.

Dennett, D. C. (1991). *Consciousness explained*. London: Allen Lane.

Dennett, D. C. (1992). The self as a center of narrative gravity. In F. Kessel, P. Cole, and D. Johnson, eds., *Self and consciousness: Multiple perspectives*, pp. 103–115. Hillsdale, N.J.: Erlbaum.

Derbyshire, S. W., Jones, A. K., Devani, P., Friston, K. J., Feinmann, C., Harris, M., Pearce, S., Watson, J. D., and Frackowiak, R. S. (1994). Cerebral responses to pain in patients with atypical facial pain measured by positron emission tomography. *Journal of Neurology, Neurosurgery and Psychiatry*, 57, 1166–1172.

Desimone, R. (1996). Neural mechanisms for visual memory and their role in attention. *Proceedings of the National Academy of Sciences of the USA*, 93, 13494–13499.

Desimone, R., Miller, E., and Chelazzi, L. (1994). The interaction of neural systems for attention and memory. In C. Koch and J. L. Davis, eds., *Large-scale neuronal theories of the brain*, pp. 75–91. Cambridge, Mass.: MIT Press.

Desmedt, J. E., and Debecker, J. (1979). Wave form and neural mechanism of the decision P350 elicited without pre-stimulus CNV or readiness potential. *Electroencephalography and Clinical Neurophysiology*, 47, 648–670.

Desmedt, J. E., and Tomberg, C. (1994). Transient phase-locking of 40 Hz electrical oscillations in prefrontal and parietal human cortex reflects the process of conscious somatic perception. *Neuroscience Letters*, 168, 126–129.

Devine, J. (1985). The versatility of human locomotion. *American Anthropologist*, 87, 550–567.

Devinsky, O., Morrell, M. J., and Vogt, B. A. (1995). Contributions of anterior cingulate cortex to behaviour. *Brain*, 118, 279–306.

Dimitrov, M., Granetz, J., Peterson, M., Hollnagel, C., Alexander, G., and Grafman, J. (1999). Associative learning impairments in patients with frontal lobe damage. *Brain and Cognition*, 41, 213–230.

Dinse, H. R., and Krüger, K. (1994). The timing of processing along the visual pathway in the cat. *NeuroReport*, 5, 893–897.

Divale, W. (1984). *Matrilocal residence in pre-literate society*. Ann Arbor, Mich.: UMI Research Press.

Dominy, N. J., and Lucas, P. W. (2001). Ecological importance of trichromatic vision to primates. *Nature*, 410, 363–364.

Doron, N., and Wollberg, Z. (1994). Cross-modal neuroplasticity in the blind mole rat *Spalaz ehrenbergi*. *NeuroReport*, 5, 2697–2701.

Dove, A., Pollmann, S., Schubert, T., Wiggins, C. J., and von Cramon, D. Y. (2000). Prefrontal cortex activation in task switching. *Cognitive Brain Research*, 9, 103–109.

Drevets, W. C., Burton, H., Videen, T. O., Snyder, A. Z., Simpson, J. R., and Raichle, M. E. (1995). Blood flow changes in human somatosensory cortex during anticipated stimulation. *Nature*, 373, 249–252.

Drewe, E. A. (1974). The effect of type and area of brain lesion on Wisconsin Card Sorting Test performance. *Cortex*, 10, 159–170.

Dunbar, R. I. (1992). Why gossip is good for you. *New Scientist*, Nov. 28, 28–31.

Duncan, J., Burgess, P., and Emslie, H. (1995). Fluid intelligence after frontal lobe lesions. *Neuropsychologia*, 33, 261–268.

Duncan, J., Emslie, H., Williams, P., Johnson, R., and Freer, C. (1996). Intelligence and the frontal lobe: The organization of goal-directed behavior. *Cognitive Psychology*, 30, 257–303.

Duncan, J., Johnson, R., Swales, M., and Freer, C. (1997). Frontal lobe deficits after head injury: Unity and diversity of function. *Cognitive Neuropsychology*, 14, 713–741.

Dwight, T. (1892). Fusion of hands. *Memoirs of the Boston Society of Natural History*, 4, 473–486.

Eacott, M., and Gaffan, D. (1992). Inferotemporal-frontal disconnection: The uncinate fascicle and visual associative learning in monkeys. *European Journal of Neuroscience*, 4, 1320–1332.

Edelman, G. (1987). *Neural Darwinism: A theory of neuronal group selection*. New York: Basic Books.

Eichenbaum, H., and Cohen, N. (1988). Representation in the hippocampus: What do hippocampal neurons code? *Trends in Neurosciences*, 11, 244–248.

Ekman, P., Davidson, R., and Friesen, W. (1990). The Duchenne smile: Emotional expression and brain physiology. II. *Journal of Personality and Social Psychology*, 58, 342–353.

Ekman, P., and Davidson, R. J. (1993). Voluntary smiling changes regional brain activity. *Psychological Science*, 4, 342–345.

Elbert, T., Pantev, C., Wienbruch, C., Rockstroh, B., and Taub, E. (1995). Increased cortical representation of the fingers of the left hand in string players. *Science*, 270, 305–307.

Ellis, H., and Young, A. (1990). Accounting for delusional misidentifications. *British Journal of Psychiatry*, 157, 239–248.

Elston, G. N. (2000). Pyramidal cells of the frontal lobe. *Journal of Neuroscience*, 20:RC95, 1–4.

Endo, T., Roth, C., Landolt, H. P., Werth, E., Aeschbach, D., Achermann, P., and Borbely, A. A. (1998). Selective REM sleep deprivation in humans: effects on sleep and sleep EEG. *American Journal of Physiology*, 274, R1186–R1194.

Engel, A. K., König, P., Kreiter, A. K., Schillen, T. B., and Singer, W. (1992). Temporal coding in the visual cortex: New vistas on integration in the nervous system. *Trends in Neurosciences*, 15, 218–226.

Engel, A. K., König, P., Kreiter, A. K., and Singer, W. (1991). Interhemispheric synchronization of oscillatory neuronal responses in cat visual cortex. *Science*, 252, 1177–1179.

Engel, A. K., König, P., and Singer, W. (1991). Direct physiological evidence for scene segmentation by temporal coding. *Proceedings of the National Academy of Sciences of the USA*, 88, 9136-9140.

Engel, A. K., and Singer, W. (2001). Temporal binding and the neural correlates of sensory awareness. *Trends in Cognitive Science*, 5, 16–25.

Engen, T. (1991). *Odor sensation and memory*. New York: Praeger.

Ericsson, K. A. (1985). Memory skill. *Canadian Journal of Psychology*, 39, 188–231.

Ersland, L., Rosen, G., Lundervold, A., Smievoll, A. I., Tillung, T., Sundberg, H., and Hugdahl, K. (1997). Phantom limb imaginary fingertapping causes primary motor cortex activation: An fMRI study. *NeuroReport*, 8, 207–210.

Eslinger, P., and Grattan, L. (1993). Frontal lobe and frontal-striatal substrates for different forms of human cognitive flexibility. *Neuropsychologia*, 31, 17–28.

Eslinger, P., Grattan, L., Damasio, H., and Damasio, A. (1992). Developmental consequences of childhood frontal lobe damage. *Archives of Neurology*, 49, 764–769.

Espy, K. A., Kaufmann, P. M., McDiarmid, M. D., and Glisky, M. L. (1999). Executive functioning in preschool children: Performance on A-not-B and other delayed response format tasks. *Brain and Cognition*, 41(2), 178–199.

Etcoff, N. L. (1994). Beauty and the beholder. *Nature*, 368, 186–187.

Ettlin, T. M., Seiler, W., and Kaeser, H. E. (1980). Phantom and amputation illusions in paraplegic patients. *European Journal of Neurology*, 19, 12–19.

Evans, A., and Tomesello, M. (1986). Evidence for social referencing in young chimpanzees (*Pan troglodytes*). *Folia Primatologia*, 47, 49–54.

Evarts, E. V. (1987). Role of motor cortex in voluntary movements in primates. In F. Plum, ed., *Handbook of physiology, nervous system* vol. 5, part II, pp. 1083–1120. Bethesda, Md.: American Physiology Society.

Everitt, B. J., Cador, M., and Robbins, T. W. (1989). Interactions between the amygdala and ventral striatum in stimulus-reward associations: Studies using a second-order schedule of sexual reinforcement. *Neuroscience*, 30, 63–75.

Fadiga, L., Fogassi, L., Pavesi, G., and Rizzolatti, G. (1995). Motor facilitation during action observation: A magnetic stimulation study. *Journal of Neurophysiology*, 73, 2608–2611.

Fagen, R. (1981). *Animal play behavior*. New York: Oxford University Press.

Fahn, S. (1993). Motor and vocal tics. In R. Kurlan, ed., *Handbook of Tourette's syndrome and related tic and behavioral disorders*, pp. 3–16. New York: Dekker.

Fairbanks, L. A. (1988). Vervet monkey grandmothers: Interactions with infant grandoffspring. *International Journal of Primatology*, 9, 425–441.

Farah, M., Péronnet, F., Gonon, M., and Giard, M. (1988). Electrophysiological evidence for a shared representational medium for visual images and visual precepts. *Journal of Experimental Psychology: General*, 117, 248–257.

Farah, M. J. (1998). Why does the somatosensory homunculus have hands next to face and feet next to genitals? A hypothesis. *Neural Computation*, 10, 1983–1985.

Feeney, D., and Baron, J.-C. (1986). Diaschisis. *Stroke*, 17, 817–830.

Feinberg, T. E., Schindler, R. J., Flanagan, G. N., and Haber, L. D. (1992). Two alien hand syndromes. *Neurology*, 42, 19–24.

Feinman, S. (1992). Social referencing and conformity. In S. Feinman, ed., *Social referencing and the social construction of reality in infancy*, pp. 229–267. New York: Plenum.

Felleman, D., and Van Essen, D. (1991). Distributed hierarchical processing in the primate cerebral cortex. *Cerebral Cortex*, 1, 1–47.

Field, T. (2001). Massage therapy facilitates weight gain in preterm infants. *Current Directions in Psychological Science*, 10, 51–54.

Fillmore, L. W. (1979). Individual differences in second language acquisition. In C. J. Fillmore, D. Kempler, and W. S.-Y. Wang, eds., *Individual differences in language ability and language behavior*, pp. 203–228. New York: Academic Press.

Fisher, C. (1982). Disorientation for place. *Archives of Neurology*, 39, 33–36.

Fisher, C. M. (1999). Phantom erection after amputation of penis: Case description and review of the relevant literature on phantoms. *Canadian Journal of Neurological Sciences*, 26, 53–56.

Fisher, M., Ryan, S. B., and Dobyns, W. B. (1992). Mechanism of interhemispheric transfer and patterns of cognitive function in acallosal patients of normal intelligence. *Archives of Neurology*, 49, 271–277.

Fiske, A. (1992). The four elementary forms of sociality: Framework for a unified theory of social relations. *Psychological Bulletin*, 99, 689–723.

Flament, D., Onstott, D., Fu, Q.-G., and Ebner, T. J. (1993). Distance- and error-related discharge of cells in premotor cortex of rhesus monkeys. *Neuroscience Letters*, 153, 144–148.

Fletcher, P. C., Frith, C. D., Grasby, P. M., Shallice, T., Frackowiak, R. S., and Dolan, R. J. (1995). Brain systems for encoding and retrieval of auditory-verbal memory. *Brain*, 118, 401–415.

Fletcher, P. C., Shallice, T., and Dolan, R. J. (2000). "Sculpting the response space"—an account of left prefrontal activation at encoding. *NeuroImage*, 12, 404–417.

Flor, H., Braun, C., Elbert, T., and Birbaumer, N. (1997). Extensive reorganization of primary somatosensory cortex in chronic back pain patients. *Neuroscience Letters*, 224, 5–8.

Flor, H., Elbert, T., Knecht, S., Wienbruch, C., Pantev, C., Birbaumer, N., Larbig, W., and Taub, E. (1995). Phantom limb pain as a perceptual correlate of massive cortical reorganization in upper extremity amputees. *Nature*, 375, 482–484.

Forsyth, D. (1990). *Group dynamics*, 2nd ed. Pacific Grove, Calif.: Brooks/Cole.

Fox, N. A., Bell, M. A., and Jones, N. A. (1992). Individual differences in response to stress and cerebral asymmetry. *Developmental Neuropsychology*, 8, 161–184.

Frauenglass, M. H., and Diaz, R. M. (1985). Self-regulatory functions of children's private speech: A critical analysis of recent challenges to Vygotsky's theory. *Developmental Psychology*, 21, 357–364.

Freedman, D. J., Riesenhuber, M., Poggio, T., and Miller, E. K. (2001). Categorical representation of visual stimuli in the primate prefrontal cortex. *Science*, 291, 312–315.

Frison, G. (1987). Prehistoric Plains-Mountains, larger mammals and communal hunting strategies. In M. Nitecki and D. Nitecki, eds., *Evolution of human hunting*, pp. 177–223. New York: Plenum.

Frith, C. D., Friston, K., Liddle, P. E., and Frackowiak, R. S. (1991). Willed action and the prefrontal cortex in man: A study with PET. *Proceedings of the Royal Society of London, Series B*, 244, 241–246.

Frost, D. O. (1999). Functional organization of surgically created visual circuits. *Restorative Neurology and Neuroscience*, 15, 107–113.

Frost, H., Lamb, S. E., Klaber Moffett, J. A, Fairbank, J. C., and Moser, J. S. (1998). A fitness programme for patients with chronic low back pain: 2-year follow-up of a randomised controlled trial. *Pain*, 75, 273–279.

Fuster, J. M. (1980). *The prefrontal cortex: Anatomy, physiology, and neuropsychology of the frontal lobe*. New York: Raven.

Fuster, J. (1989). *Prefrontal cortex: Anatomy, physiology, and neuropsychology*, 2nd ed. New York: Raven.

Fuster, J. M. (1997). *The prefrontal cortex: anatomy, physiology, and neuropsychology of the frontal lobe*, 3rd ed. New York: Lippincott-Raven.

Fuster, J. M. (2000). Prefrontal neurons in networks of executive memory. *Brain Research Bulletin*, 52, 331–336.

Fuster, J. M., Bodner, M., and Kroger, J. K. (2000). Cross-modal and cross-temporal association in neurons of frontal cortex. *Nature*, 405, 347–351.

Gabilondo, A. M., Meana, J. J., and García-Servilla, J. A. (1995). Increased density of m-opioid receptors in the postmortem brains of suicide victims. *Brain Research*, 682, 245–250.

Gaffan, D., Murray, E., and Fabre-Thorpe, M. (1993). Interaction of the amygdala with the frontal lobe in reward memory. *European Journal of Neuroscience*, 5, 968–975.

Gafman, J., Pascual-Leone, A., Alway, D., Nichelli, P., Gomez-Tortosa, E., and Hallett, M. (1994). Induction of a recall deficit by rapid-rate transcranial magnetic stimulation. *NeuroReport*, 5, 1157–1160.

Galea, M. P.,and Darian-Smith, I. (1994). Multiple corticospinal neuron populations in the macaque monkey are specified by their unique cortical origins, spinal terminations, and connections. *Cerebral Cortex*, 4, 166–194.

Galkin, E. W. (1988). *A history of orchestral conducting in theory and practice*. New York: Pendragon Press.

Galliot, B., and Miller, D. (2000). Origin of anterior patterning: how old is our head? *Trends in Genetics*, 16, 1–5.

Gamble, C. (1982). Culture and society in the Upper Palaeolithic of Europe. In G. Bailey, ed., *Hunter-gatherer economy in prehistory*, pp. 201–219 . Cambridge: Cambridge University Press.

Gao, W.-J., and Pallas, S. L. (1999). Cross-modal reorganization of horizontal connectivity in auditory cortex without altering thalamocortical projections. *Journal of Neuroscience*, 19, 7940–7950,

Garofalo, R., Chheda, S., Mei, F., Palkowetz, K. H., Rudloff, H. E., Schmalstieg, F. C., Rassin, D. K., and Goldman, A. S. (1995). Interleukin-10 in human milk. *Pediatric Research*, 37, 444–449.

Garrett, A. S., Wood, F. B., and Keyes, J. W. (1994). Anxiety and social inhibition are reflected in regional glucose metabolism as measured by PET. *Society for Neuroscience Abstracts*, 20, 368.

Gastaut, H., Naquet, R., and Gastaut, Y. (1965). A study of the mu rhythm in subjects lacking one or more limbs. *Electroencephalography and Clinical Neurophysiology*, 18, 720–721.

Gathercole, S. E., Adams, A.-M., and Hitch, G. J. (1994). Do young children rehearse? An individual-differences analysis. *Memory and Cognition*, 22, 201–207.

George, M. S., Wassermann, E. M., Williams, W. A., Callahan, A., Ketter, T. A., Basser, P., Hallett, M., and Post, R. M. (1995). Daily repetitive transcranial magnetic stimulation (rTMS) improves mood in depression. *NeuroReport*, 6, 1853–1856.

Gevins, A., Le, J., Martin, N. K., Brickett, P., Desmond, J., and Reutter, B. (1994). High resolution EEG: 124-channel recording, spatial deblurring and MRI integration methods. *Electroencephalography and Clinical Neurophysiology*, 90, 337–358.

Ghiglieri, M. P. (1984). *The chimpanzees of Kibale Forest*, New York: Columbia University Press.

Giedd, J. N., Snell, J. W., Lange, N., Rajapakse, J. C., Casey, B. J., Kozuch, P. L., Vaituzis, A. C., Vauss, Y. C., Hamburger, S. D., Kaysen, D., and Rapoport, J. L. (1996). Quantitative magnetic resonance imaging of human brain development: Ages 4–18. *Cerebral Cortex*, 6, 551–560.

Giedd, J. N., Vaituzis, C., Hamburger, S. D., Lange, N., Rajapakse, J., Kaysen, D., Vauss, Y. C., and Rapoport, J. L. (1996). Quantitative MRI of the temporal lobe, amygdala, and hippocampus in normal human development: Ages 4–18 years. *Journal of Comparative Neurology*, 366, 223–230.

Gilbert, S. F. (2000). *Developmental biology*, 6th ed. Sunderland, Mass.: Sinauer.

Giraud, A. L., Garnier, S., Micheyl, C., Lina, G., Chays, A., and Chéry-Croze, S. (1997). Auditory efferents involved in speech-in-noise intelligibility. *NeuroReport*, 8, 1779–1783.

Glass, D. C., Singer, J. E., and Friedman, L. N. (1969). Psychic cost of adaptation to an environmental stressor. *Journal of Personality and Social Psychology*, 12, 200–210.

Godefroy, O., and Rousseaux, M. (1997). Novel decision making in patients with prefrontal or posterior brain damage. *Neurology*, 49, 695–701.

Goel, V., and Grafman, J. (2000). Role of the right prefrontal cortex in ill-structured planning. *Cognitive Neuropsychology*, 17, 417–436.

Goldberg, E., and Podell, K. (2000). Adaptive decision making, ecological validity and the frontal lobes. *Journal of Clinical and Experimental Neuropsychology*, 22, 56–68.

Goldberg, G. (1985). Supplementary motor area structure and function: Review and hypotheses. *Behavioural and Brain Sciences*, 8, 567–616.

Goldberg, T. E., Randolph, C., Gold, J. M., Berman, K. F., and Weinberger, D. R. (1994). Spatial delayed response in normal humans. *Society for Neuroscience Abstracts*, 20, 7.

Goldenberg, G., Podreka, I., Steiner, M., Willmes, K., Suess, E., and Deecke, L. (1989). Regional cerebral blood flow patterns in visual imagery. *Neuropsychologia*, 27, 641–664.

Goldin-Meadow, S., and Feldman, H. (1977). The development of language-like communication without a language model. *Science*, 197, 401–403.

Goldman-Rakic, P. (1984). Modular organization of prefrontal cortex. *Trends in Neurosciences*, 7, 419–424.

Goldman-Rakic, P. S. (1994). The issue of memory in the study of prefrontal function. In A. M. Thierry, J. Glowinski, P. S. Goldman-Rakic, and Y. Chisten, eds., *Motor and cognitive functions of the prefrontal cortex*, pp. 112–121, Berlin: Springer-Verlag.

Goldman-Rakic, P. S. (1996). Regional and cellular fractionation of working memory. *Proceedings of the National Academy of Sciences of the USA*, 93, 13473–13480.

Goldman-Rakic, P. S., Bates, J. F., and Chafee, M. V. (1992). The prefrontal cortex and internally generated motor acts. *Current Opinions in Neurobiology*, 2, 830–835.

Goldman-Rakic, P. S., Selemon, L. D., and Schwartz, M. L. (1984). Dual pathways connecting the dorsolateral prefrontal cortex with the hippocampal formation and parahippocampal cortex in the rhesus monkey. *NeuroScience*, 2, 719–743.

Gonzalez, J., Gamundi, A., Rial, R., Nicolau, M. C., de Vera, L., and Pereda, E. (1999). Nonlinear, fractal, and spectral analysis of the EEG of lizard, *Gallotia galloti*. *American Journal of Physiology*, 277, R86–R93.

Goodall, J. (1986). *The chimpanzees of Gombe: Patterns of behavior*. Cambridge, Mass.: Harvard University Press.

Goodall, J. (1990). *Through a window*. London: Weidenfeld and Nicolson.

Goodall, J. (1992). Unusual violence in the overthrow of an alpha male chimpanzee at Gombe. In T. Nishida, W. C. McGrew, P. Marler, M. Pickford, and B. de Waal, eds., *Topics in primatology:* vol. 1: *Human origins*, pp. 131–142. Tokyo: University of Tokyo Press.

Goodwyn, S. W., and Acredolo, L. P. (1993). Symbolic gesture versus word. *Child Development*, 64, 688–701.

Gopnik, A. (1993). How we know our minds: The illusion of first-person knowledge of intentionality. *Behavioral and Brain Sciences*, 16, 1–14.

Gorman, D. G., and Unützer, J. (1993). Brodmann's "missing" numbers. *Neurology*, 43, 226–227.

Grafman, J., and Christen, Y. (1999). *Neural plasticity: Building a bridge from the laboratory to the clinic.* Berlin: Springer-Verlag.

Grafman, J., Vance, S. C., Weingartner, H., Salazar, A. M., and Amin, D. (1986). The effects of lateralised frontal lesions on mood regulation. *Brain*, 109, 1127–1148.

Grasby, P. M., Frith, C. D., Friston, K., Frackowiak, R. S., and Dolan, R. J. (1993). Activation of the human hippocampal formation during auditory-verbal long-term memory function. *Neuroscience Letters*, 163, 185–188.

Gratton, G., and Fabiani, M. (2001) The event-related optical signal: A new tool for studying brain function. *International Journal of Psychophysiology*, 42, 109–121.

Gratton, G., Maier, J. S., Fabiani, M., Mantulin, W. W., and Gratton, E. (1994). Feasibility of intracranial near-infrared optical scanning. *Psychophysiology*, 31, 211–215.

Gray, C. M. (1999). The temporal correlation hypothesis of visual feature integration. *Neuron*, 24, 31–47.

Gray, C. M., and McCormick, D. A. (1996). Chattering cells: Superficial pyramidal neurons contributing to the generation of synchronous oscillations in the visual cortex. *Science*, 274, 109–113.

Grazino, M. S., Yap, G. S., and Gross, C. G. (1994). Coding of visual space by premotor neurons. *Science*, 266, 1054–1057.

Green, M. F., and Kinsbourne, M. (1990). Subvocal activity and auditory hallucinations: Clues for behavioral treatments? *Schizophrenia Bulletin*, 16, 617–625.

Griffith, H., and Davidson, M. (1966). Long-term changes in intellect and behaviour after hemispherectomy. *Journal of Neurology, Neurosurgery and Psychiatry*, 29, 271–576.

Grinvald, A., Frostig, R., Siegel, R., and Bartfeld, E. (1991). High-resolution optical imaging of functional brain architecture in the awake monkey. *Proceedings of the National Academy of Sciences of the USA*, 88, 11559–11563.

Gruzelier, J. (1998). A working model of the neurophysiology of hypnosis. *Contemporary Hypnosis*, 15, 3–15.

Gruzelier, J., Allison, J., and Conway, A. (1984). Hypnotic susceptibility: A lateral predisposition and altered cerebral asymmetry under hypnosis. *International Journal of Psychophysiology*, 2, 131–139.

Gruzelier, J., and Warren, K. (1993). Neuropsychological evidence of reductions on left frontal tests with hypnosis. *Psychological Medicine*, 23, 93–101.

Gunnar-Vongnechten, M. R. (1972). Changing a frightening toy into a pleasant toy by allowing the infant to control its actions. *Developmental Psychology*, 14, 157–162.

Gurney, K. (1997). *An introduction to neural networks*. London: UCL Press.

Gustafson, G. E., Green, J. A., and Cleland, J. W. (1994). Robustness of individual identity in the cries of human infants. *Developmental Psychobiology*, 27, 1–9.

Habel, W. B. (1956). Observations on phantom-limb phenomena. *Archives of Neurology and Psychiatry*, 75, 624–636.

Haig, A. R., Gordon, E., De Pascalis, V., Meares, R. A., Bahramali, H., and Harris, A. (2000). Gamma activity in schizophrenia: Evidence of impaired network binding. *Clinical Neurophysiology*, 111, 1461–1468.

Halgren, E., Baudena, P., and Biraben, A. (1995). Intracerebral potentials to rare target and distractor auditory and visual stimuli. II. Medial, lateral and posterior temporal lobe. *Electroencephalography and Clinical Neurophysiology*, 94, 229–250.

Halligan, P. W., Marshall, J. C., and Wade, D. T. (1995). Unilateral somatoparaphrenia after right hemisphere stroke. *Cortex*, 31, 173–182.

Halligan, P. W., Marshall, J. C., Wade, D., Davey, J., and Morrison, D. (1993). Thumb in cheek? Sensory reorganization and perceptual plasticity after limb amputation. *NeuroReport*, 4, 233–236.

Hamilton, R., Keenan, J. P., Catalá, M., and Pascual-Leone, A. (2000). Alexia for Braille following bilateral occipital stroke in an early blind woman. *NeuroReport*, 11, 237–240.

Hamilton, R., and Pascual-Leone, A. (1998). Cortical plasticity associated with Braille learning. *Trends in Cognitive Science*, 2, 168–174.

Hamilton, W. D. (1984). Significance of paternal investment by primates to the evolution of adult male-female associations. In D. Taub, ed., *Primate paternalism*, pp. 309–335. New York: Van Nostrand Reinhold.

Hamilton, W. D., and Zuk, M. (1982). Heritable true fitness and bright birds: A role for parasites? *Science*, 218, 384–387.

Hanowell, S. T., and Kennedy, S. T. (1979). Phantom tongue pain and causalgia. *Anaesthesia and Analgesia*, 58, 436–438.

Hansen, M. H. (1991). *Athenian democracy in the age of Demosthenes*. Oxford: Blackwell.

Hari, R. (1993). Magnetoencephalography as a tool of clinical neurophysiology. In E. Niedermeyer and F. Lopes da Silva, eds., *Electroencephalography: Basic principles, clinical approaches and related fields*, 3rd ed., pp. 1035–1061. Baltimore: Williams and Wilkins.

Harlow, J., and Roll, S. (1992). Frequency of day residue in dreams of young adults. *Perception and Motor Skills*, 74, 832–834.

Harris, L. J. (1990). Cultural influences on handedness: Historical and contemporary theory and evidence. In S. Coren, ed., *Left-handedness: Behavioral implications and anomalies*, pp. 195–258. Amsterdam: Elsevier.

Harrison, Y., Horne, J. A., and Rothwell, A. (2000). Prefrontal neuropsychological effects of sleep deprivation in young adults—a model for healthy aging? *Sleep*, 23, 1067–1073.

Harvey, P. H., and Arnold, S. J. (1982). Female mate choice and runaway sexual selection. *Nature*, 297, 533–534.

Hashimoto, C., and Furuichi, T. (1994). Social role and development of noncopulatory sexual behavior of wild bonobos. In R. W. Wrangham, W. C. McGrew, F. B. W. de Waal, and P. Heltine, eds., *Chimpanzee culture*, pp. 155–168. Cambridge, Mass.: Harvard University Press.

Hashimoto, R., and Tanaka, Y. (1998). Contribution of the supplementary motor area and anterior cingulate gyrus to pathological grasping phenomena. *European Neurology*, 40, 151–158.

Hasselmo, M. E. (1999). Neuromodulation: acetylcholine and memory consolidation. *Trends in Cognitive Science*, 3, 351–359.

Hatano, G., and Osawa, K. (1983). Digit memory of grand experts in abacus-derived mental calculation. *Cognition*, 15, 95–110.

Hatfield, E., Cocioppo, J. T., and Rapson, R. L. (1994). *Emotional contagion*. Cambridge: Cambridge University Press.

Haug, H., Kühl, S., Mecke, E., Sass, N.-L., and Wasner, K. (1984). The significance of morphometric procedures in the investigation of age changes in cytoarchitectonic structures of the human brain. *Journal für Hirnforschung*, 25, 353–374.

Hebb, D. O. (1939). Intelligence in man after large removals of cerebral tissue: Report of four left frontal lobe cases. *Journal of General Psychology*, 21, 73–87.

Heinen, F., Glocker, F. X., Fietzek, U., Meyer, B. U., Lucking, C. H., and Korinthenberg, R. (1998). Absence of transcallosal inhibition following focal magnetic stimulation in preschool children. *Annals of Neurology*, 43(5), 608–612.

Heinrich, B. (2001). *Racing the antelope: What animals can teach us about running and life*. New York: Cliff Street Books.

Henry, D. O. (1989). *From foraging to agriculture: The Levant at the end of the Ice Age*. Philadelphia: University of Pennsylvania Press.

Hinke, R. M., Hu, X., Stillman, A. E., Kim, S.-G., Merkle, H., Salmi, R., and Ugurbil, K. (1993). Functional magnetic resonance imaging of Broca's area during internal speech. *NeuroReport*, 4, 675–678.

Hinton, G. E. (1992). How neural networks learn from experience. *Scientific American*, 267 (3) 104–109.

Hirstein, W., and Ramachandran, V. S. (1997). Capgras syndrome: A novel probe for understanding the neural representation of the identity and familiarity of persons. *Proceedings of the Royal Society of London, Series B*, 264, 437–444.

Hof, P. R., Nimchinsky, E. A., Perl, D. P., and Erwin, J. M. (2001) An unusual population of pyramidal neurons in the anterior cingulate cortex of hominids contains the calcium-binding protein calretinin. *Neuroscience Letters*, 307, 139–142.

Hofer, M. A. (1984). Relationships as regulators: A psychobiologic perspective on bereavement. *Psychosomatic Medicine*, 46, 183–196.

Hofer, M. A. (1987). Early social relationships: A psychobiologist's view. *Child Development*, 58, 633–647.

Hofer, M. A. (1996). On the nature and consequences of early loss. *Psychosomatic Medicine*, 58, 570–581.

Hofman, M. (1983). Encephalization in hominids. *Brain Behavior and Evolution*, 22, 102–117.

Højer–Pedersen, E., and Petersen, O. (1989). Changes of blood flow in the cerebral cortex after subcortical ischemic infarction. *Stroke*, 20, 211–216.

Holland, N. D., and Chen, J. (2001). Origins and early evolution of the vertebrates: New insights from advances in molecular biology, anatomy, and palaeontology, *BioEssays*, 23, 142–151.

Hollaway, R. (1983). Cerebral brain endocast patterns of *Australopithecus afarensis*. *Nature*, 303, 420–422.

Hong, C. C., Potkin, S. G., Antrobus, J. S., Dow, B. M., Callaghan, G. M., and Gillin, J. C. (1997). REM sleep eye movement counts correlate with visual imagery in dreaming: A pilot study. *Psychophysiology*, 34, 377–381.

Hopkins, W., and Savage-Rumbaugh, E. S. (1991). Vocal communication as a function of differential rearing experience in *Pan paniscus*: A preliminary report. *International Journal of Primatology*, 12, 559–583.

Horne, J. A. (1993). Human sleep, sleep loss and behaviour: Implications for the prefrontal cortex and psychiatric disorder. *British Journal of Psychiatry*, 162, 413–419.

Horwitz, B., Maisog, J., Kirschner, P., Haxby, J. V., McIntosh, A. R., Schapiro, M. B., Friston, K. J., Ungerleider, L. G., and Grady, C. L. (1993). Functional pathways in the brain during object and spatial visual processing. *Journal of Cerebral Blood Flow and Metabolism*, 13, S527.

Hoshi, Y., and Tamura, M. (1993a). Detection of dynamic changes in cerebral oxygenation coupled to neuronal function during mental work in man. *Neuroscience Letters*, 150, 5–8.

Hoshi, Y., and Tamura, M. (1993b). Dynamic multichannel near-infrared optical imaging of human brain activity. *Journal of Applied Physiology*, 75, 1842–1846.

Houghton, P. (1993). Neanderthal supralaryngeal vocal tract. *American Journal of Physical Anthropology*, 90, 139–149.

House, J. S., Landis, K. R., and Umberson, D. (1988). Social relationships and health. *Science*, 241, 540–545.

Howard, D., Patterson, K., Wise, R., Brown, W. D., Friston, K., Weiller, C., and Frackowiak, R. (1992). The cortical localization of the lexicons. *Brain*, 115, 1769–1782.

Hrdy, S. (1981). *The woman that never evolved*. Cambridge, Mass: Harvard University Press.

Hrdy, S. B. (1999). *Mother nature*. New York: Pantheon.

Hsu, F. (1971). Psychosocial homeostasis and jen: Conceptual tools for advancing psychological anthropology. *American Anthropologist*, 73, 23–44.

Huang, H. S., and Hanley, J. R. (1994). Phonological awareness and visual skills in learning to read Chinese and English. *Cognition*, 54, 73–98.

Hudspeth, W. J., and Pribram, K. H. (1990). Stages of brain and cognitive maturation. *Journal of Educational Psychology*, 82, 881–884.

Humphrey, N. K. (1976). The social function of intellect. In P. P. G. Bateson and R. A. Hinde, eds., *Growing points in ethology*, pp. 303–321. Cambridge: Cambridge University Press.

Hunt, K. D. (1994). The evolution of human bipedality. *Journal of Human Evolution*, 26, 183–202.

Hurtado, A. M. (2000). Origins of trade-offs in maternal care (review of S. B. Hrdy, *Mother nature*). *Science*, 287, 433–434.

Huttenlocker, P. R. (1979). Synaptic density in human frontal cortex— Developmental changes and effects of aging. *Brain Research*, 163, 195–205.

Illingworth, V. (1986). *Dictionary of computing*, 2nd ed. Oxford: Oxford University Press.

Incisa Della Rocchetta, A., and Milner, B. (1993). Strategic search and retrieval inhibition: The role of the frontal lobes. *Neuropsychologia*, 31, 503–524.

Ingram, F., Levin, H. S., Guinto, F. C., and Eisenberg, H. M. (1994). Case report of a massive congenital left hemisphere lesion. *Developmental Neuropsychology*, 10, 443–453.

Ingvar, D. H. (1993). Language functions related to prefrontal cortical activity: Neurolinguistic implications. *Annals of the New York Academy of Sciences*, 682, 240–247.

Insel, T. R. (1993). Oxytocin and the neuroendocrine basis of affiliation, In J. Schulkin, ed., *Hormonally induced changes in mind and brain*, pp. 225–252, San Diego: Academic Press.

Insel, T. R., and Shapiro, L. E. (1992). Oxytocin receptor distribution reflects social organization in monogamous and polygamous voles. *Proceedings of the National Academy of Sciences of the USA*, 89, 5981–5985.

Insel, T. R., Wang, Z.-X., and Ferris, C. F. (1994). Patterns of brain vasopressin receptor distribution associated with social organization in monogamous rodents. *Journal of Neuroscience*, 14, 5381–5392.

Insel, T. R., and Young, L. J. (2000). Neuropeptides and the evolution of social behavior. *Current Opinion in Neurobiology*, 10(6), 784–789.

Insel, T. R., and Young, L. J. (2001). The neurobiology of attachment. *Nature Reviews Neuroscience*, 2(2), 129–136.

Jablonski, N. G., and Chaplin, G. (2000). The evolution of human skin coloration. *Journal of Human Evolution*, 39, 57–106.

Jack, A. I., and Shallice, T. (2001). Introspective physicalism as an approach to the science of consciousness. *Cognition*, 79, 161–196.

Jackson, H. J. (1958). *Selected writing of John Hughlings Jackson*, vol. 2. London: Staples Press.

Jacobs, R. A. (1999). Computational studies of the development of functionally specialized neural modules. *Trends in Cognitive Science*, 3, 31–38.

Jacobsen, T. (1998). Delay behavior at age six: Links to maternal expressed emotion. *Journal of Genetic Psychology*, 159, 117–120.

Jacobson, R. (1986). Disorders of facial recognition, social behaviour and affect after combined bilateral amygdalotomy and subcaudate tractotomy. *Psychological Medicine*, 16, 437–450.

Jahanshahi, M., Dirnberger, G., Fuller, R., and Frith, C. D. (2000). The role of dorsolateral prefrontal cortex in random number generation: A study with PET. *NeuroImage*, 12, 713–725.

Jahanshahi, M., Jenkins, I. H., Brown, R. G., Marsden, C. D., Passingham, R. E., and Brooks, D. J. (1995). Self-initiated versus externally triggered movements. *Brain*, 118, 913–933.

James, W. (1891). *Principles of psychology*, vol. 1. London: MacMillan.

Jameson, K. A., Highnote, S. M., and Wasserman, L. M. (2001). Richer color experience in observers with multiple photopigment opsin genes. *Psychonomic Bulletin and Review*, 8, 244–261.

Janowsky, J. S., Shimamura, A. P., and Squire, L. R. (1989). Source memory impairments in patients with frontal lobe lesions. *Neuropsychologia*, 27, 1043–1056.

Jaynes. J. (1979) *The origins of consciousness in the breakdown of the bicameral mind*. London: Allen Lane.

Jernigan, T. L., Trauner, D. A., Hesselink, J. R., and Tallal, P. A. (1991). Maturation of human cerebrum observed *in vivo* during adolescence. *Brain*, 114, 2037–2049.

Jetter, W., Poser, U., Freeman, R. B., and Markowitsch, H. J. (1986). A verbal long term memory deficit in frontal lobe damaged patients. *Cortex*, 22, 229–242.

Johanson, D., and Edey, M. (1981). *Lucy: The beginnings of humankind*. London: Granada.

Johnson, M. (1991). Reality monitoring: Evidence from confabulation in organic brain disease patients. In G. Prigatano and D. Schacter, eds., *Awareness of deficit after brain injury*, pp. 176–197. Oxford: Oxford University Press.

Johnson, S. C., Saykin, A. J., Flashman, L. A., McAllister, T. W., and Sparling, M. B. (2001). Brain activation on fMRI and verbal memory ability: functional neuroanatomic correlates of CVLT performance. *Journal of the International Neuropsychological Society*, 7, 55–62.

Jokeit, H., and Makeig, S. (1994). Different event-related patterns of g-band power in brain waves of fast- and slow-reacting subjects. *Proceedings of the National Academy of Sciences of the USA*, 91, 6339–6343.

Joliot, M., Ribary, U., and Llinás, R. (1994). Human oscillatory brain activity near 40-Hz coexists with cognitive temporal binding. *Proceedings of the National Academy of Sciences of the USA*, 91, 11748–11751.

Jones, A. K., Brown, W., Friston, K., Qi, L., and Frackowiak, R. (1991a). Cortical and subcortical localization of response to pain in man using positron emission tomography. *Proceedings of the Royal Society of London, Series B*, 244, 39–44.

Jones, A. K., Friston, K. J. Qi, L. Y., Harris, M., Cunningham, V. J., Jones, T., Feinman, C., and Frackowiak, R. S. (1991b). Sites of action of morphine in the brain. *Lancet*, 338, 825.

Jones, E. G. (1985). *The thalamus*. New York: Plenum.

Joutsiniemi, S.-L., and Hari, R. (1989). Omissions of auditory stimuli may activate frontal cortex. *European Journal of Neuroscience*, 1, 524–528.

Just, M. A., and Carpenter, P. A. (1993). The intensity dimension of thought: Pupillometric indices of sentence processing. *Canadian Journal of Psychology*, 47, 310–339.

Kaas, J., Krubitzer, L., Chino, Y., Langston, A., Polley, E., and Blair, N. (1990). Reorganization of retinotopic cortical maps in adult mammals after lesions of the retina. *Science*, 248, 229–231.

Kalin, N. (1993). The neurobiology of fear. *Scientific American*, 263, (5), 54–60.

Kang, D.-H., Davidson, R. J., Coe, C. L., Wheeler, R. E., Tomarken, A. J., and Ershler, W. B. (1991). Frontal brain asymmetry and immune function. *Behavioral Neuroscience*, 105, 860–869.

Kano, T. (1992). *The last ape: Pygmy chimpanzee behavior and ecology*. Stanford, Calif.: Stanford University Press.

Kansu, T., and Zacks, S. I. (1979). "Agenesis" of the temporal lobe. *Archives of Neurology*, 36, 590–591.

Kapur, N., Turner, A., and King, C. (1988). Reduplicative paramnesia: Possible anatomical and neuropsychological mechanisms. *Journal of Neurology, Neurosurgery and Psychiatry*, 51, 579–581.

Karni, A., Tanne, D., Rubenstein, B. S., Askenasy, J. J., and Sagi, D. (1995). Dependence on REM sleep of overnight improvement of a perceptual skill. *Science*, 265, 679–682.

Kasamatsu, A., and Hirai, T. (1966). An electroencephalographic study on Zen meditation (Zazen). *Folia Psychiatrica et Neurologica Japonica*, 20, 315–336.

Kato, T., Kamei, A., Takashima, S. and Ozaki, T. (1993). Human visual cortical function during photic stimulation monitoring by means of near-infrared spectroscopy. *Journal of Cerebral Blood Flow and Metabolism*, 13, 516–520.

Kato, T., Zhu, Z.-H., Strupp, J., Ugurbil, K., and Chen, W. (1996). Activation of hypothalamus and reticular formation induced by external and internal stimulation: Functional brain stem study. *NeuroImage*, 3, *(2, ii)*, S187.

Kawashima, R., O'Sullivan, B. T., and Roland, P. (1993). A PET study of selective attention in man: Cross-modality decreases in activity in somatosensory and visual tasks. *Journal of Cerebral Blood Flow and Metabolism*, 13, S502.

Kawashima, R., Roland, P. E., and O'Sullivan, B. T. (1994). Activity in the human primary motor cortex related to ipsilateral hand movements. *Brain Research*, 663, 251–256.

Kawashima, R., Satoh, K., Itoh, H., Ono, S., Furumoto, S., Gotoh, R., Koyama, M., Yoshioka, S., Takahashi, T., Takahashi, K., Yanagisawa, T., and Furkud, H. (1996). Functional anatomy of GO/NO-GO discrimination and response selection—a PET study in man. *Brain Research*, 728, 79–89.

Keller, H. (1903). *Story of my life*, 6th ed. London: Hodder and Stoughton.

Kellog, W. N. (1962). Sonar system of the blind. *Science*, 137, 399–404.

Kempermann, G., Kuhn, H. G., and Gage, F. H. (1998). Experience-induced neurogenesis in the senescent dentate gyrus. *Journal of Neuroscience*, 18, 3206–3212.

Kendon, A. (1970). Movement coordination in social interaction. *Acta Psychologia*, 32, 100–125.

Kerr, R. A. (1999). Early life thrived despite earthly travails. *Science*, 284, 2111–2113.

Kesner, R., Farnsworth, G., and DiMattia, B. (1989). Double dissociation of egocentric and allocentric space following medial prefrontal and parietal cortex lesions in the rat. *Behavioral Neuroscience*, 103, 956–961.

Kesner, R. P., Hopkins, R. O., and Fineman, B. (1994). Item and order dissociation in humans with prefrontal cortex damage. *Neuropsychologia*, 32, 881–891.

Keverne, E. B., Martensz, N. D., and Tuite, B. (1989). Beta-endorphin concentrations in cerebrospinal fluid of monkeys are influenced by grooming relationships. *Psychoneuroendocrinology*, 14, 155–161.

Kiehl, K. A., Smith, A. M., Hare, R. D., and Liddle, P. F. (2000). An event-related potential investigation of response inhibition in schizophrenia and psychopathy. *Biological Psychiatry*, 48(3), 210–221.

Kievit, J., and Kuypers, H. G. (1975). Basal forebrain and hypothalamic connections to frontal and parietal cortex in the rhesus monkey. *Science*, 187, 660–662.

King, A., and Moore, D. (1991). Plasticity of auditory maps in the brain. *Trends in Neurosciences*, 14, 31–37.

Kinsbourne, M. (1974). Mechanism of hemispheric interaction in man. In M. Kinsbourne and W. L. Smith, eds., *Hemisphere disconnection and cerebral function*. Springfield, Ill.: Thomas.

Kirchhoff, B. A., Wagner, A. D., Mari, I. A., and Stern, C. E. (2000). Prefrontal-temporal circuitry for episodic encoding and subsequent memory. *Journal of Neuroscience*, 20, 6173–6180.

Kirkpatrick, L. A. (1992). An attachment-theory approach to the psychology of religion. *International Journal for the Psychology of Religion*, 2, 3–28.

Klein, R. (1983). The stone age prehistory of Southern Africa. *Annual Review of Anthropology*, 12, 25–48.

Klein, R. (1987). Reconstructing how early people exploited animals: Problems and prospects. In M. Nitecki and D. Nitecki, eds., *Evolution of human hunting*, pp. 11–45. New York: Plenum.

Klein, R. (1989). *The human career: Human biological and cultural origins.* Chicago: University of Chicago Press.

Klinnert, M., Campos, J., Emde, R., and Svejda, M. (1983). Social referencing: Emotional expressions as behavior regulators. In R. Plutchik and H. Kellerman, eds., *Emotion: Theory, research and experience,* vol. 2. *Emotions in early development,* pp. 57–86. Orlando: Academic Press.

Klostermann, F., Funk, T., Vesper, J., Siedenberg, R., and Curio, G. (2000). Double-pulse stimulation dissociates intrathalamic and cortical high-frequency (>400Hz) SEP components in man. *NeuroReport,* 11, 1295–1299.

Knight, R. T. (1984). Decreased response to novel stimuli after prefrontal lesions in man. *Electroencephalography and Clinical Neurophysiology,* 59, 9–20.

Knight, R. T. (1991). Evoked potential studies of attention capacity in human frontal lobe lesions. In H. Levin., H. Eisenberg, and A. Benton, eds., *Frontal lobe function and dysfunction,* pp. 139–153. New York: Oxford University Press.

Knight, R. T. (1994). Attention regulation and human prefrontal cortex. In A. Thierry, J. Glowinski, P. S. Goldman-Rakic, and Y. Christen, eds., *Motor and cognitive functions of the prefrontal cortex,* pp. 160–173, Berlin: Springer-Verlag.

Knight, R. T., Scabini, D., and Woods, D. L. (1989). Prefrontal cortex gating of auditory transmission in humans. *Brain Research,* 504, 338–341.

Knight, S. (1983). *The brotherhood.* London: Granada.

Knoll, A. H. (1999). A new molecular window on early life. *Science,* 285, 1025–1026.

Kobayashi, H, and Kohshima, S. (2001). Unique morphology of the human eye and its adaptive meaning: Comparative studies on external morphology of the primate eye. *Journal of Human Evolution,* 40, 419–435.

Koch, C. (1997). Computation and the single neuron. *Nature,* 385, 207–210.

Koch, C., and Crick, F. (1994). Some further ideas regarding the neuronal basis of awareness. In C. Koch and J. L. Davis, eds., *Large-scale neuronal theories of the brain,* pp. 93–109. Cambridge, Mass.: MIT Press.

Koechlin, E., Corrado, G., Pietrini, P., and Grafman. J. (2000). Dissociating the role of the medial and lateral anterior prefrontal cortex in human planning. *Proceedings of the National Academy of Sciences of the USA,* 97, 7651–7656.

Kojima, K., Ogomori, K., Mori, Y., Hirata, K., Kinukawa, N., and Tashiro, N. (1996). Relationship of emotional behaviors induced by electrical stimulation of the hypothalamus to changes in EKG, heart, stomach, adrenal glands, and thymus. *Psychosomatic Medicine*, 58, 383–391.

König, P., and Engel, A. E. (1995). Correlated firing in sensory-motor systems. *Current Opinion in Neurobiology*, 5, 511–519.

König, P., Engel, A. E., and Singer, W. (1995). Relation between oscillatory activity and long-range synchronization in cat visual cortex. *Proceedings of the National Academy of Sciences of the USA*, 92, 290–294.

Konishi, S., Wheeler, M. E., Donaldson, D. I., and Buckner, R. L. (2000). Neural correlates of episodic retrieval success. *NeuroImage*, 12, 276–286.

Korte, M., and Rauschecker, J. P. (1993). Auditory spatial tuning of cortical neurons is sharpened in cats with early blindness. *Journal of Neurophysiology*, 70, 1717–1721.

Kosslyn, S. M. Alpert, N. M., Thompson, W. L., Maljkovic, V., Weise, S. B., Chabris, C. F., Hamilton, S. E., Rauch, S. L., and Buonanno, F. S. (1993). Visual mental imagery activates topographically organized visual cortex: PET investigations. *Journal of Cognitive Neuroscience*, 5, 263–287.

Kosslyn, S. M., Daly, P. F., McPeek, R. M., Alpert, N. M., Kennedy, D. N., and Caviness, V. S. (1993). Using locations to store shape: An indirect effect of a lesion. *Cerebral Cortex*, 3, 567–582.

Kosslyn, S. M., Thompson, W. L., Kim, I. J., and Alpert, N. M. (1995). Topographical representations of mental images in primary visual cortex. *Nature*, 378, 496–498.

Kosslyn, S. M., Thompson, W. L., Kim, I. J., Rauch, S. L., and Alpert, N. M. (1996). Individual differences in cerebral blood flow in area 17 predict the time to evaluate visualized letters. *Journal of Cognitive Neuroscience*, 8, 78–82.

Kostovic, I. (1990). Structure and histochemical reorganisation of the human prefrontal cortex during perinatal and postnatal life. *Progress in Brain Research*, 85, 223–240.

Koyama, T., Kato, K., Tanaka, Y. Z., and Mikami, A. (2001). Anterior cingulate activity during pain-avoidance and reward tasks in monkeys. *Neuroscience Research*, 39(4), 421–430.

Krause, C. M., Korpilahti, P., Pörn, B., Jäntti, J., and Lang A. H. (1998). Automatic auditory word perception as measued by 40 Hz EEG responses. *Electroencephalography and Clinical Neurophysiology*, 107, 84–87.

Krauss, L. M. (2001). *Atom: An odyssey from the Big Bang to life on earth*. New York: Little Brown.

Kreiman, J., and Van Lancker, D. (1988). Hemispheric specialization for voice recognition: Evidence from dichotic listening. *Brain and Language*, 34, 246–252.

Kristeva-Feige, R., Feige, B., Makeig, S., Ross, B., and Elbert, T. (1993). Oscillatory brain activity during a motor task. *NeuroReport*, 4, 1291–1294.

Kubin, L., Davis, R. O., and Pack, A. I. (1998). Control of upper airway motoneurons during REM sleep. *News in Physiology*, 13, 91–95.

Kujala, T., Alho, K., and Näätänen, R. (2000). Cross-modal reorganization of human cortical functions. *Trends in Neurosciences*, 23, 115–120.

Kujala, T., Alho, K. Paavilainen, P., Summala, H., and Näätänen, R. (1992). Neural plasticity in processing of sound location by the early blind: An event-related potential study. *Electroencephalography and Clinical Neurophysiology*, 84, 467–472.

Kujala, T., Lehtokoski, A., Alho, K., Kekoni, J., and Näätänen, R. (1997). Faster reaction times in the blind than sighted during bimodal divided attention. *Acta Psychologia*, 96, 75–82.

Kulli, J., and Koch, C. (1991). Does anaesthesia cause loss of consciousness? *Trends in Neurosciences*, 14, 6–10.

Kurth, J. M., and Le Quesne, P. M. (1992). Toxic disorders. In J. H. Adams and L. W. Duchen, eds., *Greenfield's neuropathology*, 5th ed. London: Arnold.

Kuzawa, C. W. (1998). Adipose tissue in human infancy and childhood: An evolutionary perspective. *Yearbook of Physical Anthropology*, 41, 177–209.

Laird, D. J., De Tomaso, A. W., Cooper, M. D., and Weissman, I. L. (2000). 50 million years of chordate evolution: Seeking the origins of adaptive immunity. *Proceedings of the National Academy of Sciences of the USA*, 97, 6924–6926.

Lancaster, J. B., and Lancaster, C. S. (1983). Parental investment: The hominid adaptation. In D. J. Ortner, ed., *How humans adapt: A biocultural odyssey*, pp. 33–65. Washington, D.C.: Smithsonian Institution Press.

Landauer, T. K., and Dumais, S. T. (1997). A solution of Plato's problem: The latent semantic analysis theory of acquisition, induction, and representation of knowledge. *Psychological Review*, 104, 211–240.

Landis, T., Regard, M., Bliestle, A., and Kleihues, P. (1988). Prosopagnosia and agnosia for noncanonical views: An autopsied case. *Brain*, 111, 1287–1297.

Lang, W., Lang, M., Podreka, I., Steiner, M., Uhl, F., Suess, E., Müller, C., and Deecke, L. (1988). DC-potential shifts and regional cerebral blood flow reveal frontal cortex involvement in human visuomotor learning. *Experimental Brain Research*, 71, 353–364.

Lapierre, D., Braun, C. M., and Hodgins, S. (1995). Ventral frontal deficits in psychopathy. *Neuropsychologia*, 33, 139–151.

Le Bihan, D., Turner, R., Zeffiro, T. A., Cuenod, C. A., Jezzard, P., and Bonnerot, V. (1993). Activation of human primary visual cortex during visual recall: A magnetic resonance imaging study. *Proceedings of the National Academy of Sciences of the USA*, 90, 11802–11805.

Leakey, M. G., Spoor, F., Brown, F. H., Gathogo, P. N., Kiarie, C., Leakey, L. N., and McDougall, I. (2001). New hominid genus from eastern Africa shows diverse middle Pliocene lineages. *Nature*, 410, 433–440.

Leakey, R., and Walker, A. (1993). Introduction. In A. Walker and R. Leakey, eds., *Nariokotome* Homo erectus *skeleton*, pp. 1–5. Cambridge, Mass.: Harvard University Press.

Leary, M. R., Tamber, E. S., Terdal, S. K., and Downs, D. L. (1995). Self-esteem as an interpersonal monitor: The sociometer hypothesis. *Journal of Personality and Social Psychology*, 68, 518–530.

Leavens, D. A., Hopkins, W. D., and Bard, K. A. (1996). Indexical and referential pointing in chimpanzees (*Pan troglodytes*). *Journal of Comparative Psychology*, 110, 346–353.

Leberge, D. (1995). Computational and analytical models of selective attention in object identification. In M. Gazzaniga, ed., *Cognitive neuroscience*, pp. 649–663. Cambridge, Mass.: MIT Press.

Leclerc, C., Saint-Amour, D., Lavoie, M. E., Lassonde, M., and Lepore, F. (2000). Brain functional reorganization in early blind humans revealed by auditory event-related potentials. *NeuroReport*, 11, 545–550.

Lee, A. C., Robbins, T. W., Pickard, J. D., and Owen, A. M. (2000). Asymmetric frontal activation during episodic memory: The effects of stimulus type on encoding and retrieval. *Neuropsychologia*, 38, 677–692.

Lee, R. B., and DeVore, I. (1968). Problems in the study of hunters and gatherers. In R. B. Lee and I. DeVore, eds, *Man the hunter*, pp. 3–12. New York: Aldine.

Leimkuhler, M. E., and Mesulam, M.-M. (1985). Reversible go–no go deficits in a case of frontal lobe tumor. *Annals of Neurology*, 18, 617–618.

Lepage, M., and Richer, F. (2000). Frontal brain lesions affect the use of advance information during response planning. *Behavioral Neuroscience*, 114, 1034–1040.

LaPlante, E. (1993). *Seized*. New York: HarperCollins.

Lessard, N., Paré, M., Lepore, F., and Lassonde, M. (1998). Early-blind human subjects localise sound sources better than sighted subjects. *Nature*, 395, 278–280.

Leung, H. C., Skudlarski, P., Gatenby, J. C., Peterson, B. S., and Gore, J. C. (2000). An event-related functional MRI study of the Stroop color word interference task. *Cerebral Cortex*, 10(6), 552–560.

Levänen, S., and Hamdorf, D. (2001). Feeling vibrations: enhanced tactile sensitivity in congenitally deaf humans. *Neuroscience Letters*, 301, 75–77.

LeVay, S. (1991). A difference in hypothalamic structure between heterosexual and homosexual men. *Science*, 253, 1034–1037.

Levenson, R. W., Ekman, P., and Friesen, W. V. (1990). Voluntary facial action generates emotion-specific autonomic nervous system activity. *Psychophysiology*, 27, 363–384.

Levenson, R. W., and Ruef, A. M. (1992). Empathy: A physiological substrate. *Journal of Personality and Social Psychology*, 63, 234–246.

Levin, H., Elsenberg, H. ,and Benton, A., eds. (1991). *Frontal lobe function and dysfunction*, New York: Oxford University Press.

Levine, D., Calvanio, R., and Popovics, A. (1982). Language in the absence of inner speech. *Neuropsychologia*, 20, 391–409.

Levy, J., and Trevarthen, C. (1976). Metacontrol of hemispheric function in human split-brain patients. *Journal of Experimental Psychology: Human Perception and Performance*, 2, 299–312.

Lévy-Bruhl, L. (1985). *How natives think*. Princeton, N.J.: Princeton University Press.

Lewin, R. (1993). *The origins of modern humans*. New York: Scientific American Library.

Lewis, M., Sullivan, M. W., Ramsay, D. S., and Alessandri, S. M. (1992). Individual differences in anger and sad expressions during extinction: Antecedents and consequences. *Infant Behavior and Development*, 15, 443–452.

Lhermitte, F. (1986). Human autonomy and the frontal lobes. Part II: Patient behavior in complex and social situations: The "environmental dependency syndrome." *Annals of Neurology*, 19, 335–343.

Lhermitte, F., Pillon, B., and Serdaru, M. (1986). Human autonomy and the frontal lobes. Part I: Imitation and utilization behavior: A neuropsychological study of 75 patients. *Annals of Neurology*, 19, 326–334.

Lhermitte, J. (1951). Visual hallucination of the self. *British Medical Journal*, 1, 403–434.

Libet, B. (1985). Unconscious cerebral initiative and the role of conscious will in voluntary action. *Behavioral and Brain Sciences*, 8, 529–566.

Liddle, P. F., Kiehl, K. A., and Smith, A. M. (2001). Event-related fMRI study of response inhibition. *Human Brain Mapping*, 12(2), 100–109.

Lieberman, P. (1984). *The biology and evolution of language*. Cambridge, Mass: Harvard University Press.

Livingstone, M. S., and Hubel, D. H. (1987). Psychophysical evidence for separate channels for the perception of form, colour, movement and depth. *Journal of Neuroscience*, 7, 4316–3416.

Llinás, R. R., Grace, A. A., and Yarom, Y. (1991). *In vitro* neurons in mammalian cortical layer 4 exhibit intrinsic oscillatory activity in the 10- to 50-Hz frequency range. *Proceedings of the National Academy of Sciences of the USA*, 88, 897–901.

Llinás, R. R., and Paré, D. (1991). Of dreaming and wakefulness. *NeuroScience*, 44, 521–535.

Llinás, R., and Ribary, U. (1993). Coherent 40-Hz oscillation characterizes dream state in humans. *Proceedings of the National Academy of Science of the USA*, 90, 2078–2081.

Locke, J. (1689/1975). *An essay concerning human understanding*. Oxford: Oxford University Press.

Lord, T., and Kasprzak, M. (1989). Identification of self through olfaction. *Perceptual and Motor Skills*, 69, 219–224.

Lovejoy, C. (1981). The origin of man. *Science*, 211, 341–350.

Luciana, M., and Nelson, C. A. (1998). The functional emergence of pre-frontally-guided working memory systems in four- to eight-year-old children. *Neuropsychologia*, 36, 273–293.

Luria, A. R. (1973). *Working brain*. London: Allan Lane.

Luria, A. R. (1980). *Higher cortical functions in man*, 2nd rev. ed. New York: Basic Books.

Lutzenberger, W., Pulvermüller, F., and Birbaumer, N. (1994). Words and pseudowords elicit distinct patterns of 30-Hz EEG responses in humans. *Neuroscience Letters*, 176, 115–118.

Lutzenberger, W., Pulvermüller, F., Elbert, T., and Birbaumer, N. (1995). Visual stimulation alters local 40-Hz responses in humans: An EEG-study. *Neuroscience Letters*, 183, 39–42.

Lykken, D., McGue, M., Tellegen, A., and Bouchard, T. (1992). Emergenesis: Genetic traits that may not run in families. *American Psychologist*, 47, 1565–1577.

Lykken, D., Tellegen, A., and Thorkelson, K. (1974). Genetic determination of EEG frequency spectra. *Biological Psychology*, 1, 245–259.

Lynch, D. R., and Dawson, T. M. (1994). Secondary mechanisms in neuronal trauma. *Current Opinion in Neurology*, 7, 510–516.

MacDonald, A. W., Cohen, J. D., Stenger, V. A., and Carter, C. S. (2000). Dissociating the role of the dorsolateral prefrontal and anterior cingulate cortex in cognitive control. *Science*, 288(5472), 1835–1838.

MacDonald, C. R., Crosson, B., Valenstein, E., and Bowers, D. (2001). Verbal encoding deficits in a patient with a left retrosplenial lesion. *Neurocase*, 7, 407–417.

Mackowiak, P. A., Wasserman, S. S., and Levine, M. M. (1992). A critical appraisal of 98.8°F, the upper limit of the normal body temperature, and other legacies of Carl Reinhold August Wunderlich. *JAMA*, 268, 1578–1580.

MacLarnon, A. M., and Hewitt, G. P. (1999). The evolution of human speech: The role of enhanced breathing control. *American Journal of Physical Anthropology*, 109, 341–363.

MacLean, P. (1990). *The triune brain in evolution: Role in paleocerebral functions*. New York: Plenum Press.

MacLeod, C. M. (1991). John Ridley Stroop: Creator of a landmark cognitive task. *Canadian Psychology*, 32, 521–524

MacLeod, C. M., and MacDonald, P. A. (2000). Inter-dimensional interference in the Stroop effect: Uncovering the cognitive and neural anatomy of attention. *Trends in Cognitive Science*, 4, 383–391

MacLeod-Morgan, C., and Lack, L. (1982). Hemispheric specificity: A physiological concomitant of hypnotizability. *Psychophysiology*, 19, 687–690.

Maguire, E. A., Burgess, N., Donnett, J. G., Frackowiak, R. S., Frith, C. D., and O'Keefe, J. (1998). Knowing where and getting there: A human navigation network. *Science*, 280, 921–944.

Maguire, E. A., Frackowiak, R. S., and Frith, C. D. (1997). Recalling routes around London: Activation of the right hippocampus in taxi drivers. *Journal of Neuroscience*, 17, 7103–7110.

Main, M. (1990). Parental aversion to infant-initiated contact is correlated with the parent's own rejection during childhood. In K. E. Barnard and T. B. Brazelton, eds., *Touch*, pp. 461–495. Madison, Conn.: International University Press.

Majkowski, J. (1988). Kindling: A model for epilepsy and memory. *Acta Neurologica Scandinavica*, 74(suppl. 109), 97–108.

Maloy, P., Bihrle, A., Duffy, J., and Cimino, C. (1993). The orbitomedial frontal syndrome. *Archives of Clinical Neuropsychology*, 8, 185–201.

Mano, Y., Nakamuro, T., Tamura, R., Takayanagi, T., Kawanishi, K., Tamai, S., and Mayer, R. F. (1995). Central motor reorganization after anastomosis of the musculocutaneous and intercostal nerves following cervical root avulsion. *Annals of Neurology*, 38, 15–20.

Maquet, P., Peters, J., Aerts, J., Delfiore, G., Degueldre, C., Luxen, A., and Franck, G. (1996). Functional neuroanatomy of human rapid-eye-movement sleep and dreaming. *Nature*, 383, 163–166.

Marais, D. J. (2000). When did photosynthesis emerge on Earth? *Science*, 289, 1703–1705

Margulis, L. (1982). *Early life*. Boston: Jones and Bartlett.

Markand, O. N. (1990). Alpha rhythms. *Journal of Clinical Neurophysiology*, 7, 163–189.

Markowitsch, H. J. (1992). *Intellectual functions and the brain: An historical perspective*. Seattle: Hogrefe & Huber.

Markowitsch, H. J., Calabrese, P., Haupts, M., Durwen, H. F., Liess, J., and Gehlen, W. (1993). Searching for the anatomical basis of retrograde amnesia. *Journal of Clinical and Experimental Neuropsychology*, 15, 947–967.

Markus, H. R., and Kitayama, S. (1991). Culture and the self: Implications for cognition, emotion, and motivation. *Psychological Review*, 98, 224–253.

Marosi, E., Harmony, T., Sánchez, L., Becker, J., Bernal, J., Reyes, A., Díaz de León, A. E., Rodríguez, M., and Fernández, T. (1992). Maturation of the coherence of EEG activity in normal and learning-disabled children. *Electroencephalography and Clinical Neurophysiology*, 83, 350–357.

Marshack, A. (1976). Implications of the paleolithic symbolic evidence for the origins of language. *American Scientist*, 64, (2), 136–145.

Marshall, L. (1976). *The !Kung of Nyae Nyae*. Cambridge, Mass.: Harvard University Press.

Martin, A., Haxby, J. V., Lalonde, F. M., Wiggs, C. L., and Ungerleider, L. G. (1995). Discrete cortical regions associated with knowledge of color and knowledge of action. *Science*, 270, 102–105.

Martin, A., Wiggs, C. L., Ungerleider, L. G., and Haxby, J. V. (1996). Neural correlates of category-specific knowledge. *Nature*, 379, 649–652.

Martin, L., Spicer, D., Lewis, M., Gluck, J., and Cork, L. (1991). Social deprivation of infant rhesus monkeys alters the chemoarchitecture of the brain: I. Subcortical regions. *Journal of Neuroscience*, 11, 3344–3358.

Martin, R. D. (1982). *Human brain evolution in an ecological context: Fifty-second James Arthur lecture on the evolution of the human brain*. New York: American Museum of Natural History.

Martin, R. D. (1993). Primate origins: Plugging the gaps. *Nature*, 363, 223–234.

Masserman, J., Wechkin, S., and Terris, W. (1964). "Altruistic" behavior in rhesus monkeys. *American Journal of Psychiatry*, 121, 584–585.

Matsuzawa, T. (1996). Chimpanzee intelligence in nature and in captivity: Isomorphism of symbol use and tool use. In W. C. McGrew, L. F. Marchant, and T. Nishida, eds., *Great ape societies*, pp. 196–209. Cambridge: Cambridge University Press.

Matthews, P. (1995). *The Guinness book of records*. London: Guinness.

Maurice, D. M. (1998). The Von Sallmann Lecture 1996: an ophthalmological explanation of REM sleep. *Experimental Eye Research*, 66, 139–145.

Max, L. W. (1937). Experimental study of the motor theory of consciousness. IV. Action-current responses in the deaf during awakening, kinaesthetic imagery and abstract thinking. *Journal of Comparative Psychology*, 24, 301–338.

Mazziotta, J. C., Phelps, M. E., Carson, R. E., and Kuhl, D. E. (1982). Tomographic mapping of human cerebral metabolism: Sensory deprivation. *Annals of Neurology*, 12, 435–444.

McCarty, B., and Worchel, P. (1954). Rate of motion and object perception in the blind. *New Outlook*, 48, 316–322.

McClelland, J. L., McNaughton, B. I.., and O'Reilly, R. C. (1995). Why there are complementary learning systems in the hippocampus and neocortex. *Psychological Review*, 102, 419–457.

McClelland, J. L., and Rumelhart, D. E. (1988). *Explorations in parallel distributed processing: A handbook of models, programs and exercises*. Cambridge, Mass.: MIT Press.

McClintock, M. K. (1971). Menstrual synchrony and suppression. *Nature*, 229, 244–245.

McConnell, S. (1992). The control of neuronal identity in the developing cerebral cortex. *Current Opinion in Neurobiology*, 2, 23–27.

McCormick, P. W., Stewart, M., Lewis, G., Dujovny, E., and Ausman, J. I. (1992). Intracerebral penetration of infrared light. *Journal of Neurosurgery*, 76, 315–318.

McGrew, W. C. (1992). *Chimpanzee material culture*. Cambridge, Mass.: Cambridge University Press.

McGuire, P. K., Robertson, D., Thacker, A., David, A. S., Kitson, N., Frackowiak, R. S., and Frith, C. D. (1997). Neural correlates of thinking in sign language. *NeuroReport*, 8, 678–698.

McGuire, P. K., Silbersweig, D. A., Murray, R. M., David, A. S., Frackowiak, R. S., and Frith, C. D. (1996). Functional anatomy of inner speech and auditory imagery. *Psychological Medicine*, 26, 29–38.

McGurk, H., and MacDonald, J. (1976). Hearing lips and seeing voices. *Nature*, 264, 746–748.

McIntosh, A. R., Sekuler, A. B., Penpeci, C., Rajah, M. N., Grady, C. L., Sekuler, R., and Bennett, P. J. (1999). Recruitment of unique neural systems to support visual memory in normal aging. *Current Biology*, 9, 1275–1278.

Meador, K., Loring, D., Bowers, D., and Heilman, K. (1987). Remote memory and neglect syndrome. *Neurology*, 37, 522–526.

Medalia, A., Merriam, A., and Ehrenreich, J. (1991). The neuropsychological sequelae of attempted hanging. *Journal of Neurology, Neurosurgery, and Psychiatry*, 54, 546–548.

Meiser-Koll, A. (1992). Ultradian behavior cycles in humans: Developmental and social aspects. In D. Lloyd and L. Rossi, eds., *Ultradian rhythms in life processes*, pp. 243–281. London: Springer-Verlag.

Mellet, E., Briscogne, S., Tzourio-Mazoyer, N., Ghaem, O., Petit, L., Zago, L., Etard, O., Berthoz, A., Mazoyer, B., and Denis, M. (2000). Neural correlates of topographic mental exploration: The impact of route versus survey perspective learning. *NeuroImage* 12, 588–600.

Meltzoff, A., and Gopnik, A. (1993). The role of imitation in understanding persons and developinga theory of mind . In S. Baron-Cohen, H. Tager-Flusberg, and D. J. Cohen, eds., *Understanding other minds*, pp. 335–366. Oxford: Oxford University Press.

Melzack, R. (1990). Phantom limbs and the concept of a neuromatrix. *Trends in Neurosciences*, 13, 88–92.

Melzack, R. (1992). Phantom limbs. *Scientific American*, 265, (4), 90–96.

Merzenich, M., and Kaas, J. (1980). Principles of organization of sensory-perceptual systems in mammals. *Progress in Psychology and Physiological Psychology*, 9, 1–40.

Merzenich, M. M., Recanzone, G., Jenkins, W. M., Allard, T. T., and Nudo, R. J. (1988). Cortical representational plasticity. In P. Rakic and W. Singer, eds., *Neurobiology of Neocortex*, pp. 41–67, London: Wiley.

Mestel, R. (1994). Exercise machine works out the perfect work-out. *New Scientist*, 26 March, 6.

Metcalfe, J. (1993). Novelty monitoring, metacognition, and control in a composite holographic associative recall model: Implications for Korsakoff amnesia. *Psychological Review*, 100, 3–22.

Metherate, R., Cox, C. L., and Ashe, J. H. (1992). Cellular bases of neocortical activation: Modulation of neural oscillations by the nucleus basalis and endogenous acetylcholine. *Journal of Neuroscience*, 12, 4701–4711.

Metter, E. J., Riege, W. H., Hanson, W. R., Phelps, M. E., and Kuhl, D. E. (1984). Local cerebral metabolic rates of glucose in movement and language disorders from positron tomography. *American Journal of Physiology*, 246, R897–R900.

Milgram, S. (1963). Behavioral study of obedience. *Journal of Abnormal and Social Psychology*, 67, 371–378.

Milgram, S. (1974). *Obedience to authority: An experimental view*. London: Tavistock.

Milgram, S., and Sabini, J. (1978). On maintaining urban norms: A field experiment in the subway. In A. Baum, J. E. Singer, and S. Valins, eds., *Advances in environmental psychology*, vol. 1, pp. 31–40. Hillsdale, N.J.: Erlbaum.

Miller, A. G. (1986). *The obedience experiment*. New York: Praeger.

Miller, E. K., and Desimone, R. (1994). Parallel neuronal mechanisms for short-term memory. *Science*, 263, 520–522.

Miller, E. K., Erickson, C. A., and Desimone, R. (1996). Neural mechanisms of visual working memory in prefrontal cortex of the macaque. *Journal of Neuroscience*, 16, 5154–5167.

Miller, G. F. (2000). *The mating mind*. New York: Doubleday.

Mills, A. E. (1987). The development of phonology in the blind child. In B. Dodds and R. Campbell, eds., *Hearing by eye: The psychology of lip reading*, pp. 145–161. London: Erlbaum.

Miltner, W. H., Braun, C., Arnold, M., Witte, H., and Taub, E. (1999). Coherence of gamma-band EEG activity as a basis for associative learning. *Nature*, 397, 434–436.

Minsky, M. (1987). *The society of mind*. London: Heinemann.

Mischel, W., Shoda, Y., and Peake, P. (1988). The nature of adolescent competencies predicted by preschool delay of gratification. *Journal of Personality and Social Psychology*, 54, 687–696.

Mischel, W., Shoda, Y., and Rodriguez, M. (1989). Delay of gratification in children. *Science*, 244, 933–938.

Mitani, J. C., Hasegawa, T., Gros-Louis, J., Marler, P., and Byrne, R. (1992). Dialects in wild chimpanzees? *American Journal of Primatology*, 27, 233–243.

Mitchell, S. W. (1872/1965). *Injuries of nerves and their consequences*. New York: Dover .

Mitrofanis, J., and Guillery, R. W. (1993). New views of the thalamic reticular nucleus in the adult and the developing brain. *Trends in Neurosciences*, 16, 240–245.

Miyashita, Y. (1995). How the brain creates imagery: Projection to primary visual cortex. *Science*, 268, 1719–1720.

Mock, D., and Fujioka, M. (1990). Monogamy and long-term pair bonding in vertebrates. *Trends in Ecology and Evolution*, 5, 39–43.

Mogilner, A., Grossman, J. A, Ribary, U., Joliot, M., Volkmann, J., Rapaport, D., Beasley, R. W., and Llinás, R. R. (1993). Somatosensory cortical plasticity in adult humans revealed by magnetoencephalography. *Proceedings of the National Academy of Science of the USA*, 90, 3593–3597.

Molchan, S. E., Sunderland, R., McIntosh, A. R., Herscovitch, P., and Schreurs, B. G. (1994). A functional anatomical study of associative learning in humans. *Proceedings of the National Academy of Sciences of the USA*, 91, 8122–8126.

Molleson, T. (1989). Seed preparation in the Mesolithic: The osteological evidence. *Antiquity*, 63, 356–362.

Molleson, T. (1994). The eloquent bones of Abu Hureyra. *Scientific American*, 271(2),70–75.

Money, J. (1960). Phantom orgasm in the dreams of paraplegic men and women. *Archives of General Psychiatry*, 3, 373–382.

Montagu, A. (1978). *Touching: The human significance of the skin*. New York: Harper and Row.

Morais, J. (1987). Phonetic awareness and reading acquisition. *Psychological Research*, 49, 147–152.

Morais, J., Cary, L., Alegria, J., and Bertelson, P. (1979). Does awareness of speech as a sequence of phonemes arise spontaneously? *Cognition*, 7, 323–331.

Morecraft, R. J., Geula, C., and Mesulam, M.-M. (1992). Cytoarchitecture and neural afferents of orbitofrontal cortex in the brain of the monkey. *Journal of Comparative Neurology*, 323, 341–358.

Morgan, J. M., Wenzl, M., Lang, W., Lindinger, G., and Deecke, L. (1992). Frontocentral DC-potential shifts predicting behavior with or without a motor task. *Electroencephalography and Clinical Neurophysiology*, 83, 378–388.

Morgan, M. A., Romanski, L. M., and LeDoux, J. E. (1993). Extinction of emotional learning: Contribution of medial prefrontal cortex. *Neuroscience Letters*, 163, 109–113.

Morton, J., Hammersley, R., and Bekerian, D. (1985). Headed records: A model for memory and its failure. *Cognition*, 20, 1–23.

Motluk, A. (1997). What colour is innocence? *New Scientist*, 22 March, 16–17.

Mühlnickel, W., Elbert, T., Taub, E., and Flor, H. (1998). Reorganization of auditory cortex in tinnitus. *Proceedings of the National Academy of Sciences of the USA.*, 95, 10340–10343.

Müller-Preuss, P., Newman, J. D., and Jürgens, U. (1980). Anatomical and physiological evidence for a relationship between the "cingular" vocalization area and the auditory cortex in the squirrel monkey. *Brain Research*, 202, 307–315.

Munk, M. H., Roelfsema, P. R., König, P., Engel, A. K., and Singer, W. (1996). Role of reticular activation in the modulation of intracortical synchronization. *Science*, 272, 271–274.

Murphy, M. R., Checkley, S. A., Seckl, S. R., and Lightman, S. L. (1990). Naloxone inhibits oxytocin release at orgasm in man. *Journal of Clinical Endocrinology and Metabolism*, 71, 1056–1058.

Murphy, V. N., and Fetz, E. E. (1992). Coherent 25- to 35-Hz oscillations in the sensorimotor cortex of awake behaving monkeys. *Proceedings of the National Academy of Sciences of the USA*, 89, 5670–5674.

Murray, E. A., Bussey, T. J., and Wise, S. P. (2000). Role of prefrontal cortex in a network for arbitrary visuomotor mapping. *Experimental Brain Research*, 133, 114–129.

Murtha, S., Chertkow, H., Beauregard, M., Dixon, R., and Evans, A. (1996). Anticipation causes increased blood flow to the anterior cingulate cortex. *Human Brain Mapping*, 4, 103–112.

Nabokov, P., and MacLean, M. (1980). Ways of Native American running. *CoEvolution Quarterly (Summer)*, 4–21.

Nadel, L., Samsonovich, A., Ryan, L., and Moscovitch, M. (2000). Multiple trace theory of human memory: Computational, neuroimaging, and neuropsychological results. *Hippocampus*, 10, 352–368.

Nauta, W. (1971). The problem of the frontal lobe: A reinterpretation. *Journal of Psychiatric Research*, 8, 167–187.

Neafsey, E. J. (1990). Prefrontal cortical control of the autonomic nervous system: Anatomical and physiological observations. *Progress in Brain Research*, 85, 147–161.

Negroponte, N. (1995). *Being digital*. London: Hodder and Stoughton.

Nelson, E. E., and Panksepp, J. (1998). Brain substrates of infant-mother attachment: Contributions of opioids, oxytocin, and norepinephrine. *Neuroscience and Biobehavioral Review*, 22, 437–452.

Neville, H. J., and Lawson, D. (1987). Attention to central and peripheral visual space in a movement detection task: An event-related potential and behavioral study. II. Congenitally deaf adults. *Brain Research*, 405, 268–283.

Nicol, S. C., Andersen, N. A., Phillips, N. H., Berger, R. J. (2000). The echidna manifests typical characteristics of rapid eye movement sleep. *Neuroscience Letters*, 283, 49–52.

Nicolau, M. C., Akaarir, M., Gamundi, A., Gonzalez, L., and Rial, R. (2000). Why we sleep: The evolutionary pathway to the mammalian sleep. *Progress in Neurobiology*, 62, 379–406.

Nielsen, J. M. (1938). Gerstmann syndrome. *Archives of Neurology and Psychiatry*, 39, 536–559.

Nielsen, T. I. (1963). Volition: A new experimental approach. *Scandinavian Journal of Psychology*, 4, 225–230.

Niemeyer, W., and Starlinger, I. (1981). Do the blind hear better? Investigations on auditory processing in congenital or early acquired blindness. II. Central functions. *Audiology*, 20, 510–515.

Niiyama, Y., Fushimi, M., Sekine, A., and Hishikawa, Y. (1995). K-complex evoked in NREM sleep is accompanied by a slow negative potential related to cognitive process. *Electroencephalography and Clinical Neurophysiology*, 95, 27–33.

Nisbett, R. E., and Wilson, T. D. (1977). Telling more than we can know: Verbal reports on mental processes. *Psychological Review*, 84, 231–259.

Nishida, T. (1994). Review of recent finding on Mahake chimpanzees. In R. W. Wrangham, W. C. McGrew, F. B. W. de Waal, and P. Heltine, eds., *Chimpanzee culture*, pp. 373–393. Cambridge, Mass.: Harvard University Press.

Nishida, T., Hasegawa, T., Hayaki, H., Takahata, Y., and Uehara, S. (1992). Meat-sharing as a coalition strategy by an alpha male chimpanzee. In T. Nishida, W. C. McGrew, P. Marler, M. Pickford, and B. de Waal, eds., *Topics in primatology: vol. 1: Human origins*, pp. 159–174. Tokyo: University of Tokyo Press.

Nishida, T., and Hiraiwa-Hasegawa, M. (1987). Chimpanzees and bonobos: Cooperative relationships among males. In B. Smuts, D. Cheng, R. Seyfarth, R. Wrangham, and T. Struhsaker, eds., *Primate societies*, pp. 165–177. Chicago: University of Chicago Press.

Nishida, T., Takasaki, H. , and Takahata, Y. (1990). Demography and reproductive profiles. In T. Nishida, *The chimpanzees of the Mahale mountains*, pp. 63–97. Tokyo: University of Tokyo Press.

Nishimura, H., Hashikawa. K., Doi, K., Iwaki, T., Watanabe, Y., Kusuoka, H., Nishimura, T., and Kubo, T. (1999). Sign language "heard" in the auditory cortex. *Nature*, 397, 116.

Noble, W., and Davidson, I. (1989). The archeology of depiction and language. *Current Anthropology*, 30, 125–156.

Núñez, A., Amzica, F., and Steriade, M. (1992). Voltage-dependent fast (20–40 Hz) oscillations in long-axoned neocortical neurons. *NeuroScience*, 51, 7–10.

Obler, L. K., Nicholas, M., Albert, M. L., and Woodward, S. (1985). On comprehension across the adult lifespan. *Cortex*, 21, 273–280.

O'Keefe, J., and Nadel, L. (1978). *The hippocampus as a cognitive map*. Oxford: Oxford University Press.

O'Keefe, J., and Speakman, A. (1987). Single unit activity in the rat hippocampus during a spatial memory task. *Experimental Brain Research*, 68, 1–27.

O'Leary, D. D., Schlaggar, B. L., and Tuttle, R. (1994). Specification of neocortical areas and thalamocortical connections. *Annual Review of Neuroscience*, 17, 419–439.

Oliphant, M. (1992). *Atlas of the ancient world*. London: Ebury Press.

Olivero, W. C., Lister, J. R., and Elwood, P. E. (1995). The natural history and growth rate of asymptomatic meningiomas. *Journal of Neurosurgery*, 83, 222–224.

Olshausen, B. A., Anderson, C. H., and Van Essen, D. C. (1993). A neurobiological model of visual attention and invariant pattern recognition based on dynamic routing of information. *Journal of Neuroscience*, 13, 4706–4719.

Olson, C. R., Musil, S. Y., and Goldberg, M. E. (1993). Posterior cingulate cortex and visuospatial cognition. In B. A. Vogt and A. Gabriel, eds., *Neurobiology of the cingulate cortex and limbic thalamus*, pp. 366–380. Boston: Birkhäuser.

Olson, E. J., Boeve, B. F., and Silber, M. H. (2000). Rapid eye movement sleep behaviour disorder: demographic, clinical and laboratory findings in 93 cases. *Brain*, 123, 331–339.

Ongur, D., An, X., and Price, J. L. (1998). Prefrontal cortical projections to the hypothalamus in macaque monkeys. *Journal of Comparative Neurology*, 401, 480–505.

Opitz, B., Mecklinger, A., and Friederici, A. D. (2000). Functional asymmetry of human prefrontal cortex: encoding and retrieval of verbally and nonverbally coded information. *Learning and Memory*, 7, 85–96.

Otten, E., and Sacks, O. (1992). Letter: "Phantom limbs." *New York Review of Books*, January 30, 45–46.

Owen, A. M., Milner, B., Petrides, M., and Evans, A. C. (1996) Memory for object features versus memory for object location: A positron-emission tomography study of encoding and retrieval processes. *Proceedings of the National Academy of Sciences of the USA*, 93, 9212–9217.

Owen, A. M., Schneider, W. X., and Duncan, J. (2000). Executive control and the frontal lobe: Current issues. *Experimental Brain Research*, 133, 1–2.

Panksepp, J., Nelson, E., and Siviy, S. (1994). Brain opioids and mother-infant social motivation. *Acta Paediatrica Supplement*, 397, 40–46.

Panksepp, J., Siviy, S. M., and Normansell, L. A. (1985). Brain opioids and social emotions. In M. Reite and T. Field, eds., *Psychobiology of attachment*, pp. 3–49. Orlando, Fla.: Academic Press.

Pantev, C., Makeig, S., Hoke, M., Galambos, R., Hampson, S., and Gallen, C. (1991). Human auditory evoked gamma-band magnetic fields. *Proceedings of the National Academy of Sciences of the USA*, 88, 8996–9000.

Paradiso, S., Johnson, D. L., Andreasen, N. C., O'Leary, D. S., Watkins, G. L., Ponto, L. L., and Hichwa, R. D. (1999). Cerebral blood flow changes associated with attribution of emotional valence to pleasant, unpleasant, and neutral visual stimuli in a PET study of normal subjects. *American Journal of Psychiatry*, 156, 1618–1629.

Pardo, J., Pardo, P., Janer, K., and Raichle, M. (1990). The anterior cingulate cortex mediates processing selection in the Stroop attentional conflict paradigm. *Proceedings of the National Academy of Science of the USA*, 87, 256–259.

Pardo, J., Pardo, P., and Raichle, M. (1993). Neural correlates of self-induced dysphoria. *American Journal of Psychiatry*, 150, 713–719.

Parent, A., and Hazrati, L.-N. (1995). Functional anatomy of the basal ganglia. I. The cortico-basal ganglia-thalamo-cortical loop. *Brain Research Reviews*, 20, 91–127.

Parish, A. R. (1994). Sex and food control in the "uncommon chimpanzee": How bonobo females overcome a phylogenic legacy of male dominance. *Ethology and Sociobiology*, 15, 157–179.

Parsons, L. M., Fox, P. T., Downs, J. H., Glass, T., Hirsh, T. B., Martin, C. C., Jerabek, P. A., and Lancaster, J. L. (1995). Use of implicit motor imagery for visual shape discrimination as revealed by PET. *Nature*, 375, 54–58.

Pascual-Leone, A., Blanco, T., et al. (1994). Mental movement exercises prevent motor disturbances after immobilization of a limb. *Society for Neuroscience, Abstracts*, 20, 1294.

Pascual-Leone, A., Gates, J. R., and Dhuna, A. (1991). Induction of speech arrest and counting errors with rapid-rate transcranial magnetic stimulation. *Neurology*, 41, 697–702.

Pascual-Leone, A., and Hallett, M. (1994). Induction of errors in a delayed response task by repetitive transcranial magnetic stimulation of the dorsolateral prefrontal cortex. *NeuroReport*, 5, 2517–2520.

Pascual-Leone, A, Hamilton, R., Tormes, J. M., Keenan, J. P., and Catalá, M. D. (1999). Neuroplasticity in the adjustment to blindness. In J. Grafman and J. Christen, eds., *Neuronal plasticity*, pp. 93–109. Berlin: Spring-Verlag.

Pascual-Leone, A., and Torres, F. (1993). Plasticity of the sensorimotor cortex representation of the reading finger in Braille readers. *Brain*, 116, 39–52.

Pascual-Leone, A., Wassermann, E. M., Sadato, N., and Hallett, M. (1993). Modulation of motor cortical outputs to the reading hand of Braille readers. *Annals of Neurology*, 34, 33–37.

Pasman, R. H., and Weisberg, P. (1975). Mothers and blankets as agents for promoting play and exploration by young children in a novel environment. *Developmental Psychology*, 11, 170–177.

Passingham, R. E. (1973). Anatomical differences between the neocortex of man and other primates. *Brain, Behaviour and Evolution*, 7, 337–359.

Passingham, R. E. (1985). Cortical mechanisms and cues for action. *Philosophical Transactions of the Royal Society of London*, B 308, 101–111.

Passingham, R. E. (1993). *Frontal lobes and voluntary action*. Oxford: Oxford University Press.

Pate, R. R., Pratt, M., Blair, S. N., Haskell, W. L., Macera, C. A., Blouchard, C., Buchner, D., Ettinger, W., Health, G. W., King, A. C., Kriska, A., Leon, A. S., Marcus, B. H., Morris, J., Paffenbarger, R. S., Patrick, K., Pollock, M. L., Rippe, J. M., Sallis, J., and Wilmore, J. H. (1995). Physical activity and public health. *JAMA*, 273, 402–407.

Paternoster, R., Saltzman, L. E., Waldo, G. P., and Chiricos, T. G. (1983). Perceived risk and social control: Do sanctions really deter? *Law and Society Review*, 17, 457–479.

Paul, G. (1989). *Predatory dinosaurs of the world*. New York: Simon and Schuster.

Paulesu, E., Frith, C. D., and Frackowiak, R. S. (1993). The neural correlates of the verbal component of working memory. *Nature*, 362, 342–345.

Paus, T., Petrides, M., Evans, A., and Meyer, E. (1993). Role of the human anterior cingulate cortex in the control of oculomotor, manual, and speech responses. *Journal of Neurophysiology*, 70, 453–469.

Paus, T., Zijdenbos, A., Worsley, K., Collins, D. L., Blumenthal, J., Giedd, J. N., Rapoport, J. L., and Evans, A. C. (1999). Structural maturation of neural pathways in children and adolescents: In vivo study. *Science*, 283(5409), 1908–1911.

Pegelow, C. H., Wang, W., Granger, S., et al., (2001). Silent infarcts in children with sickle cell anemia and abnormal cerebral artery velocity. *Archives of Neurology*, 58, 2017–2021.

Peirce, C. S. (1932). *Collected papers of Charles Sanders Peirce*, vol. 2. Cambridge, Mass.: Harvard University Press.

Perez-Garci, E., del-Rio-Portilla, Y., Guevara, M. A., Arce, C., and Corsi-Cabrera, M. (2001). Paradoxical sleep is characterized by uncoupled gamma activity between frontal and perceptual cortical regions. *Sleep*, 24, 118–126.

Peterhans, J. C., Wrangham, R. W., Carter, M. L., and Hauser, M. D. (1993). A contribution to tropical rain forest taphonomy. *Journal of Human Evolution*, 25, 485–514.

Peterson, B. S., Skudlarski, P., Gatenby, J. C., Zhang, H., Anderson, A. W., and Gore, J. C. (1999). An fMRI study of Stroop word-color interference: Evidence for cingulate subregions subserving multiple distributed attentional systems. *Biological Psychiatry*, 45(10), 1237–1258.

Petersson, K. M., Reis, A., Askelof, S., Castro-Caldas, A., and Ingvar, M. (2000). Language processing modulated by literacy: A network analysis of verbal repetition in literate and illiterate subjects. *Journal of Cognitive Neuroscience*, 12, 364–382.

Petrides, M. (1990). Nonspatial conditional learning impaired in patients with unilateral frontal but not unilateral temporal lobe excisions. *Neuropsychologia*, 28, 137–149.

Petrides, M., Alivisatos, B., Meyer, E., and Evans, A. (1993). Functional activation of the human frontal cortex during the performance of verbal working memory tasks. *Proceedings of the National Academy of Sciences of the USA*, 90, 878–882.

Petrie, M. (1994). Improved growth and survival of offspring of peacocks with more elaborate trains. *Nature*, 371, 598–599.

Petsche, H. (1996). Approaches to verbal, visual and musical creativity by EEG coherence analysis. *International Journal of Psychophysiology*, 24, 145–159.

Petsche, H., Richter, P., von Stein, A., Etlinger, S. C., and Fitz, O. (1993). EEG coherence and musical thinking. *Music Perception*, 11, 117–151.

Petsche, H., von Stein, A., and Filz, O. (1996). EEG aspects of mentally playing an instrument. *Cognitive Brain Research*, 3, 115–123.

Pfurtscheller, G., Neuper, C., and Kalcher, J. (1993). 40-Hz oscillations during motor behavior in man. *Neuroscience Letters*, 164, 179–182.

Phillis, N. (1987). Symbolism. In M. Eliade, ed., *The encyclopedia of religion*, pp. 198–208. New York: Macmillan.

Pinault, D., and Deschênes, M. (1992). Voltage-dependent 40-Hz oscillation in rat reticular thalamic neurons *in vivo*. *NeuroScience*, 51, 245–258.

Pitkow, L. Sharer, C. A, Ren, X., Insel, T. R., Terwilliger, E. F., and Young, L. J. (2001). Facilitation of affiliation and pair-bond formation by vasopressin receptor gene transfer into the ventral forebrain of a monogamous vole. *Journal of Neuroscience*, 21, 7392–7396.

Plihal, W., and Born, J. (1997). Effects of early and late nocturnal sleep on declarative and procedural memory. *Journal of Cognitive Neuroscience*, 9, 534–547.

Plihal, W., and Born, J. (1999). Effects of early and late nocturnal sleep on priming and spatial memory. *Psychophysiology*, 36, 571–582.

Plourde, G. (1993). The clinical use of the 40Hz auditory steady state response. *International Anaesthesiology Clinic*, 31, 107–120.

Pochon, J. B., Levy, R., Poline, J. B., Crozier, S., Lehericy, S., Pillon, B., Deweer, B., Le Bihan, D., and Dubois, B. (2001). The role of dorsolateral prefrontal cortex in the preparation of forthcoming actions: An fMRI study. *Cerebral Cortex*, 11, 260–266.

Polan, H. J., and Ward, M. J. (1994). Role of the mother's touch in failure to thrive: A preliminary investigation. *Journal of the American Academy of Child and Adolescent Psychiatry*, 33, 1098–1105.

Polish, J., and Burns, T. (1987). P300 from identical twins. *Neuropsychologia*, 25, 299–304.

Pons, T. (1994). Response to Lund, Sun, and Lamarre. *Science*, 265, 548.

Pons, T., Garraghty, P., Ommaya, A., Kaas, J., Taub, E., and Mishkin, M. (1991). Massive cortical reorganization after sensory deafferentation in adult macaques. *Science*, 252, 1857–1860.

Pöppel, E. (1989). Taxonomy of the subjective: An evolutionary perspective. In J. N. Brown, ed., *Neuropsychology of visual perception*, pp. 219–232. Hillsdale, N.J.: Erlbaum.

Popper, K. R. (1972). *Objective knowledge: An evolutionary approach.* Oxford: Oxford University Press.

Popper, K. R. (1976). *Unended quest.* London: Fontana/Collins.

Porter, A. M. (1993). Sweat and thermoregulation in hominids: Comments prompted by the publications of P. E. Wheeler 1984–1993. *Journal of Human Evolution,* 25, 417–423.

Posner, M. I. (1994). Attention: The mechanisms of consciousness. *Proceedings of the National Academy of Sciences of the USA,* 91, 7398–7403.

Postle, B. R., Berger, J. S., and D'Esposito, M. (1999). Functional neuroanatomical double dissociation of mnemonic and executive control processes contributing to working memory performance. *Proceedings of the National Academy of Science of the USA,* 96, 12959–12964.

Potts, W. K. (1984). The chorus-line hypothesis of manoeuvre coordination in avian flocks. *Nature,* 309, 344–345.

Prabhakaran, V., Narayanan, K., Zhao, Z., and Gabrieli, J. D. (2000). Integration of diverse information in working memory within the frontal lobe. *Nature Neuroscience,* 3, 85–90.

Preissl, H., Pulvermüller, F., Lutzenberger, W., and Birbaumer, N. (1995). Evoked potentials distinguish between nouns and verbs. *Neuroscience Letters,* 197, 81–83.

Pribram, K. H. (1986). The hippocampus system and recombinant processing. In R. L. Isaacson and K. H. Pribram, eds., *The hippocampus,* vol. 4, pp. 329–370. New York: Plenum Press.

Price, B., Daffner, K., Stowe, R., and Mesulam, M. (1990). The compartmental learning disabilities of early frontal lobe damage. *Brain,* 113, 1383–1393.

Prior, A., Bertolasi, L., Rothwell, B., Day., B., and Marsden, C. (1993). Some saccadic movements can be delayed by transcranial magnetic stimulation of the cerebral cortex in man. *Brain,* 116, 355–347.

Proust, M. (1982). *Remembrance of things past:* vol 1. *Swann's way.* London: Chatto and Windus.

Pulvermüller, F., Harle, M., and Hummel, F. (2000). Neurophysiological distinction of verb categories. *NeuroReport,* 11, 2789–2793.

Pulvermüller, F., Preissl, H., Lutzenberger, W., and Birbaumer, N. (1996). Brain rhythms of language: Nouns from verbs. *European Journal of Neuroscience,* 8, 937–941.

Pütz, B., Miyauchi, S., Sasaki, Y., Takino, R., Ohki, M., and Okamoto, J. (1996). Activation of the visual cortex by imagery used in mental calculations. *NeuroImage*, 3, (2, ii), S215.

Raab, J., and Gruzelier, J. (1994). A controlled investigation of right hemispheric processing enhancement after restricted environmental stimulation (REST) with floatation. *Psychological Medicine*, 24, 457–452.

Radcliffe-Brown, A. (1965). *Structure and function in primitive society*. New York: Free Press.

Raeva, S., and Lukashev, A. (1993). Unit activity in human thalamic reticularis neurons. II. Activity evoked by significant and non-significant verbal or sensory stimuli. *Electroencephalography and Clinical Neurophysiology*, 86, 110–122.

Raichle, M. E., Fiez, J. A., Videen, T. O., MacLeod, A. K., Pardo, J. V., Fox, P. T., and Peterson, S. E. (1994). Practice-related changes in human brain functional anatomy during nonmotor learning. *Cerebral Cortex*, 4, 8–26.

Raij, T., Uutela, K., and Hari, R. (2000). Audiovisual integration of letters in the human brain. *Neuron*, 28, 617–625.

Raine, A. (1993). *Psychopathology of crime*. San Diego: Academic Press.

Raine, A., Brennam, P., and Mednick, S. A. (1994). Birth complications combined with early maternal rejection at age 1 year predispose to violent crime at age 18 years. *Archives of General Psychiatry*, 51, 984–988.

Raine, A., Buchsbaum, M. S., Stanley, J., Lottenberg, S., Abel, L., and Stoddard, J. (1994). Selective reductions in prefrontal glucose metabolism in murderers. *Biological Psychiatry*, 36, 365–373.

Raine, A., and Venables, P. H. (1992). Evolution, genetics, neuropsychology and psychophysiology of antisocial behaviour. In A. Gale and M. W. Eysenck, eds., *Handbook of individual differences: Biological perspective*, pp. 287–321. Chichester: Wiley.

Raleigh, M., McGuire, M., Melega, W., Cherry, S., Huang, S.-C., and Phelps, M. (1996). Neural mechanism supporting successful social decisions in simians. In A. R. Damasio, M. Damasio, and Y. Christen, eds., *Neurobiology of decision-making*, pp. 63–82. Berlin: Springer.

Ramachandran, V. S. (1993). Behavioral and magnetoencephalographic correlates of plasticity in the adult human brain. *Proceedings of the National Academy of Sciences of the USA*, 90, 10413–10420.

Ramachandran, V. S., and Blakeslee, S. (1999). *Phantoms in the brain.* London: Fourth Estate.

Ramachandran, V. S., and Rogers-Ramachandran, D. (1996). Synaesthesia in phantom limbs induced with mirrors. *Proceedings of the Royal Society, London B, Biological Sciences,* 263, 377–386.

Ramachandran, V. S., Rogers-Ramachandran, D., and Stewart, M. (1992). Perceptual correlates of massive cortical reorganization. *Science,* 258, 1159–1160.

Ranganath, C., Johnson, M. K., and D'Esposito, M. (2000). Left anterior prefrontal activation increases with demands to recall specific perceptual information. *Journal of Neuroscience,* 20, RC108.

Rao, S. M., Binder, J. R., Bandettini, P. A., Hammeke, T. A., Yetkin, Y. Z, Jesmanowicz, A., Lisk, L. M., Morris, G. L., Mueller, W. M., Estkowski, L. D., Wong, E. C., Haughton, V. M., and Hyde, J. S. (1993). Functional magnetic resonance imaging of complex human movements. *Neurology,* 43, 2311–2318.

Rapcsak, S. Z., Polster, M. R., Glisky, M. L., and Comer, J. F. (1996). False ecognition of unfamiliar faces following right hemisphere damage: Neuropsychological and anatomical observations. *Cortex,* 32, 593–611.

Rauschecker, J. P. (1995). Compensatory plasticity and sensory substitution in the cerebral cortex. *Trends in Neurosciences,* 18, 36–43.

Rawlins, J. (1985). Association across time: The hippocampus as a temporary memory store. *Behavioral and Brain Sciences,* 8, 479–496.

Rayner, K., and Pollatsek, A. (1981). Eye-movement control during reading: Evidence for direct control. *Quarterly Journal of Experimental Psychology,* 33A, 351–373.

Read, C., Zhang, Y., Nie, H., and Ding, B. (1986). The ability of manipulate speech sounds depends on knowing alphabetic writing. *Cognition,* 24, 31–44.

Rebillard, G., Carlier, E., Rebillard, M., and Pujol, R. (1977). Enhancement of visual response on the primary auditory cortex of the cat after an early destruction of cochlear receptors. *Brain Research,* 129, 162–164.

Reid, I., Young, A. W., and Hellawell, D. J. (1993). Voice recognition impairment in a blind Capgras patient. *Behavioral Neurology,* 6, 225–228.

Reinhardt, V., Reinhardt, A., Bercovitch, F., and Goy, R. (1986). Does intermale mounting function as dominance demonstration in rhesus monkeys? *Folia Primatologica,* 47, 55–60.

Reiter, R. J. (2000). Melatonin: Lowering the high price of free radicals. *News in Physiological Science*, 15, 246–250.

Rempp, K. A., Brix, G., Wenz, F., Becker, C. R., Gückel, F., and Lorenz, W. J. (1994). Quantification of regional cerebral blood flow and volume with dynamic susceptibility contrast-enhanced MR imaging. *Radiology*, 193, 637–641.

Reynolds, J. H., and Desimone, R. (1999). The role of neural mechanisms of attention in solving the binding problem. *Neuron*, 24, 19–29.

Ribary, U., Ioannides, A. A., Singh, K. D., Hasson, R., Bolton, J. P., Lado, F., Mogilner, A., and Llinás, R. (1991). Magnetic field tomography of coherent thalamocortical 40-Hz oscillations in humans. *Proceedings of the National Academy of Sciences of the USA*, 88, 11037–11041.

Richardson, M. K., Hanken, J., Gooneratne, M. L., Pieau, C., Raynaud, A., Selwood, L., and Wright, G. M. (1997). There is no highly conserved embryonic stage in the vertebrates: implications for current theories of evolution and development. *Anatomy and Embryology*, 196, 91–106.

Riddoch, G. (1941). Phantom limbs and body shape. *Brain*, 64, 197–222.

Rizzolatti, G., Fadiga, L., Gallese, V., and Fogassi, L. (1996). Premotor cortex and the recognition of motor actions. *Cognitive Brain Research*, 3, 131–141.

Robertson, P. (1974). *The Shell book of firsts*. London: Michael Joseph.

Robins, A. H. (1991). *Biological perspectives on human pigmentation*. Cambridge: Cambridge University Press.

Robinson, D. L., and Peterson, S. E. (1992). The pulvinar and visual salience. *Trends in Neurosciences*, 15, 127–132.

Rockstroh, B., Elbert, T., Birbaumer, N. ,and Lutzenberger, W. (1983). *Slow brain potentials and behavior*. Baltimore: Urban and Schwarzenberg.

Rockstroh, B., Elbert, T., Canavan, A., Lutzenberger, W., and Birbaumer, N. (1990). *Slow cortical potentials and behavior*. Baltimore: Urban and Schwarzenberg.

Rockstroh, B., Müller, M., Cohen, R., and Elbert, T. (1992). Probing the functional brain state during P300-evocation. *Journal of Psychophysiology*, 6, 175–184.

Rodseth, L., Wrangham, R. W., Harrigan, A. M., and Smuts, B. B. (1991). The human community as a primate society. *Current Anthropology*, 32, 221–254.

Roe, A., Pallas, S., Kwon, Y., and Sur, M. (1991). Visual projections routed to the auditory pathway in ferrets: Receptive fields of visual neurons in primary auditory cortex. *Journal of Neuroscience*, 12, 3651–3664.

Roelfsema, P. R., König, P., Engel, A. K., Sireteanu, R., and Singer, W. (1994). Reduced synchronization in the visual cortex of cats with strabismic amblyopia. *European Journal of Neuroscience*, 6, 1645–1655.

Rolak, L. A. (1991). Literary neurologic syndromes: Alice in Wonderland. *Archives of Neurology*, 48, 649–651.

Roland, P. R. (1984). Metabolic measurements of the working frontal cortex in man. *Trends in Neurosciences*, 7, 430–435.

Roland, P. R. (1993). *Brain activation*. New York: Wiley-Liss.

Roland, P. R., Larsen, B., Lassen, N.,and Skinhøj, E. (1980). Supplementary motor area and other cortical areas in organization of voluntary movements in man. *Journal of Neurophysiology*, 43, 118–136.

Rolls, E. T., and O'Mara, S. (1993). Neurophysiological and theoretical analysis of how the primate hippocampus functions in memory. In T. Ono, L. R. Squire, M. R. Raichle, D. I. Perrett, and M. Fukuda, eds., *Brain mechanisms of perception and memory*. pp. 276–300. New York: Oxford University Press.

Rolls, E. T., and Tovee, M. J. (1994) Processing speed in the cerebral cortex and the neurophysiology of visual masking. *Proceedings of the Royal Society of London*, Series B, 257, 9–15.

Ronne-Engström, E., Hillered, L., Flink, R., Spannare, B., Ungerstedt, U., and Carlson, H. (1992). Intracerebral microdialysis of extracellular amino acids in the human epileptic focus. *Journal of Cerebral Blood Flow and Metabolism*, 12, 873–876.

Rose, M. R., and Moore, C. (1993). A Darwinian function for the orbital cortex. *Journal of Theoretical Biology*, 161, 119–129.

Rosen, M., and Lunn, J.,eds. (1987). *Consciousness, awareness, and pain in general anesthesia*. London: Butterworth.

Rosenberg, K. (1992). The evolution of modern human childbirth. *Yearbook of Physical Anthropology*, 35, 89–124.

Rosenblum, L. A., Coplan, J. D., Friedman, S., Bassoff, T., Gorman, J. M., and Andrews, M. W. (1994). Adverse early experiences affect noradrenergic and serotonergic functioning in adult primates. *Biological Psychiatry*, 35, 221–227.

Ross, L., and Nisbett, R. E. (1991). *The person and the situation*. Philadelphia: Temple University Press.

Rossi, L. N., Candini, G., Scarlatti, G., Rossi, G., Prina, E., and Alberti, S. (1987). Autosomal dominant microcephaly without mental retardation. *American Journal of Diseases of Children*, 141, 655–659.

Ruff, C. B., and Walker, A. (1993). Body size and body shape. In A. Walker and R. Leakey, eds., *Nariokotome* Homo erectus *skeleton*, pp. 234–265. Cambridge, Mass.: Harvard University Press.

Rugg, M. D., Fletcher, P. C., Frith, C. D., Frackowiak, R. S., and Dolan, R. J. (1996). Differential activation of the prefrontal cortex in successful and unsuccessful memory retrieval. *Brain*, 119, 2073–2083.

Rumelhart, D. E., McClelland, J. L., and the PDP Research Group (1986). *Parallel distributed processing*, vols. 1 and 2. Cambridge, Mass.: MIT Press.

Russek, L. G., and Schwartz, G. A. (1997). Perceptions of parental caring predict health status in midlife: A 35-year follow-up of the Harvard Mastery of Stress study. *Psychosomatic Medicine*, 59, 144–149.

Russell, J. G. (1969). Moulding of the pelvic outlet. *Journal of Obstetrics and Gynaecology of The British Commonwealth*, 76, 817–820.

Russell, M., Switz, G., and Thompson, K. (1980). Olfactory influences on the human menstrual cycle. *Pharmacology, Biochemistry and Behaviour*, 13, 737–738.

Ruvolo, M., Pan, D., Zehr, S., Goldberg, T., Distrell, T. R., and von Dornum, M. (1994). Gene trees and hominoid phylogeny. *Proceedings of the National Academy of Sciences of the USA*, 91, 8900–8904.

Sacks, O. (1984). *A leg to stand on*. London: Duckworth.

Sadato, N., Pascual-Leone, A., Grafman, J., Deiber, M.-P., Ibañez, V., and Hallett, M. (1998). Neural networks for Braille reading by the blind. *Brain*, 121, 1213–1229.

Sadato, N., Pascual-Leone, A., Grafman, J., Ibañez, V., Deiber, M.-P., Dold, G., and Hallett, M. (1996). Activation of the primary visual cortex by Braille reading in blind subjects. *Nature*, 380, 526–528.

Sagan, C., (1977). *The dragons of Eden: Speculations on the evolution of human intelligence*. New York: Ballantine Books.

Sagi, A., and Hoffman, M. L. (1976). Empathic distress in newborns. *Developmental Psychology*, 12, 175–176.

Saito, S., Yoshokawa, D., Nishihara, F., Morita, T., Kitani, Y., Amaya, T., and Fujita, T. (1995). The cerebral hemodynamic responses to electrically induced seizures in man. *Brain Research*, 673, 93–100.

Sakurai, Y., Momose, T., Iwata, M., Watanabe, T., Ishikawa, T., Takeda, K., and Kanazawa, I. (1992). *Kanji* word reading process analyzed by positron emission tomography. *NeuroReport*, 3, 445–448.

Salamé, P., and Baddeley, A. (1982). Disruption of short-term memory by unattended speech. *Journal of Verbal Learning and Verbal Behavior*, 21, 150–164.

Sams, M. Aulanko, R., Hämäläinen, M., Hari, R., Lounasmaa, O. V., Lu., S.-T., and Simola, J. (1991). Seeing speech: Visual information from lip movements modifies activity in the human auditory cortex. *Neuroscience Letters*, 127, 141–145.

Sandrew, B. B., Stamm, J. S., and Rosen, S. C. (1977). Steady potential shifts and facilitated learning of delayed response in monkeys. *Experimental Neurology*, 55, 43–55.

Sanes, J. N., and Donoghue, J. P. (1993). Oscillations in local field potentials of the primate motor cortex during voluntary movement. *Proceedings of the National Academy of Sciences of the USA*, 90, 4470–4474.

Santamaria, J., and Chiappa, K. H. (1987). The EEG of drowsiness in normal adults. *Journal of Clinical Neurophysiology*, 4, 327–382.

Sarter, M., and Markowitsch, H. (1985). The amygdala's role in human mnemonic processing. *Cortex*, 21, 7–24.

Sasaki, K., and Gemba, H. (1984). Compensatory motor function of the somatosensory cortex for the motor cortex temporarily impaired by cooling in the monkey. *Experimental Brain Research*, 55, 60–68.

Sassaman, E. A., and Zartler, A. S. (1982). Mental retardation and head growth abnormalities. *Journal of Pediatric Psychology*, 7, 149–156.

Sauve, K. (1999). Gamma-band synchronous oscillations: Recent evidence regarding their functional significance. *Consciousness and Cognition*, 8, 213–224.

Savage, C. R., Deckersbach, T., Heckers, S., Wagner, A. D., Schacter, D. L., Alpert, N. M., Fischman, A. J., and Rauch, S. L. (2001). Prefrontal regions supporting spontaneous and directed application of verbal learning strategies: Evidence from PET. *Brain*, 124, 219–231.

Savage-Rumbaugh, E. S., and Lewin, R. (1994). *Kanzi*. London: Doubleday.

Savage-Rumbaugh E. S., McDonald, K., Sevcik, R. A., Hopkins, W. D., and Rubert, E. (1986). Spontaneous symbol acquisition and communicative use by pygmy chimpanzees (*Pan paniscus*). *Journal of Experimental Psychology: General*, 115, 211–235.

Savage-Rumbaugh, E. S., and Rumbaugh, D. M. (1993). The emergence of language. In K. R. Gibson and T. Ingold, eds., *Tools, language and cognition in human evolution*, pp. 86–108. Cambridge: Cambridge University Press.

Savage-Rumbaugh, E. S., Rumbaugh, D., Smith, S., and Lawson, J. (1980). Reference: The linguistic essential. *Science*, 210, 922–925.

Savage-Rumbaugh, E. S., Williams, S. L., Furuichi, T., and Kano, T. (1996). Language perceived, *paniscus* branches out. In W. C. McGrew, L. F. Marchant, and T. Nishida, eds., *Great ape societies*, pp. 173–184. Cambridge: Cambridge University Press.

Sawaguchi, T., and Kudo, H. (1991). *Human Evolution*, 6, 201–212.

Sawamoto, N., Honda, M., Okada, T., Hanakawa, T., Kanda, M., Fukuyama, H., Konishi, J., and Shibasaki, H. (2000). Expectation of pain enhances responses to nonpainful somatosensory stimulation in the anterior cingulate cortex and parietal operculum/posterior insula: An event-related functional magnetic resonance imaging study. *Journal of Neuroscience*, 20(19), 7438–7445.

Schanberg, S., and Field, T. (1987). Sensory deprivation stress and supplemental stimulation in the rat pup and preterm human neonate. *Child Development*, 58, 1431–1447.

Schenck, C. H., Pareja, J. A., Patterson, A. L., and Mahowald, M. W. (1998). Analysis of polysomnographic events surrounding 252 slow-wave sleep arousals in thirty-eight adults with injurious sleepwalking and sleep terrors. *Journal of Clinical Neurophysiology*, 15, 159–166.

Schieber, M. H. (1990). How might the motor cortex individuate movements? *Trends in Neuroscience*, 13, 440–445.

Schieber, M. H., and Hibbard, L. S. (1993). How somatotopic is the motor cortex hand area? *Science*, 261, 489–492.

Schiff, B. B., and Lamon, M. (1989). Inducing emotions by unilateral contraction of facial muscles: A new look at hemispheric specialization and the experience of emotion. *Neuropsychologia*, 27, 923–935.

Schlaug, G., Jäncke, L., Huang, Y., Staiger, J. F., and Steinmetz, H. (1995). Increased corpus callosum size in musicians. *Neuropsychologia*, 33, 1047–1055.

Schmandt-Besserat, D. (1992). *Before writing*, vol. 1. *From counting to cuneiform*. Austin: University of Texas Press.

Schneider, R. J., Friedman, D. P., and Mishkin, M. (1993). A modality-specific somatosensory area within the insula of the rhesus monkey. *Brain Research*, 621, 116–120.

Schore, A. N. (1994). *Affect regulation and the origin of the self: The neurobiology of emotional development*. Hillsdale, N.J.: Erlbaum.

Schore, A. N. (1996). The experience-dependent maturation of a regulatory system in the orbital prefrontal cortex and the origin of developmental psychopathology. *Development and Psychopathology*, 8, 59–87.

Schupp, H. T., Lutzenberger, W., Birbaumer, N., Miltner, W., and Braun, C. (1994). Neurophysiological differences between perception and imagery. *Cognitive Brain Research*, 2, 77–86.

Schwartz, J. H., and Tattersall, I., (2000). The human chin revisited: What is it and who has it? *Journal of Human Evolution*, 38 (3), 367–409

Schwender, D., Madler, C., Klasing, S., Peter, K., and Pöppel, E. (1994). Anaesthetic control of 40-Hz brain activity and implicit memory. *Consciousness and Cognition*, 3, 129–147.

Sebeka, A., and Ringo, J. L. (1993). Investigation of long term recognition and association memory in unit responses from infer temporal cortex. *Experimental Brain Research*, 96, 28–38.

Seeck, M., Mainwaring, N., Ives, J., Blume, H., Dubuisson, D., Cosgrove, R., Mesulam, M. M., and Schomer, D. L. (1993). Differential neural activity in the human temporal lobe evoked by faces of family members and friends. *Annals of Neurology*, 34, 369–372.

Seigel, B. V., Buchsbaum, M. S., Bunney, W. E., Gottschalk, L. A., Haier, R. J., Lohr, J. B., Lottenberg, S., Najafi, A., Nuechterlein, K. H., Porkin, S. G., and Wu, J. C. (1993). Cortical-striatal-thalamic circuits and brain glucose metabolic activity in 70 unmedicated male schizophrenic patients. *American Journal of Psychiatry*, 150, 1325–1336.

Siegel, J. M. (2001). The REM sleep-memory consolidation hypothesis. *Science*, 294, 1058–1063.

Seitz, R. J., Huang, Y., Knorr, U., Tellmann, L., Herzog, H., and Freund, H.-J. (1995). Large-scale plasticity of the human motor cortex. *NeuroReport*, 6, 742–744.

Seitz, R. J., and Roland, P. E. (1992). Vibratory simulation increases and decreases the regional cerebral blood flow and oxidative metabolism. *Acta Neurologica Scandivanica*, 86, 60–67.

Sells, C. J. (1977). Microcephaly in a normal school population. *Pediatrics*, 59, 262–265.

Semendeferi, K., Armstrong, E., Schleicher, A., Zilles, K., and Van Hoesen, G. W. (2001) Prefrontal cortex in humans and apes: A comparative study of area 10. *American Journal of Physical Anthropology*, 114, 224–241.

Semendeferi, K., Damasio, H., Frank, R., and Van Hoesen, G. W. (1997). The evolution of the frontal lobes: a volumetric analysis based on three-dimensional reconstructions of magnetic resonance scans of human and ape brains. *Journal of Human Evolution*, 32, 375–388.

Semmes, J., Weinstein, S., Ghent, L., and Teuber, H.-L. (1963). Correlates of impaired orientation in personal and extrapersonal space. *Brain*, 86, 747–772.

Sergent, J., Zuck, E., Terriah, S., and MacDonald, B. (1992). Distributed neural network underlying musical sight-reading and keyboard performance. *Science*, 257, 106–109.

Seyfarth, R., Cheney, D. L., and Marler, P. (1980). Monkey responses to three different alarm calls. *Science*, 210, 801–804.

Shackelford, T. K., and Larsen, R. J. (1999). Facial attractiveness and physical health. *Evolution and Human Behavior*, 20, 71–76.

Shadlen, M. N., and Movshon, J. A. (1999). Synchrony unbound: A critical evaluation of the temporal binding hypothesis. *Neuron* , 24, 67–77.

Shah N. J., Marshall, J. C., Zafiris, O., Schwab, A., Zilles, K., Markowitsch, H. J., and Fink, G. R. (2001). The neural correlates of person familiarity: A functional magnetic resonance imaging study with clinical implications. *Brain*, 124, 804–815.

Shallice, T. (1988). *From neuropsychology to mental structure*. Cambridge: Cambridge University Press.

Shallice, T., and Burgess, P. (1991). Higher-order cognitive impairments and frontal lobe lesions. In H. Levin, H. Eisenberg, and A. Benton, eds., *Frontal lobe function and dysfunction*, pp. 125–138. New York: Oxford University Press.

Shallice, T., and Evans, M. (1978). The involvement of the frontal lobes in cognitive estimation. *Cortex*, 14, 294–303.

Shallice, T., Fletcher, P., Frith, C., Grasby, P., Frackowiak, R., and Dolan, R. (1994). Brain regions associated with acquisition and retrieval of verbal episodic memory. *Nature*, 368, 633–635.

Shand, M. A. (1982). Sign-based short-term coding of American Sign Language signs and printed English words by congenitally deaf signers. *Cognitive Psychology*, 14, 1–12.

Share, D. (1995). Phonological recoding and self-teaching: *Sine qua non* of reading acquisition. *Cognition*, 55, 151–216.

Sharma, J., Angelucci, A., and Sur, M. (2000). Induction of visual orientation modules in auditory cortex. *Nature*, 404, 841–847.

Shashri, L., and Ajjanagadde, V. (1993). From simple associations to systematic reasoning: A connectionist representation of rules, variables and dynamic bindings using temporal synchrony. *Behavioral and Brain Sciences*, 16, 417–494.

Shatz, C. (1992). How are specific connections formed between thalamus and cortex? *Current Opinion in Neurobiology*, 2, 78–82.

Shaver, P. R., and Clark, C. L. (1994). Psychodynamics of adult romantic attachment. In J. M. Masling and R. F. Bornstein, eds., *Empirical perspectives on object relations theory*, pp. 105–156. Washington, D.C.: American Psychological Association.

Sheer, D. E. (1976). Focused arousal and 40 Hz EEG. In R. M. Knight and D. J. Bakker, eds., *The neuropsychology of learning disorders*, pp. 71–87. Baltimore: University Park Press.

Sheldon, B. C., Merilä, J., Qvarnström, A., Gustafsson, L., and Ellegren, H. (1997). Paternal genetic contribution of offspring condition predicted by size of male secondary sexual character. *Proceedings of the Royal Society of London, Series B*, 264, 297–302.

Shima, K., Aya, K., Mushiake, H., Inase, M., Aizawa, H., and Tanki, J. (1991). Two movement-related foci in the primate cingulate cortex observed in signal-triggered and self-paced forelimb movements. *Journal of Neurophysiology*, 65, 188–202.

Shimamura, A. P. (2000). The role of the prefrontal cortex in dynamic filtering. *Psychobiology*, 28, 207–218.

Shulman, G. L., Fiez, J. A., Corbetta, M., and Buckner, R. L. (1997). Common blood flow changes across visual tasks. II. Decreases in cerebral cortex. *Journal of Cognitive Neuroscience*, 9, 648–663.

Shulman, R. G., Blamire, A. M., Rothman, D. L., and McCarthy, G. (1993). Nuclear magnetic resonance imaging and spectroscopy of human brain function. *Proceedings of the National Academy of Sciences of the USA*, 90, 3127–3133.

Shumikhina, S., and Molotchnikoff, S. (1995). Visually-triggered oscillations in the cat lateral posterior-pulvinar complex. *NeuroReport*, 6, 2341–2347.

Sigg, H., Stolba, A., Abegglen J., and Dasser, V. (1982). Life history of hamadryas baboons: Physical development, infant mortality, reproductive parameters and family relationships. *Primates*, 23, 473–487.

Sillito, A. M., Jones, H. E., Gerstein, G. L., and West, D. C. (1994). Feature-linked synchronization of thalamic relay cell firing induced by feedback from the visual cortex. *Nature*, 369, 479–482.

Simmel, M. L. (1956). On phantom limbs. *Archives of Neurology and Psychiatry*, 75, 637–647.

Singer, W. (1993). Synchronization of cortical activity and its putative role in information processing and learning. *Annual Review of Physiology*, 53, 349–374.

Singer, W. (1994). Putative functions of temporal correlations in neocortical processing. In C. Koch and J. L. Davis, eds., *Large-scale neuronal theories of the brain*, pp. 201–237. Cambridge, Mass.: MIT Press.

Singer, W. (1999). Neuronal synchrony: A versatile code for the definition of relations? *Neuron*, 24, 49–65.

Singh, J., and Knight, R. T. (1990). Prefrontal lobe contribution to voluntary movements in humans. *Brain Research*, 531, 45–54.

Skinner, J. E., and Yingling, C. D. (1977). Central gating mechanisms that regulate event-related potentials and behavior. In J. E. Desmedt, ed., *Progress in clinical neurophysiology: Attention, voluntary contraction and event-related cerebral potentials*, vol. 1, pp. 30–69. Basel: Karger.

Skotko, D. J. (1992). Structural properties of verbal commands and their effects on the regulation of motor behavior. In R. Diaz and L. Berk, eds., *Private speech: From social interaction to self-regulation*. Hillsdale, N.J.: Erlbaum.

Skoyles, J. R. (1991a). Connectionism, reading and the limits of cognition. *Psycoloquy*, 2.8.4.

Skoyles, J. R. (1991b). No free meals for reading networks—all reading networks need to be trained. *Psycoloquy*, 2.9.4.2.

Skoyles, J. R. (1997). Evolution's "missing link": A hypothesis upon neural plasticity, prefrontal working memory and the origins of modern cognition. *Medical Hypotheses*, 48, 499–501.

Skoyles, J. R. (1998). Speech phones are a replication code. *Medical Hypotheses*, 50, 167–173.

Skoyles, J. R. (1999a). Human evolution expanded brains to increase expertise capacity, not IQ. *Psycoloquy*, 10(002). http://www.cogsci.soton.ac.uk/cgi/psyc/newpsy?10.002

Skoyles, J. R. (1999b). Neural plasticity and exaptation. *American Psychologist*, 54, 438–439.

Slatter, K. H. (1960). Alpha rhythms and mental imagery. *Electroencephalography and Clinical Neurophysiology*, 12, 851–859.

Smith, A. (1974). Dominant and nondominant hemispherectomy. In M. Kinsbourne and W. Smith., eds., *Hemisphere disconnection and cerebral function*, pp. 5–33, Springfield, Ill.: Thomas.

Smith, A., and Sugar, O. (1975). Development of above normal language and intelligence 21 years after left hemispherectomy. *Neurology*, 25, 813–818.

Smith, C. (1995). Sleep states and memory processes. *Behavioral Brain Research*, 69, 137–145.

Smith, E. E., and Jonides, J. (1999). Storage and executive processes in the frontal lobes. *Science*, 283, 1657–1661.

Smuts, B. (1985). *Sex and friendship in baboons*. New York: Aldine.

Sneden, C. (2001). The age of the universe. *Nature*, 409, 673–674.

Snell, B. (1953). *The discovery of the mind: The Greek origins of European thought*. Oxford: Blackwell.

Snyder, A. Z., Abdullaev, Y. G., Posner, M. I., and Raichle, M. E. (1995). Scalp electrical potentials reflect regional cerebral blood flow responses during processing of written words. *Proceedings of the National Academy of Sciences of the USA*, 92, 1689–1693.

Sobotka, S., and Ringo, J. L. (1993) Investigation of long term recognition and association memory in unit responses from inferotemporal cortex. *Experimental Brain Research*, 26, 28–38.

Sokolov, A. N. (1972). *Inner speech and thought*. New York: Plenum.

Sorce, J. F., Emde, R. N., Campos, J., and Klinnert, M. D. (1985). Maternal emotional signaling: Its effect on the visual cliff behavior of 1-year-olds. *Developmental Psychology*, 21, 195–200.

Sowell, E. R., Delis, D., Stiles, J., and Jernigan, T. L. (2001). Improved memory functioning and frontal lobe maturation between childhood and adolescence: A structural MRI study. *Journal of the International Neuropsychological Society*, 7(3), 312–322.

Sowell, E. R., Thompson, P. M., Holmes, C. J., Jernigan, T. L., and Toga, A. W. (1999). In vivo evidence for post-adolescent brain maturation in frontal and striatal regions. *Nature Neuroscience*, 2, 859–861.

Spiegel, D. (1991). Psychosocial aspects of cancer. *Current Opinion in Psychiatry*, 4, 889–897.

Spiegel, D. (1992). Hypnosis: Brain basis. In B. Smith and O. Adelman, eds., *Neuroscience year book*, suppl. 2, pp. 75–78. Boston: Birkhäuser.

Spydell, J. D., Ford, M. R., and Sheer, D. E. (1979). Task dependent cerebral lateralization of the 40 hertz EEG rhythm. *Psychophysiology*, 16, 347–350.

Spydell, J. D., and Sheer, D. E. (1982). Effect of problem solving on right and left hemisphere 40 hertz EEG activity. *Psychophysiology*, 19, 420–425.

Squire, L. (1992). Memory and the hippocampus: A synthesis from findings with rats, monkeys, and humans. *Psychological Review*, 99, 195–231.

Stamm, J. S. (1987). The riddle of the monkey's delayed-response deficit has been solved. In E. Perecman, ed., *The frontal lobes revisited*, pp. 73–89. New York: IRBN Press.

Stanley, S. (1992). An ecological theory of the origin of *Homo*. *Paleobiology*, 18, 237–257.

Staton, R., Brumback, R., and Wilson, H. (1982). Reduplicative paramnesia: A disconnection syndrome of memory. *Cortex*, 18, 23–36.

Steen, R. G., Ogg, R. J., Reddick, W. E., and Kingsley, P. B. (1997). Age-related changes in the pediatric brain: Quantitative MR evidence of maturational changes during adolescence. *American Journal of Neuroradiology*, 18(5), 819–828.

Stein, J. (1992). The representation of egocentric space in the posterior parietal cortex. *Brain and Behavioral Sciences*, 15, 691–700.

Steriade, M., Contreras, D., Dossi, R. C., and Núñez, A. (1993). The slow (<1 Hz) oscillation in reticular thalamic and thalamocortical neurons: Scenario of sleep rhythm generation in interacting thalamic and neocortical networks. *Journal of Neuroscience*, 13, 3284–3299.

Stickgold, R., Malia, A., Maguire, D., Roddenberry, D., and O'Connor, M. (2000). Replaying the game: Hypnagogic images in normals and amnesics. *Science*, 290, 350–353.

Stigler, J. (1984). "Mental abacus": The effect of abacus training on Chinese children's mental calculation. *Cognitive Psychology*, 16, 145–176.

Strafella, A. P., Paus, T., Barrett, J., and Dagher, A. (2001). Repetitive transcranial magnetic stimulation of the human prefrontal cortex induces dopamine release in the caudate nucleus. *Journal of Neuroscience*, 21, RC157,1–4.

Straus, L. (1987). Hunting in Late Upper Paleolithic Western Europe. In M. Nitecki and D. Nitecki, eds., *Evolution of human hunting*, pp. 147–176. New York: Plenum.

Streissguth, A. P. (1993). Fetal alcohol syndrome in older patients. *Alcohol and Alcoholism (*suppl. 2*)*, 209–212.

Stringer, C., and Gamble, C. (1993). *In search of the Neanderthals*. London: Thames and Hudson.

Stroop, J. R. (1935). Studies of interference in serial verbal reactions. *Journal of Experimental Psychology*, 18, 643–662.

Sturm, R. A., Box, N. F., and Ramsay, M. (1998). Human pigmentation genetics: The difference is only skin deep. *BioEssays*, 20, 712–721.

Stuss, D. T. (1991). Disturbance of self-awareness after frontal system damage. In G. Prigatano and D. Schacter, eds., *Awareness of deficit after brain injury*. New York: Oxford University Press.

Stuss, D. T. (1992). Biological and psychological development of executive functions. *Brain and Cognition*, 20, 8–23.

Sur, M., Pallas, S., and Roe, A. (1990). Cross-model plasticity in cortical development: Differentiation and specification of sensory neocortex. *Trends in Neurosciences*, 13, 227–233.

Swanson, L. W. (2000). Cerebral hemisphere regulation of motivated behavior (1). *Brain Research*, 886(1–2), 113–164.

Swanson, L. W., and Petrovich, G. D. (1998). What is the amygdala? *Trends in Neurosciences*, 21(8), 323–331.

Swisher, C. C., Curtis, G. H., Jacob, T., Getty, A. G., Suprijo, A., and Widiasmoro, (1994). Age of the earliest known hominids in Java, Indonesia. *Science*, 263, 1118–1121.

Symons, D. (1995). Beauty is in the adaptations of the beholder: The evolutionary psychology of human female sexual attractiveness. In P. R. Abramson and S. D. Pinkerton, eds., *Sexual nature/sexual culture*, pp. 80–118. Chicago: University of Chicago Press.

Tabert, M. H., Borod, J. C., Tang, C. Y., Lange, G., Wei, T. C., Johnson, R., Nusbaum, A. O , and Buchsbaum, M. S. (2001). Differential amygdala activation during emotional decision and recognition memory tasks using unpleasant words: an fMRI study. *Neuropsychologia*, 39, 556–573.

Tajfel, H. (1970). Experiments in intergroup discrimination. *Scientific American*, 223, (11), 96–102.

Tallon, C., Bertrand, O., Bouchet, P., and Pernier, J. (1995). Gamma-range activity evoked by coherent visual stimuli in humans. *European Journal of Neuroscience*, 7, 1285–1291.

Tallon-Baudry, C., and Bertrand, O. (1999). Oscillatory gamma activity in humans and its role in object representation. *Trends in Cognitive Science*, 3, 151–162.

Tambiah, S. J. (1968). The magic power of words. *Man*, 3, 175–208.

Tamminga, C. A., Thaker, G. K., Buchanan, R., Kirkpatrick, B., Alphas, L. D., Chase, T. N., and Carpenter, W. T. (1992). Limbic system abnormalities identified in schizophrenia using positron emission tomography. *Archives of General Psychiatry*, 49, 522–530.

Tamura, R., Ono, T., Fukuda M., and Nakamura, K. (1990). Recognition of egocentric and allocentric visual and auditory space by neurons in the hippocampus of monkeys. *Neuroscience Letters*, 109, 293–298.

Tan, A.-A., and Breen, S. (1993). Radial mosaicism and tangential cell dispersion both contribute to mouse neocortical development. *Nature*, 362, 638–640.

Tanaka, K. (1993). Neuronal mechanism of object recognition. *Science*, 685–688.

Tanji, J., and Shima, K. (1994). Role for supplementary motor area cells in planning several movements ahead. *Nature*, 371, 413–416.

Tanner, N., and Zihlman, A. (1976). Women in evolution. I. Innovation and selection in human origins. *Signs*, 1, 585–608.

Taylor, S. F., Liberzon, I., and Koeppe, R. A. (2000). The effect of graded aversive stimuli on limbic and visual activation. *Neuropsychologia*, 38(10), 1415–1425.

Tennov, D. (1979). *Love and limerence*. New York: Stein and Day.

Terada, K., Ikeda, A., Negamine, T., and Shibasaki, H. (1995). Movement-related cortical potentials associated with voluntary muscle relaxation. *Electroencephalography and Clinical Neurophysiology*, 95, 335–345.

Tesche, C., and Hari, R. (1993). Independence of steady-state 40-Hz and spontaneous 10-Hz activity in the human auditory cortex. *Brain Research*, 629, 19–22.

Teyler, T., and DiScenna, P. (1986). The hippocampal memory indexing theory. *Behavioral Neuroscience*, 100, 147–154.

Thatcher, R. (1991). Maturation of the human frontal lobes: Physiological evidence for staging. *Developmental Neuropsychology*, 7, 397–419.

Thatcher, R., Walker, R., and Giudice, S. (1987). Human cerebral hemispheres develop at different rates and ages. *Science*, 236, 1110–1113.

Thieme, H. (1997). Lower Palaeolithic hunting spears from Germany. *Nature*, 387, 807–810.

Thiessen, D. (1999). Social influences on human assortative mating. In M. C. Corballis, and S. E. Lea, eds., *Descent of mind*, pp. 311–323. Oxford: Oxford University Press.

Thomas, M. L., Sing, H. C., Belenky, G., Holcomb, H. H., Dannals, R. F., Wagner, H. N., Peller, P., Mayberg, H. S., Wright, J. E., Thorne, D. R., Popp, K. A., Redmond, D. R., Zurer, J., and Balwinski, S. (1993). Cerebral glucose utilization during task performance and prolonged sleep loss. *Journal of Cerebral Blood Flow and Metabolism*, 13 (suppl.), S531.

Tidswell, T., Gibson, A., Bayford, R. H., and Holder, D. S. (2001). Three-dimensional electrical impedance tomography of human brain activity. *NeuroImage*, 13, 283–294.

Tiihonen, J., Kuikka, J., Hakola, P., Paania, J., Airaksinen, J., Eronen, M., and Hallikainen, T. (1994). Acute ethanol-induced changes in cerebral blood flow. *American Journal of Psychiatry*, 151, 1505–1508.

Tiihonen, J., Kuikka, J., Kupila, J., Partanen, K., Vainio, P., Airaksinen, J., Eronen, M., Hallikainen, T., Paanila, J., Kinnunen, I., and Huttunen, J. (1994). Increase in cerebral blood flow of right prefrontal cortex in man during orgasm. *Neuroscience Letters*, 170, 241–243.

Tiitinen, H., Sinkkonen, J., Reinikainen, K., Alho, K., Lavikainen, J., and Naatanen, R. (1993). Selective attention enhances the auditory 40-Hz transient response in humans. *Nature*, 364, 59–60.

Tomarken, A. J., and Davidson, R. J. (1994). Frontal brain activation in repressers and nonrepressers. *Journal of Abnormal Psychology*, 103, 339–349.

Tomarken, A. J., Davidson, R., Wheeler, R., and Doss, R. (1992). Individual differences in anterior brain asymmetry and fundamental dimensions of emotions. *Journal of Personality and Social Psychology*, 62, 676–687.

Tomasello, M., Kruger, A. C., and Ratner, H. H. (1993). Cultural learning. *Behavioral and Brain Sciences*, 16, 495–552.

Travis, J. (1994). Glia: The brain's other cells. *Science*, 266, 970–972.

Trivers, R. L. (1971). The evolution of reciprocal altruism. *Quarterly Review of Biology*, 46, 35–57.

Tsunoda, K., Yamane, Y., Nishizaki, M., and Tanifuji, M. (2001). Complex objects are represented in macaque inferotemporal cortex by the combination of feature columns. *Nature Neuroscience*, 4, 832–838.

Tulving, E., Kapur, S., Craik, F. I., Moscovitch, M., and Houle, S. (1994). Hemispheric encoding/retrieval asymmetry in episodic memory. *Proceedings of the National Academy of Sciences of the USA*, 91, 2016–2020.

Tulving, E., Markowitsch, H. J., Kapur, S., Habib, R., and Houle, S. (1994). Novelty encoding networks in the human brain: Positron emission tomography data. *NeuroReport*, 5, 2525–2528.

Turner, J. (1987). *Rediscovering the social group: A self-categorization theory*. Oxford: Blackwell.

Twitty, V. C. (1966). *Of scientists and salamanders*. San Francisco: Freeman.

Tzourio, N., De Schonen, S., Pietrzyk, U., Bore, A., Bruck, B., Crouzel, M., Hassan, M., Aujard, Y., and Mazoyer, B. M. (1993). Brain activation studies using PET and ^{15}]-labeled water in two-month-old alert infants. *Journal of Cerebral Blood Flow and Metabolism*, 13 (suppl.), S414.

Uhl, F., Franzen, P., Lindinger, G., Lang, W., and Deecke, L. (1991). On the functionality of the visually deprived occipital cortex in early blind persons. *Neuroscience Letters*, 124, 256–259.

Uhl, F., Kretschmer, T., Lindinger, G., Goldenberg, G., Lang, W., Oder, W., and Deecke, L. (1994). Tactile mental imagery in sighted persons and in patients suffering from peripheral blindness early in life. *Electroencephalography and Clinical Neurophysiology*, 91, 249–255.

Uhl, F., Podreka, I., and Deecke, L. (1994). Anterior frontal cortex and the effect of proactive interference in word pair learning—results of Brain-SPECT. *Neuropsychologia*, 32, 241–247.

Umiker-Sebeok, D. J., and Sebeok, T. A. (1978). *Aboriginal sign languages of the Americas and Australia*. vol. 1. New York: Plenum.

Ungerleider, L. G., and Mishkin, M. (1982). Two cortical visual systems. In D. J. Ingle, M. A. Goodale, and R. J. Mansfield, eds., *Analysis of visual behavior*, pp. 549–586. Cambridge, Mass.: MIT Press.

Uvnas-Moberg, K. (1997a). Antistress pattern induced by oxytocin. *News in Physiological Science*, 13, 22–26.

Uvnas-Moberg, K. (1997b). Oxytocin linked antistress effects—the relaxation and growth response. *Acta Physiologica Scandinavica. Supplementum*, 640, 38–42.

Uvnas-Moberg, K. (1997c). Physiological and endocrine effects of social contact. *Annals of the New York Academy of Sciences*, 807, 146–163.

Uvnas-Moberg, K. (1998). Oxytocin may mediate the benefits of positive social interaction and emotions. *Psychoneuroendocrinology*, 23(8), 819–835.

Uvnas-Moberg, K., Bjokstrand, E., Hillegaart, V., and Ahlenius, S. (1999). Oxytocin as a possible mediator of SSRI-induced antidepressant effects. *Psychopharmacology (Berlin)*, 142(1), 95–101.

Vaadia, E., Haalman, I., Abeles, M., Bergman, H., Prut, Y., Slovin, H., and Aertsen, A. (1995). Dynamics of neuronal interactions in monkey cortex in relation to behavioral events. *Nature*, 373, 616–518.

Valenstein, E. S. (1986). *Great and desperate cures*. New York: Basic Books.

Valenstein, E. S. (1990). The prefrontal area and psychosurgery. *Progress in Brain Research*, 85, 539–554.

Van den Berg, C. L., Hol, T., Van Ree, J. M., Spuijt, B. M., Everts, H., and Koolhaas, J. M. (1999). Play is indispensable for an adequate development of coping with social challenges in the rat. *Developmental Psychobiology*, 43, 129–139.

Vanderwolf, C. (1992). Hippocampal activity, olfaction, and sniffing: An olfactory input to the dentate gyrus. *Brain Research*, 593, 197–208.

Vandiver, P., Soffer, O., Klima, B., and Svoboda, J. (1989). The origins of ceramic technology at Dolni Vestonice, Czechoslovakia. *Science*, 246, 1002–1008.

Vaneechoutte, M. ,and Skoyles, J. R. (1998). The memetic origin of language: Humans as musical primates. *Journal of Memetics—Evolutionary Models of Information Transmission*, 2. http://www.cpm.mmu.ac.uk/jomemit/1998/vol2/vaneechoutte_m&skoyles_jr.html

Vaughan, T. A. (1986). *Mammalogy*, 3rd ed. Philadelphia: Saunders College Publishing.

Vaughan, T. A., Ryan, J. M., and Czaplewski, N. J. (2000). *Mammalogy*. New York: Harcourt.

Veraart, C., De Volder, A. G., Wanet-Defalque, A., Metz, R., Michel, C., Dooms, G., and Goffinet, A. (1990). Glucose utilization in human visual cortex is abnormally elevated in blindness of early onset but decreased in blindness of late onset. *Brain Research*, 510, 115–121.

Vernikos-Danellis, J. (1980). Adrenocortical responses of humans to group hierarchy, confinement, and social interaction. In S. Levine and H. Ursin, eds., *Coping and health*. New York: Plenum Press.

Villringer, A., Planck, J., Hock, C., Schleinkofer, L., and Dirnagl, U. (1993). Near infrared spectroscopy (NIRS): A new tool to study hemodynamic changes during activation of brain function in human adults. *Neuroscience Letters*, 154, 101–104.

Vogt, B. A., Finch, D., and Olson, C. (1992). Functional heterogeneity in cingulate cortex: The anterior executive and posterior evaluative regions. *Cerebral Cortex*, 2, 435–443.

Vogt, B. A., Sikes, R. W., and Vogt, L. L. (1993). Anterior cingulate cortex and the medial pain system. In B. A. Vogt and M. Gabriel, eds., *Neurobiology of cingulate cortex and limbic thalamus*, pp. 313–344. Boston: Birkhäuser.

Von der Malsburg, C. (1995). Binding in models of perception and brain function. *Current Opinion in Neurobiology*, 5, 520–526.

Von der Malsburg, C. (1999). The what and why of binding: The modeler's perspective. *Neuron*, 24, 95–104.

Von Melchner, L., Pallas, S. L., and Sur, M. (1990). Visual behaviour mediated by retinal projections directed to the auditory pathway. *Nature*, 404, 871–876.

Vygotsky, L. S. (1986). *Thought and language*, rev. ed. Cambridge, Mass: MIT Press.

Wagner, U., Gais, S., and Born, J. (2001). Emotional memory formation is enhanced across sleep intervals with high amounts of rapid eye movement sleep. *Learning and Memory*, 8, 112–119.

Walden, T. A., and Ogan, T. A. (1988). The development of social referencing. *Child Development*, 58, 1230–1240.

Walker, A. (1993). Taphonomy. In A. Walker and R. Leakey eds., *Nariokotome Homo erectus skeleton*, pp. 40–60. Cambridge, Mass.: Harvard University Press.

Walsh, C., and Cepko, C. L. (1992). Widespread dispersion of neuronal clones across functional regions of the cerebral cortex. *Science*, 255, 434–440.

Wang, L., Kakigi, R., and Hoshiyama, M. (2001). Neural activities during Wisconsin Card Sorting Test—MEG observation. *Cognitive Brain Research*, 12, 19–31.

Wang, Z., Moody, K., Newman, J. D., and Insel, T. R. (1997). Vasopressin and oxytocin immunoreactive neurons and fibers in the forebrain of male and female common marmosets. *Synapse*, 27, 14–25.

Wang, Z.-J. (1993). Ionic basis for intrinsic 40 Hz neuronal oscillations. *NeuroReport*, 5, 221–124.

Warrington, E. (1975). The selective impairment of semantic memory. *Quarterly Journal of Experimental Psychology*, 27, 635–657.

Warrington, E., and Shallice, T. (1984). Category specific semantic impairments. *Brain*, 107, 829–853.

Wassermann, E. M., Pascual-Leone, A., Vals-Sole, J., Toro, C., Cohen, L. G., and Hallett, M. (1993). Topography of the inhibitory and excitatory responses to transcranial magnetic stimulation in a hand muscle. *Electroencephalography and Clinical Neurophysiology*, 89, 424–433.

Watanabe, E., Yamashita, Y., Maki, A., Ito, Y., and Koizumi, H. (1996). Non-invasive functional mapping with multi-channel near infra-red spectroscopic topography in humans. *Neuroscience Letters*, 205, 41–44.

Watson, J. S., and Ramey, C. T. (1972). Reactions to response-contingent stimulating in early infancy. *Merrill-Palmer Quarterly*, 18, 219–227.

Weeks, R., Horwitz, B., Aziz-Sultan, A., Tian, B., Wessinger, C. M., Cohen, L. G., Hallett, M., and Rauschecker, J. P. (2000). A positron emission tomographic study of auditory localization in the congenitally blind. *Journal of Neuroscience*, 20, 2664–2672.

Weiller, C., Ramsay, S., Wise, R., Friston, K., and Frackowiak, R. (1993). Individual patterns of functional reorganization in the human cerebral cortex after capsular infarction. *Annuals of Neurology*, 33, 181–189.

Weinstein, E. A., Kahn, R. L., Malitz, S., and Rozanski, J. (1954). Delusional reduplication of parts of the body. *Brain*, 77, 45–60.

Welty, J. (1982). *The life of birds*. Philadelphia: Saunders College Publishing.

Wernig, A., Müller, S., Nanassy, A., and Cagol, E. (1995). Laufband therapy based on "rules of spinal locomotion" is effective in spinal cord injured persons. *European Journal of Neuroscience*, 7, 823–829.

Wernig, A., Nanassy, A., Muller, S. (2000). Laufband (LB) therapy in spinal cord lesioned persons. *Progress in Brain Research*, 128, 89–97.

Whitaker, H., and Selnes, O. (1976). Anatomic variations in the cortex: Individual differences and the problem of the localization of language functions. *Annals of the New York Academy of Sciences*, 280, 844–854.

White, F. J. (1988). Party composition and dynamics in *Pan paniscus*. *International Journal of Primatology*, 9, 179–193.

White, F. J. (1992). Pygmy chimpanzee social organization: Variation with party size and between study sites. *American Journal of Primatology*, 26, 203–214.

White, R. (1993). Technological and social dimensions of "Aurignacian-Age" body ornaments across Europe. In H. Knecht, A. Pike-Tay and R. White, eds., *Before Lascaux*, pp. 277–299. Boca Raton, Fla.: CRC Press.

White, T., Suwa, G., and Asfew, B. (1994). *Australopithecus ramidus*, a new species of early hominid from Aramis, Ethiopia. *Nature*, 317, 306–312.

Whitehouse, R., and Wilkins, J. (1986). *The making of civilization*. London: Collins.

Whiten, A., and Byrne, R. (1988). Tactical deception in primates. *Behavioral and Brain Sciences*, 11, 233–273.

Whitty, C. W., and Lewin, W. (1957). Vivid day-dreaming: An unusual form of confusion following anterior cingulectomy. *Brain*, 80, 72–76.

Wiertelak, E. P., Smith, K. P., Furness, L., Mooney-Heiberger, K., Mayr, T., Maier, S. F., and Watkins, L. R. (1994). Acute and conditioned hyperalgesic responses to illness. *Pain*, 56, 227–234.

Wigan, A. L. (1985/1844). *Duality of the mind*. San Francisco: Joseph Simon.

Wilder, B. G. (1911). Exhibition of, and preliminary note upon, a brain of about one-half the average size from a while man of ordinary weight and intelligence. *Journal of Nervous and Mental Diseases*, 30, 95–97.

Wilkin, M. H., and Kaplan, H. S. (1982). Sex therapy and penectomy. *Journal of Sex and Marital Therapy*, 8, 209–229.

Wilkins, A. J., Shallice, T., and McCarthy, R. (1987). Frontal lesions and sustained attention. *Neuropsychologia*, 25, 357–365.

Williams, R. J. (1967). *You are extraordinary*. New York: Random House.

Wilson, F. A., Scalaidhe, S. P., and Goldman-Rakic, P. S. (1993). Dissociation of object and spatial processing domains in primate prefrontal cortex. *Science*, 260, 1955–1958.

Wilson, M. A., and McNaughton, B. L. (1993). Dynamics of the hippocampal ensemble code for space. *Science*, 261, 1055–1058.

Wilson, M. A., and McNaughton, B. L. (1994). Reactivation of hippocampal ensemble memories during sleep. *Science*, 265, 676–679.

Winslow, J. T., Hastings, N., Carter, C. S., Harbaugh, C. R., and Insel, T. R. (1993). A role for central vasopressin in pair bonding in monogamous prairie voles. *Nature*, 365, 545–548.

Wise, R., Chollet, F., Hadar, U., Friston, K., Hoffner, E., and Frackowiak, R. (1991). Distribution of cortical neural networks involved in word comprehension and word retrieval. *Brain*, 114, 1803–1817.

Witkin, H. A., and Berry, J. W. (1975). Psychological differentiation in cross-cultural perspective. *Journal of Cross-Cultural Psychology*, 6, 4–87.

Wolfe, J. M., and Cave, K. R. (1999). The psychophysical evidence for a binding problem in human vision. *Neuron*, 24, 11–17.

Woo, T. U., Pucak, M. L., Kye, C. H., Matus, C. V., and Lewis, D. A. (1997). Peripubertal refinement of the intrinsic and associational circuitry in monkey prefrontal cortex. *NeuroScience*, 80(4), 1149–1158.

Wrangham, R. (1986). Ecology and social relationships in two species of chimpanzees. In D. Rubenstein and R. Wrangham, eds., *Ecological aspects of social evolution*, pp. 352–378. Princeton, N.J.: Princeton University Press.

Wu, M. T., Hsieh, J. C., Xiong, J., Yang, C. F., Pan, H. B., Chen, Y. C., Tsai, G., Rosen, B. R., and Kwong, K. K. (1999). Central nervous pathway for acupuncture stimulation: Localization of processing with functional MR imaging of the brain—preliminary experience. *Radiology*, 212, 133–141.

Yamagowa, J. (1987). Intra- and inter-group interactions of an all-male group of Virunga mountain gorillas (*Gorilla gorilla beringei*). *Primates*, 28, 1–30.

Yamamoto, I, Rhoton, A. L., and Peace, D. A. (1981). Microsurgery of the third ventricle: Part 1. Microsurgical anatomy. *Neurosurgery*, 8, 334–356.

Yang, T. T., Gallen, C. C., Ramachandran, V. S., Cobb, S., Schwartz, B. J., and Bloom, F. E. (1994). Noninvasive detection of cerebral plasticity in adult human somatosensory cortex. *NeuroReport*, 5, 701–704.

Yang, T. T., Gallen, C. C., Schwartz, B., Bloom, F. E., Ramachandran, V. S., and Cobb, S. (1994). Sensory maps in the human brain. *Nature*, 368, 592–593.

Yarrow, L. J., McQuiston, S., MacTurk, R. H., McCarthy, M. E., Klein, R. P., and Vietze, P. M. (1983). Assessment of mastery motivation during the first year of life. *Developmental Psychology*, 28, 159–171.

Young, A. W., Aggleton, J. P., Hellawell, D. J., Johnson, M., Broks, P., and Hanley, J. R. (1995). Face processing impairments after amygdalotomy. *Brain*, 118, 15–24.

Young, S. (1993). Human facial expressions. In S. Jones, R. D. Martin, and D. R. Pilbeam, eds., *The Cambridge encyclopedia of human evolution*, pp. 164–165. Cambridge: Cambridge University Press.

Yue, G., and Cole, K. J. (1992). Strength increases from motor program: Comparison of training with maximal voluntary and imagined muscle contractions. *Journal of Neurophysiology*, 67, 1114–1123.

Yuste, R., and Sur, M. (1999). Development and plasticity of the cerebral cortex: From molecules to maps. *Journal of Neurobiology*, 41, 1–6.

Zahavi, A., and Zahavi, A. (1997). *The handicap principle*. Oxford: Oxford University Press.

Zappoli, R., Zappoli, F., Versari, A., Arnetoli, G., Paganini, M., Arneodo, M. G., Poggiolini, D., and Thyrion, E. Z. (1995). Cognitive potentials: Ipsilateral corticocortical interconnections in prefrontal human cortex ablations. *Neuroscience Letters*, 193, 140–144.

Zatorre, R., Halpern, A. R., Perry, D. W., Meyer, E., and Evans, A. C. (1996). Hearing in the mind's ear: A PET investigation of musical imagery and perception. *Journal of Cognitive Neuroscience*, 8, 29–46.

Zeki, S. (1993). *A vision of the brain*. Oxford: Blackwell.

Zhang, J.-Z., Li, J.-Z., and He, Q.-N. (1988). Statistical brain topographic mapping analysis for EEGs recorded during Qi Gong state. *International Journal of Neuroscience*, 38, 415–425.

Zimmer, C. (2000). In search of vertebrate origins: Beyond brain and bone. *Science*, 287, 1576–1579.

Zvelebil, M. (1984). Clues to recent human evolution from specialized technologies? *Nature*, 307, 314–315.

Index

About the Authors

John Skoyles, Ph.D., obtained his higher degree from University College London and is presently a Visiting Fellow in the Centre for Philosophy of Natural and Social Science at the London School of Economics.

Dorion Sagan, son of Carl Sagan, is an award-winning science writer. He is the coauthor of several critically acclaimed books, including *Microcosmos, Slanted Truths: Essays on Gaia, Evolution and Symbiosis, Acquiring Genomes, What Is Life?*, and *Origins of Sex*. His articles have appeared in *Wired, The New York Times, Smithsonian, The Sciences*, and other leading publications.